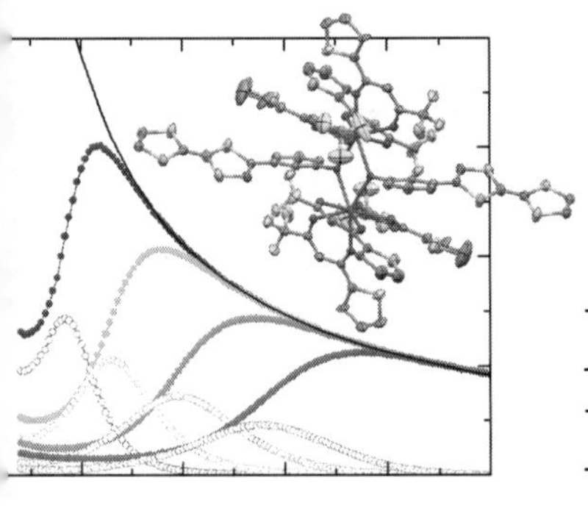

Multifunctional Molecular Materials

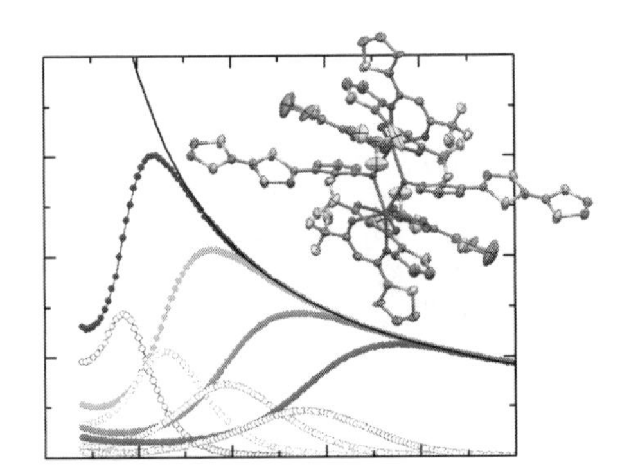

Multifunctional Molecular Materials

edited by Lahcène Ouahab

Published by

Pan Stanford Publishing Pte. Ltd.
Penthouse Level, Suntec Tower 3
8 Temasek Boulevard
Singapore 038988

Email: editorial@panstanford.com
Web: www.panstanford.com

British Library Cataloguing-in-Publication Data
A catalogue record for this book is available from the British Library.

Multifunctional Molecular Materials

ISBN 978-981-4364-29-4 (Hardcover)
ISBN 978-981-4364-30-0 (eBook)

Printed in the USA

Contents

Preface

Molecular materials are exciting and promising materials for the near-future generation of electronic devices and information-processing systems at the nanoscale level because of their chemical and physical peculiarities. In the last decade, there have been wonderful developments in molecule-based materials as regards their electrical conductivity, magnetic interactions, and optical properties. The relevance of these materials in materials science is mainly due to the almost infinite tuning of their physical properties by conventional chemical synthetic methods through soft routes, from organic chemistry to coordination chemistry to supramolecular chemistry. This opens unprecedented possibilities for the design of objects with the desired size, shape, charge, polarity, and electronic properties. Owing to important perspectives in both fundamental sciences and applications in nanotechnology or molecular electronics, for example, considerable efforts are currently being made to design and investigate new materials, named multifunctional molecular materials, which involve coexistence or interplay or synergy between multiple physical properties.

This book discusses molecular materials that combine two or more physical properties and focuses on electrical conductivity, magnetism, single-molecule magnet behavior, chirality, spin cross-over, and luminescence. It is not intended to give an exhaustive view of all possible materials—for example, materials combining electrical conductivity with chirality are not presented. The materials referred to cover transition metal and lanthanide coordination complexes as well as genuine organic materials. Their potential applications in molecule-based devices are also discussed.

This volume is divided into seven chapters. Chapter 1, a contribution of T. Sugawara *et al.*, gives a comprehensive review of the theory and experimental results related to the bifunctional properties of conductivity and magnetism, which are the most fundamental physical properties of molecular solids. In chapters 2 and 3, M. Yamashita *et al.* and J.-L. Zuo *et al.* present the recent developments in electrical conductivity versus single-molecule magnet and single-chain magnet behavior. Magnetism and chirality are presented in chapter 4 by K. Inoue *et al.*, while in chapter 5 L. Valade *et al.* review electrical conductivity and spin crossover systems. F. Pointillart *et al.* focus on magnetism and luminescence in chapter 6. Finally, L. Mercuri *et al.* conclude the book with chapter 7, which discusses multifunctional materials in molecular electronics.

Lahcène Ouahab
Winter 2012

Chapter 1

Magnetism and Conductivity

Tadashi Sugawara[a] **and Akira Miyazaki**[b]
[a]*Department of Chemistry, Kanagawa University,*
2946 Tsuchiya, Hiratsuka, Kanagawa 259-1293, Japan
[b]*Department of Environmental Applied Chemistry, University of Toyama,*
3190 Gofuku, Toyama-shi, Toyama 930-8555, Japan
sugawara-t@kanagawa-u.ac.jp

This chapter is a comprehensive review of the theory and experimental results related to the bifunctional properties of conductivity and magnetism, which are the most fundamental physical properties of molecular solids. The materials referred to range from transition metal complexes and organic conductors with inorganic magnetic ions to genuine organic materials. The plausibility of molecule-based spintronic devices is also discussed.

1.1 Introduction

The interplay between a localized spin and an itinerant electron is a key issue in modern materials science. This interaction leads to intrinsic physical phenomena, such as magnetoresistance, ferromagnetic metals, and the Kondo effect. The establishment of a strongly coupled coexisting system of conductivity and magnetism is

Multifunctional Molecular Materials
Edited by Lahcène Ouahab

ISBN 978-981-4364-29-4 (Hardcover), 978-981-4364-30-0 (eBook)
www.panstanford.com

also indispensable for the development of molecule-based spintronic devices. In particular, magnetoresistance, a phenomenon in which the resistance of a sample decreases upon application of a magnetic field, plays a crucial role in spintronic functions. Peter Grünberg and Albert Fert independently discovered this phenomenon in Fe/Cr/Fe nanolayers, for which they were awarded the Nobel Prize in Physics in 2007. This chapter is intended to provide a comprehensive overview of the cooperative coexistence of magnetism and conductivity, which is one of the most fundamental pairs of bifunctional properties in molecular materials.

This chapter first describes theoretical models of the interaction between localized spins and conduction electrons, such as the double-exchange mechanism and the RKKY (Runderman, Kittel, Kasuya, and Yoshida) mechanism, referring to some of the relevant materials, e.g., magnetic nanoparticles and carbon-based materials. Second, the coexistence of magnetism and conductivity in molecular materials, such as transition metal complexes and radical ion salts of organic donors with magnetic inorganic ions, is explained precisely. Third, the design and construction of organic molecular materials containing organic spin as a localized spin source are discussed, pointing out the importance of the spin-polarized electronic structure of a donor radical. Finally, a perspective of bifunctional materials with properties of magnetism and conductivity is mentioned in reference to plausible molecule-based spintronic devices.

Before discussing the particular individual issues, it is worthwhile to discuss the magnetic orbitals of inorganic and organic magnetic materials. In the case of inorganic spin-polarized materials, d or f atomic orbitals play the role of magnetic orbitals unexceptionally. The large positive exchange interaction is derived from the orthogonality and the effective space-sharing nature of one-centered atomic orbitals. In contrast, in the case of organic spin-polarized materials, the magnetic orbitals are p or sp^n hybridized orbitals. Since spin resides in multicentered π molecular orbitals, the exchange interaction between them is weakened by the delocalizing nature of the π molecular orbitals. On the other hand, since the spin-residing p-orbital is usually involved in a π-conjugated framework, multifunctional magnetic materials, e.g., photo-active, redox-active, protonation-active spin system, could be produced by a sophisticated molecular design.

1.2 Interaction Between Conduction Electrons and Localized Spins

For the design of molecular system in which conductivity and magnetism coexist cooperatively, it will be helpful to review how magnetic interactions are achieved in inorganic materials, such as transition or rare-earth metals. The ferromagnetism of 3*d* elements, Fe, Co, and Ni, is explained using the Stoner model.[1] In these metallic solids, the conduction electrons have a hybridized character of 3*d*, 4*s*, and 4*p* orbitals. As a result, their electronic density of state has a sharp peak around the Fermi surface. Due to the Coulomb interaction Pauli exclusion principle working between the conduction electrons, the energy bands of electrons with up- and down-spins shift energetically as illustrated in Fig. 1.1a (exchange splitting), resulting in the nonvanishing magnetic moment.

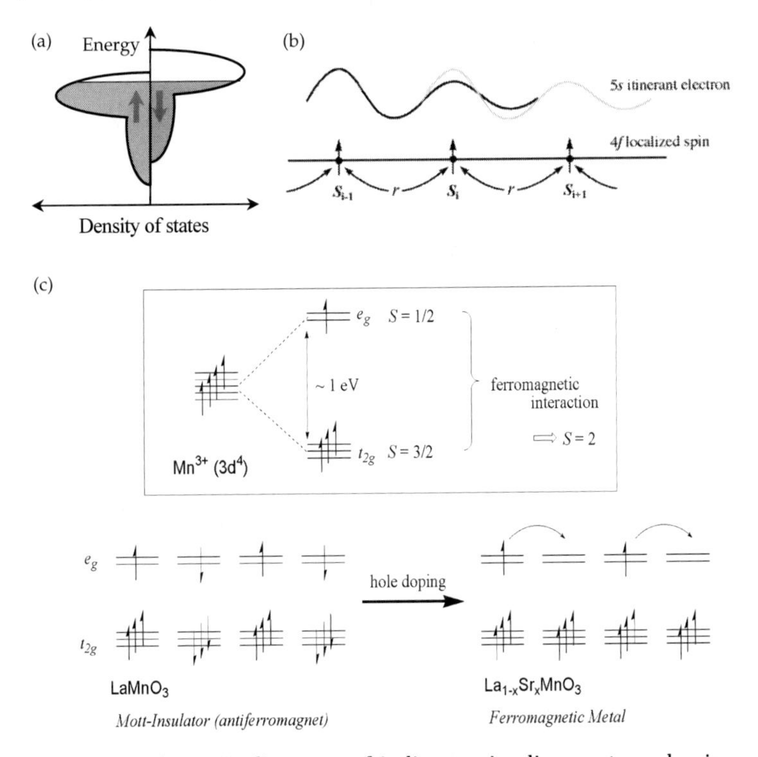

Figure 1.1 Schematic diagrams of indirect spin-alignment mechanisms: (a) Stoner mechanism, (b) RKKY and exchange enhancement mechanism, and (c) double-exchange mechanism.

Since this Stoner model is described in the reciprocal space (k-space), the design of spin alignment based on this model is difficult for chemists who design their molecules in the real space (r-space). In this section, we discuss two other indirect exchange mechanisms, namely RKKY interaction (Fig. 1.1b) and double-exchange mechanism (Fig. 1.1c). These two mechanisms have a similarity at the point that itinerant conduction electrons mediate the spin alignment of the localized electrons, and they are more realistic for making conducting magnets using molecular materials.

1.2.1 RKKY (Ruderman, Kittel, Kasuya, Yosida) Interaction

In rare-earth metals or ferromagnetic diluted alloys, itinerant conduction electrons are known to play an important role in aligning spins of localized magnetic spins. In lanthanide metals, such as Gd, Tb, and Dy, $4f$ electron spins tend to localize at the position of the nuclei, because of the less-shielded characters of $4f$ electrons. In lanthanide elements, many $4f$ electrons align their spin parallel based on the Hund's rules. When an itinerant $6s$ electron approaches the lattice point where the transition metal ion is located, it exchange-couples with $4f$ spins localized on the metal ions. As a result, the localized spin produces a wavelike spin polarization to the Fermi sea of conduction electrons, as illustrated in Fig. 1.1b, with a characteristic wavelength $\lambda \sim \pi/k_F$. Then, a second localized spin near the first one suffers a ferromagnetic or an antiferromagnetic interaction, depending on the distance between the two spins.

This basic idea of spin-ordering was formulated first by Ruderman and Kittel in connection with the interaction between nuclear spin and conduction electrons via Fermi contact,[2] and their analysis is extended by Kasuya and Yosida to the s–d and s–f interactions,[3,4] which is now known as RKKY interaction. If a pair of localized $4f$ spins $\mathbf{S}_1$ and $\mathbf{S}_2$ located at $\mathbf{R}_1$ and $\mathbf{R}_2$, respectively, inside the conduction electron system, the exchange coupling between them is

$$H_{ex} = -J_{sf}[S_1 \cdot s(R_1) + S_2 \cdot s(R_2)] \tag{1.1}$$

where $\mathbf{s}(\mathbf{R}_i)$ is the conduction electron spin having a spatial dependence.[5] Using the second-order perturbation theory, the interaction between the two localized spins is expressed with the effective Hamiltonian as

$$H_{\text{eff}} = -J(\mathbf{R}_{12})\mathbf{S}_1 \cdot \mathbf{S}_2 \tag{1.2}$$

with the indirect exchange coupling constant

$$J(\mathbf{R}_{12}) = \left(\frac{J_{\text{sf}}}{N}\right)^2 \sum_{k,q} \frac{\cos \mathbf{q} \cdot \mathbf{R}_{12}}{E(\mathbf{k}+\mathbf{q}) - E(\mathbf{k})} \tag{1.3}$$

where $E(\mathbf{k})$ is the dispersion relation of the conduction electron system. The summation should be taken under the restrictions $|\mathbf{k}| < k_F$ and $|\mathbf{k} + \mathbf{q}| < k_F$, and for a three-dimensional free-electron gas it is calculated as

$$J(R) \propto \left(\frac{J_{\text{sf}}}{N}\right)^2 \left[\frac{2k_F R \cos 2k_F R - \sin 2k_F R}{(2k_F R)^4}\right] \approx \left(\frac{J_{\text{sf}}}{N}\right)^2 \frac{\cos 2k_F R}{(2k_F R)^3} \tag{1.4}$$

where R is the distance between two localized spins, and k_F is the Fermi wavenumber that characterizes the free-electron gas. Consequently, the exchange interaction between the localized spins can either be ferromagnetic or antiferromagnetic depending on the distance R between them, and has a long-range character (proportional to R^{-3}; nonexponential decay).

Since the above derivation does not assume any specific dispersion relation $E(\mathbf{k})$ of the conduction electrons, the RKKY theory can also be applied not only to the free-electron system, but also to the tight-binding electrons, characteristic to molecular materials, just by evaluating the sum with the appropriate form of $E(\mathbf{k})$.[5] However, this theory cannot be applied for $3d$ metals in a straightforward manner because of the itinerancy of $3d$ electrons. Moreover, effects of the Coulomb interaction between conduction electrons are not taken into account in the RKKY model. In strongly correlated electron systems such as Pd and Pt, the spin polarization of conduction electrons caused by a localized moment decays not in an oscillatory but a monotonical fashion.[6] This difference may be analogous to underdamping and overdamping movements of a damped spring–mass system; the electron correlation effectively works as a friction to prevent oscillatory damping of the spin polarization of the conduction electrons.

1.2.2 Double-Exchange Mechanism in Metal Oxides

To confirm the interactive coexistence of conductive electrons and localized spins in molecule-based materials, observation of negative magnetoresistance is most indicative. Transition metal oxides, such as $La_{1-x}Sr_xMO_3$ (M = Mn, Co), which exhibit colossal negative magnetoresistance (CMR) are excellent examples.[7–9] For this system, CMR is observed near the transition temperature $T_C = 126$ K; the resistivity rapidly decreases with a relatively low magnetic field (<1 T). At 129 K, the magnetoresistance ratio (defined as $\rho(0)/\rho(B)$) reached ~200% at 0.3 T, ~3000% at 1 T, and ~20,000% at 7 T.

Although two perovskites $LaMnO_3$ and $CaMnO_3$ are antiferromagnetic Mott insulators ($T_N = 141$ K and 123 K, respectively), the solid solution $La_{1-x}Ca_xMnO_3$ exhibits ferromagnetism with metallic conductivity in the range of $0.1 < x < 0.5$.[10] The metallic conductivity is a consequence of the electron transfer from Mn^{3+} site to Mn^{4+} site, and the ferromagnetism is explained using the double-exchange mechanism.[11–13] In both Mn^{3+} and Mn^{4+} ions, the low-lying t_{2g} orbitals are occupied by three localized $3d$ electrons with parallel spins, but only in Mn^{3+} there exists an additional e_g electron. The Hund's rule yields a ferromagnetic coupling between the e_g and t_{2g} electrons, and due to hybridization with the oxygen $2p$ orbitals, the e_g electron can migrate via interatomic hopping. The electron hopping integral is expressed as $t = t_0 \cos(\theta/2)$, where θ is the relative angle between the localized spins of the neighboring Mn atoms. The hopping rate is proportional to t^2 and takes a maximum for ferromagnetic spin alignment ($\theta = 0$) and zero for antiferromagnetic alignment ($\theta = \pi$). As the interatomic hopping of the e_g electron reduces the total electronic energy, the ferromagnetic alignment of the localized moments of t_{2g} electrons are yielded. In other words, the spin-polarized electron in the e_g orbital can migrate along the metal array, aligning the localized spins ferromagnetically at each metal site (Fig. 1.1c).

The ferromagnetic coupling between the itinerant and localized electron spins is not compulsory in the double-exchange mechanism.[5] The Hamiltonian for two localized spins $\mathbf{S}_1$, $\mathbf{S}_2$, and one mobile electron that connects them is expressed as

$$H = -t \sum_{\sigma=\uparrow,\downarrow} (c_{1\sigma}^{+} c_{2\sigma} + c_{2\sigma}^{+} c_{1\sigma}) - J[\mathbf{s}_1 \cdot \mathbf{S}_1 + \mathbf{s}_2 \cdot \mathbf{S}_2] \quad (1.5)$$

The first term is equivalent to the Hückel Hamiltonian for ethylene and expresses the movement of the mobile electron, and the second term is the exchange interaction between the mobile and localized electrons. When $S = 1/2$, the ground state calculated from this Hamiltonian is $S_{tot} = 3/2$ state for ferromagnetic coupling ($J > 0$), whereas for antiferromagnetic coupling ($J < 0$), the ground state becomes $S_{tot} = 1/2$. In both cases, the two localized spins $\mathbf{S}_1$ and $\mathbf{S}_2$ are parallel. When the spin multiplicity S of the localized moments is large enough to be treated as classical vectors, some calculations show that ferromagnetic alignment of the localized spins is always preferred, regardless of the sign of J. Therefore, the most essential point in the double-exchange mechanism is the strong exchange coupling between the itinerant and the localized moments, which is prominent in one-centered exchange interaction in transition metal oxides.

1.3 Interaction Between Localized Spins and Itinerant Electrons in Metallic and Magnetic Nanoparticles

The spin polarization on conduction electrons by localized paramagnetic spin, which plays an essential role in RKKY mechanism,[2–4] has already been demonstrated in 1960s using Pd-based alloys, such as Co–Pd[14] and Fe–Pd.[15] The total amount of the spin polarization in these alloys turned out to be unexpectedly larger than the values predicted with the RKKY mechanism. Moriya dissolved this contradiction by taking the strong electron correlation of Pd into consideration.[6] In his theory, the spin polarization of the conduction electron decays monotonically as a function of the distance from the magnetic center, which is in contrast with an oscillating decay obtained in the RKKY theory. Moreover, the range of the spin polarization is estimated to be *ca.* 1 nm from the paramagnetic center, which is experimentally confirmed by a neutron scattering experiment in Co–Pd and Fe–Pd alloys.[16]

Recently, the spin-polarization effect in Pd-based alloys has been indicated more directly in their nanoparticles.[17–19] Metal nanoparticles are considered to be essential building blocks for the construction of nano-sized devices, because of their unconventional physical and chemical properties. The popularity of the transition metal nanoparticles strongly depends on the presence of the Brust–Schiffrin method,[20,21] which enables us to obtain stable transition metal nanoparticles in high yields. In this method, particle size can be controlled by carefully choosing the amount of alkanethiol molecules used as capping molecules, and the resulting particles are nearly uniform in size. It is therefore possible to reduce the size of Co–Pd and Fe–Pd alloy particles to several nanometers, which is comparable to the effective radius of the spin polarization in the RKKY and Moriya's exchange enhancement mechanisms.

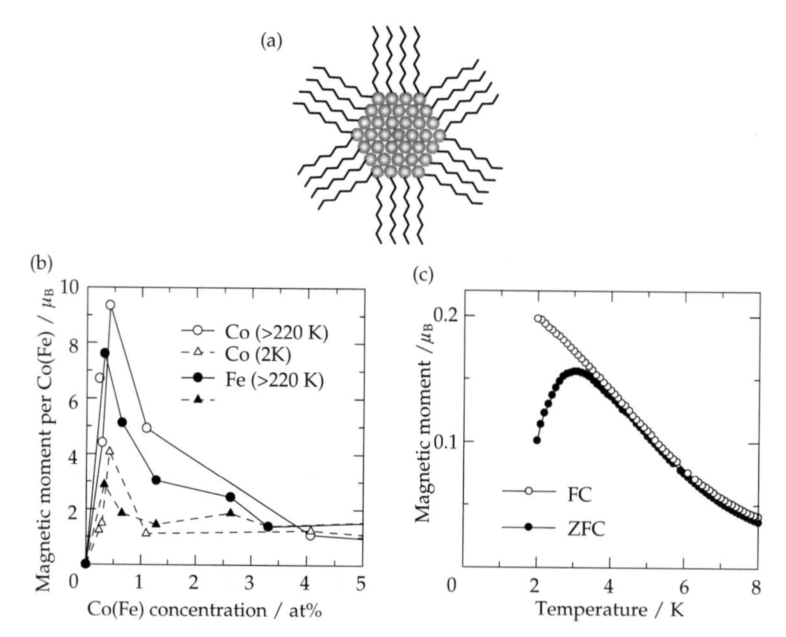

Figure 1.2 (a) Schematic structure of a Co–Pd alloy nanoparticle. (b) Magnetic moments of Co–Pd and Fe–Pd alloy nanoparticles as functions of Co and Fe concentrations.[19] (c) The magnetization per Co atom of the Co–Pd alloy nanoparticle containing 0.42 at% Co atom after field-cooled (FC) and zero-field-cooled (ZFC) processes.[18]

The high-resolution TEM images of Co–Pd and Fe–Pd nanoparticles show the lattice image of these particles, and the lattice constant is close to the interplanar spacing of bulk fcc Pd metal. Using the mean radius of the nanoparticles, the average number of atoms per particle was estimated to be 300~400. Despite the fact that the presence of Co or Fe in the nanoparticles confirmed with a bulk elemental analysis (ICP–OES), no Co atoms were detected during X-ray photoelectron spectroscopy (XPS). Since the XPS is a surface-sensitive analytical method, this result suggests that Co atoms are located inside the Pd particles whose surfaces are covered with long alkyl chains. It should also be mentioned that the concentration of the magnetic transition metal could be reduced to 0.42 atomic % (at%) and 0.33 at% for Co–Pd and Fe–Pd nanoparticles, respectively. Since the average diameter of these nanoparticles is estimated to be 2 nm, these nanoparticles of Pd-based alloy contain only one paramagnetic transition metal atoms per particle as illustrated in Fig. 1.2a. In this point, these nanoparticles can be regarded as good model system to evaluate the effect of spin polarization of localized spins of Co or Fe on the conduction electron system of Pd.

Interestingly, the magnetic moments of Co–Pd and Fe–Pd systems per transition metal atom show sharp peaks when only a single paramagnetic atom is included in a host Pd nanoparticle as the magnetic center. This result shows that the presence of only one paramagnetic atom can magnetize the whole nanoparticle with the exchange enhancement mechanism. Above 220 K, the magnetic moment of the nanoparticles of this concentration is 9.4 μ_B (Fig. 1.2b), which is nearly the same as that of bulk Co–Pd alloy.[14] This result suggests that a single Co atom magnetizes all Pd atoms inside a particle. On the other hand, the magnetic moment at 2 K is smaller than the high-temperature magnetic moment above 220 K. This difference can be explained in terms of the quantum size effect.[22] In the low-temperature region, the energy discreteness surpasses the thermal energy, resulting in the partial localization of the conduction electrons of Pd nanoparticles and the reduction of the exchange enhancement effect.

The large magnetic moment of Co–Pd alloy nanoparticles indicates that the single Co atom magnetizes all Pd atoms involved in the particle. All of the spins inside a particle are ferromagnetically coupled to form a single large spin with high multiplicity. As a result, these nanoparticles show superparamagnetic nature[23] characterized as follows. In the high-temperature range, this large spin can rotate freely to the direction of the applied external magnetic field, similar to conventional paramagnetic species. At low temperatures, however, this spin reversal is prohibited by an energy barrier produced by the axial magnetic anisotropy. This effect appears as the evident deviation of the magnetization after field-cooling (FC) process from the magnetization after zero-field cooling (ZFC) below a temperature called as blocking temperature T_B. In the present case of Co–Pd nanoparticles, the blocking temperature T_B = 3 K (Fig. 1.2c) and the particle volume gives the estimation of the effective uniaxial anisotropy as 100 kJ m^{-3}, which is comparable to the anisotropy of bulk Co (250 kJ m^{-3}).[23]

The comparison of Fe–Pd and Co–Pd nanoparticles also gives an insight into the difference in the spin polarization and the magnetic anisotropy of these two systems. The magnitude of the magnetization of the Fe–Pd nanoparticles is smaller than the corresponding Co–Pd nanoparticles, which is opposite to the trend in the bulk Fe–Cd[15] and Co–Pd[6] alloys. Moreover, the temperature dependence of the magnetization of the Fe–Pd nanoparticles shows no difference between FC and ZFC processes in the whole temperature range. The opposite trend in the magnetization can be explained with the stability of these two alloys. While the Co–Pd system has a stable single-alloy phase,[24] the phase diagram of the Fe–Pd system is complicated because of the presence of an intermetallic compound Pd_3Fe.[25] Therefore, the Fe atoms in Pd are sensitively affected by the structural instability, and then the magnetic order in the Fe–Pd alloy is partly broken down, especially at lower concentrations. This structural instability is more profound in nanoparticles than in their bulk form due to the large specific surface area, resulting in the shrinkage of the spatial distribution of the spin polarization. The absence of the blocking nature, on the other hand, can be explained from the difference in the magnetic anisotropy of Co and Fe. In their bulk form, the magnetic anisotropy of Fe is less than 20% of that of Co. Assuming that the blocking temperature of the alloy nanoparticle is proportional to the magnetic anisotropy of the magnetic transition metal in the bulk form, the blocking temperature of the Fe–Pd nanoparticle is expected to be around 0.6 K, which is below the temperature range available with usual SQUID magnetometer. In Co–Pd alloy nanoparticles, the magnetic anisotropy is large enough to evidence single-particle magnetic behavior involving superparamagnetic and blocking properties. A single-particle magnet based on metal nanoparticles can potentially become a target of molecular magnetism and spintronics applications.[26–28] It is an intriguing building block in designing unconventional magnetic properties that cannot be obtained in either molecular or bulk systems.

1.4 Magnetic Structures of Edge-State Spins in Nanographene

Recent great interest in nanocarbon materials, such as fullerenes,[29] carbon nanotubes,[30] and especially nanographenes,[31,32] yielded the Nobel Prize in Physics 2010 awarded for the discovery of graphenes. This new material of a single sheet of graphite, which can be prepared merely by repeated peeling using Scotch tapes, has a number of unique properties such as quantum Hall effect[33,34] and future applications like transparent transistors.[35]

Graphenes are also interesting material for magnetism,[36,37] as the π-electrons of graphene bear both itinerant and localized natures. The π-electron system of graphite without boundary is described in tight-binding model that is equivalent to the Hückel approximation.[38–40] The band structure calculation of graphite yields zero-gap semiconductor with linear conduction and valence bands near the Fermi level, which are the characteristics of massless Dirac Fermions. When the size of the graphite is limited to a size less than ~10 nm, the edges of the graphene significantly contributes to its electronic structure. The boundary of the nano-graphene is composed of zigzag and armchair edges, the structure corresponding to the *trans*- and *cis*-polyacetylene, respectively. While the armchair edges have closed-shell character due to bond alternation, the zigzag edges have localized spins to influence the electronic and magnetic properties of nanographene.[41–44]

In the mathematical field of graph theory, the π-electron network of condensed aromatic hydrocarbons or graphenes is described as bipartite graphs, where carbon atoms can be divided into two disjoint sets, starred and unstarred, such that each C–C bond connects a starred carbon atom only to unstarred carbon atoms. Lieb's theorem,[45] which is equivalent to the Longuet-Higgins conjecture known in the field of physical organic chemistry,[46] claims that the number of degenerate nonbonding states in a bipartite graph of the carbon network is equal to $|N^* - N|$, where N^* and N are the numbers of the starred and unstarred carbon atoms in the network, respectively. Using this rule, it is easy to show that phenalenyl and triangulene, molecules consisting of 3 and 6 rings, have 1 and 2 nonbonding states, respectively, and their electronic ground state have $S = 1/2$ and 1, respectively (Fig. 1.3a). Here it should be emphasized that these nonbonding states with unpaired electrons are populated in the peripherals of these molecules composed of zigzag edges, which has been proved experimentally.[47,48] For example, in a crystal, triangulene derivative molecules are in a monomer–dimer equilibrium, wherein a covalent bond is formed and broken between the carbon atoms on their zigzag edges.[49] The zigzag edge structure is also found in a high-spin molecule poly(*meta*-phenylenemethylene) (Fig. 1.3b).[50,51] This molecule has 4 nonbonding states and thus $S = 2$ π-electron spin on the zigzag backbone, which ferromagnetically couples with 4 unpaired sp^2 electrons yielding $S = 4$ ground state. Due to the strong electron correlation in the π-electron system, the electronic state of this molecule is described better by using Hubbard–Kondo model[52] rather than one-dimensional RKKY model.

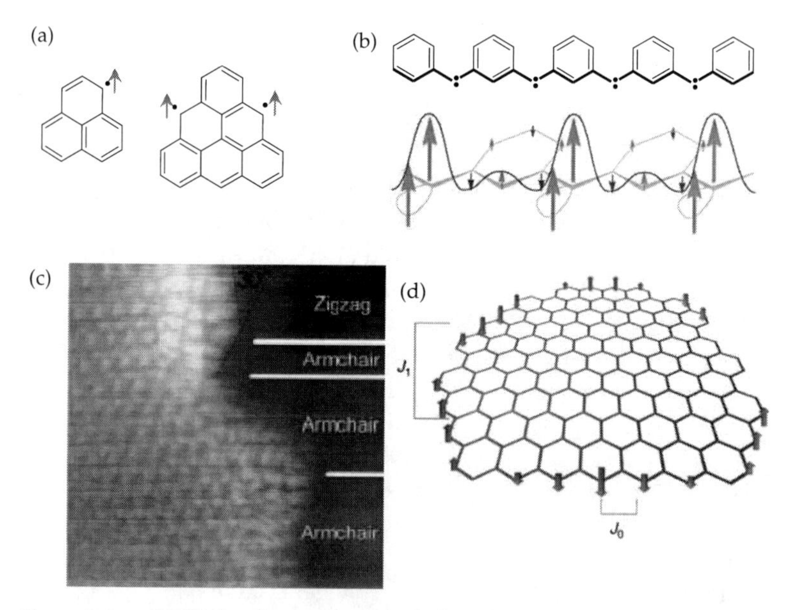

Figure 1.3 (a) Molecular structure of phenalenyl (left) and triangulene (right), indicating unpaired π-electron spin(s). (b) Spin alignment of poly(*meta*-phenylenemethylene). (c) Atomically resolved ultra-high vacuum STM lattice image of the edge region of a graphene.[54] (d) Schematic spin structure of a graphene with the edge-state spins. J_0 and J_1 are intra- and interzigzag state interaction, respectively.[36]

The nonbonding zigzag edge state exists not only in these non-Kekulé molecules, it is also found in condensed aromatic hydrocarbons. The presence of the zigzag edge states in large condensed aromatic system has been pointed out theoretically for semi-infinite graphite ribbons[44] and experimentally evidenced by scanning tunneling microscopy/spectroscopy (STM/STS) observations of hydrogen-terminated graphenes.[53–55] In the atomically resolved STM image of graphene, the edge state appears as bright spots on the zigzag edge region (Fig. 1.3c), and the STS spectrum taken around the zigzag edge has a sharp peak at the Fermi energy.

The magnetic properties of nanographenes are of great interest, as they can be regarded as an extension of polycarbene-based high-spin molecules. The boundary of a nanographene is composed of randomly distributed several zigzag edges. The localized spins on a single zigzag edge ferromagnetically couples with an exchange interaction J_0. The spins on the neighboring zigzag edges then interact with each other via RKKY interaction[2–4] as illustrated in Fig. 1.3d, whose magnitude J_1 is in the range $J_1/J_0 = 10^{-1}$–10^{-3}.[56] Activated carbon fibers (ACFs) are good model system for nanographenes, as they are characterized as porous carbon network of nanographites composed of several nanographene sheets.[57,58] The Curie–Weiss behavior of the magnetic susceptibility in pristine ACFs is explained on the basis of the present magnetic structure model. The magnetic behavior of the ACF edge spins is sensitive to the structural modification induced by heat treatment,[57,58] adsorption of guest molecules,[59–61] and chemical modification by halogens.[62,63] For example, heat treatment changes the structure of nanographene network to modify interdomain interaction, whereas adsorption of water in the nanoporous network reduces the interplanar distance inside the nanographites and increases the intersheet interaction.

Here it should be mentioned that the massless zero-gap Dirac Fermions system of graphenes is also found in a bulk molecular conductor α-BEDT-TTF$_2$I$_3$ under pressure[64,65] (BEDT-TTF = bis (ethylenedithio)tetrathiafulvalene). The theoretical band structure calculations[66–68] show that the bottom of the conduction band and the top of the valence band contact each other at two points and that the Fermi energy locates at these points. There are differences in the electronic structure of this salt from that of graphene; while graphene is a monolayer system having isotropic Dirac-cone, α-BEDT-TTF$_2$I$_3$ is a multilayered anisotropic Dirac Fermions system.

One of the characteristic features of transport in the multilayered massless Dirac Fermion system is clearly observed in the interlayer magnetoresistance, where zeroth Landau level carriers, including the Zeeman effect, is clearly observed.[69,70]

1.5 Coexistence and Cooperation of Conductivity and Magnetism in Transition Metal Complexes

In order to realize the coexistence and cooperation of conductivity and magnetism in molecular materials, we should design the conductive part and magnetic part separately, and then combine these two parts together. The electrical conductivity can be relatively easily achieved by adopting TTF and its derivatives,[71–73] of which molecular structures are summarized in Fig. 1.4. These π-electron-based flat molecules tend to form one- or two-dimensional arrangements, and the π-electron carriers give a large variety of electronic phases such as charge density wave (CDW) and spin density wave (SDW),[74] charge localization and charge disproportionation,[75,76] and metallic or superconducting phases. The great versatility of the TTF derivatives is also due to the activity of synthetic organic chemistry.[77] The magnetic part can be introduced by using either transition metal complexes or stable organic radicals. In this section, the molecular conductor with transition metal complexes are introduced, and the interplay between organic conductor and organic stable radical will be discussed in the following sections.

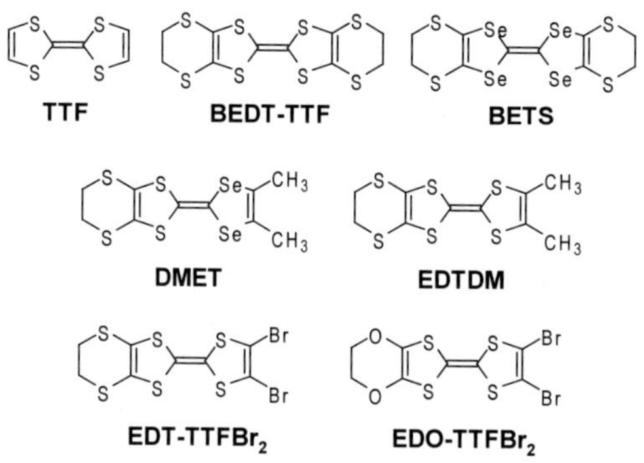

TTF BEDT-TTF BETS

DMET EDTDM

EDT-TTFBr$_2$ EDO-TTFBr$_2$

Figure 1.4 Molecular structures of TTF derivatives discussed in Section 1.5.

The hybridization of molecular metals and transition metal complexes having localized magnetic moments of unpaired d-electrons yields molecular conducting magnets, where an exchange interaction between the organic donor π-electron and anion d-electron systems, referred as the π–d exchange interaction, plays an important role in their physical properties. If the ground state of the π-electron layer is metallic, these materials can be regarded as the molecular version of the RKKY system.[2–4] If the organic π-electron systems have insulating ground states, the spins of the localized π-electrons can be coupled with the magnetic moments of d-electrons through the π–d exchange interaction to produce novel magnetic systems. Due to the high degree of freedom for molecular design, it is possible to place the electronic ground state of molecular conductors in the vicinity of metallic and insulating phases. If magnetic counterparts are introduced in such marginal conductors, even a small perturbation to the conduction electrons by localized spins can produce large modification in the physical properties of molecular materials as illustrated in the following examples.

The way of combining conduction and magnetic parts is also to be considered. The simplest way is to introduce magnetic anion (cation) as the counter part of conducting cation (anion) radical ion salt. Since the conduction part and magnetic part can be designed individually, the combination between them leads to large variety of molecular systems, which may increase the possibility of finding an optimum combination between them. The π–d exchange interaction is realized through intermolecular contacts such as van der Waals contacts, which are in general difficult to design and control. Because these contacts rely on nonbonding interaction, the magnitude of the π–d exchange interaction is generally weak. Another approach is to connect conduction and magnetic counterparts with chemical bonds. The introduction of chemical bonds between conduction and magnetic part is favorable for stronger interaction between them. The most important drawback is that multistep chemical synthesis is required for each combination.

1.5.1 Radical Ion Salts Containing Monomeric Magnetic Counter Ions

In order to achieve the π–d interaction between the conduction and magnetic layers, there are several important issues to be considered.

The localized *d*-electrons of the counter anion should be sufficiently delocalized to the ligand atoms. There should also be close contacts between the ligand of the anion and the donor molecule.

Transition metal halide complexes as magnetic anions are popularly employed as magnetic counterparts. Since the Néel temperature of MBr_2 ($M = Mn^{2+}$, Fe^{2+}, Co^{2+}, Ni^{2+}, and Cu^{2+}) monotonically increases as the number of *d*-electron increases,[78] it is preferable in principle to use $CuBr_4^{2-}$ as a magnetic counter anion to obtain large exchange interaction. Unfortunately, this anion stably exists in the solution only under high concentration of bromide anion,[79] and therefore only a couple of cation salts with this anion are reported. On the other hand, FeX_4^- (X = Cl, Br) ions are relatively stable and are easily handled in the preparation of complexes. In addition, because of the similar ionic radii of Fe^{3+} and Ga^{3+},[80] there is a general tendency that the radical cation salts of magnetic FeX_4^- and nonmagnetic GaX_4^- give the same crystal structure, which enables us to investigate the role of the π–d interaction systematically.

Many radical ion salts with transition metal halide anions showing metallic conductivity have so far been reported.[81] However, π–d interactions in most of these salts are weak, resulting in the absence of magnetic ordering. For example, $(BEDT\text{-}TTF)_3CuCl_4{\cdot}H_2O$[82] is the first molecular conductor with magnetic transition metal complex that shows a metallic behavior down to 200 mK. The localized *d*-electron spins on the counter anions are paramagnetic with no magnetic transition, hence the donors and magnetic anions behave independently. The most important and celebrated exceptions in this category are the $(BETS)_2FeX_4$ family (BETS = bis(ethylenedithio) tetraselenafulvalene),[83] in which the coexistence of magnetically ordered phase and superconductivity has been successfully obtained. In the crystal of λ-$(BETS)_2FeCl_4$ as a typical example, the conduction electronic system becomes an insulator below T_{MI} = 8.5 K, where an antiferromagnetic transition of the counter anion *d*-spins also takes place. On applying an external magnetic field at low temperatures, this antiferromagnetic ordered phase is suppressed and a field-induced superconducting state appears between 18 and 42 T when the field is applied exactly parallel to the two-dimensional conduction plane as shown in Fig. 1.5a.[84,85] The origin of the field-induced superconductivity can be explained from the Jaccarino–Peter effect,[86] which was firstly confirmed in Eu–Sn Chevrel-phase compounds.[87] In this mechanism, the localized

d-electron spin S produces the effective field $\mathbf{B}_{\text{eff}} = (J_{\pi d}/g\mu_{\text{B}})\langle\mathbf{S}\rangle$ working on the conduction electron spins to cancel out the external field to recover the superconducting state. Experimental results give the estimate of the π–d interaction as $|J_{\pi d}| \sim 18$ K, which is in good agreement with the calculated value based on the extended Hückel approximation.[88]

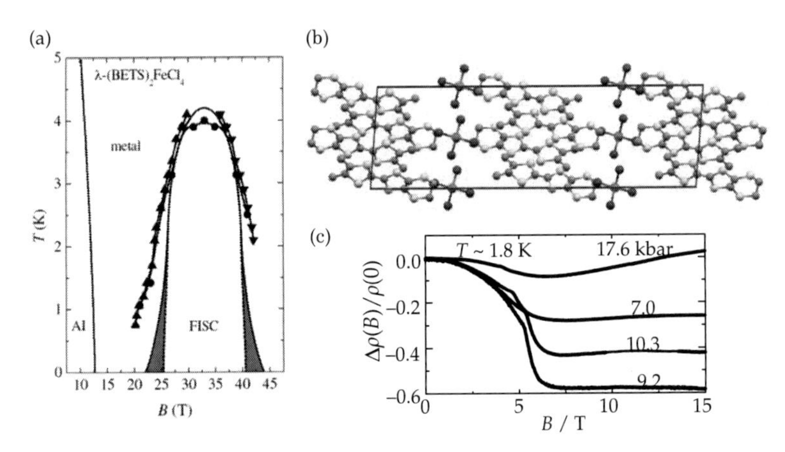

Figure 1.5 (a) Phase diagram of λ-$(BETS)_2FeCl_4$ under the applied field parallel to the conducting plane. AI: antiferromagnetic insulator, FISC: field-induced superconductor[85] (b) Crystal structure of $(EDTDM)_2FeBr_4$. (c) Field dependence of the in-plane magnetoresistance of $(EDTDM)_2FeBr_4$ at ~1.8 K under pressures.[90]

Except this exceptional case, the magnitude of the π–d interaction is in general not strong enough to perturb stable metallic electronic system. However, if a metallic electronic system has an instability coming from low dimensionality and/or electron–electron repulsion, even a small perturbation caused by the π–d interaction can drastically change its electronic ground state. Since quasi-one-dimensional (Q1D) π-electron systems have a large variety of electronic phases including nonconducting but magnetic ground state, such as spin–density wave (SDW) state, it is promising to utilize Q1D system as conducting parts. $(DMET)_2FeBr_4$[89] and $(EDTDM)_2FeBr_4$[90] are Q1D metals having the same crystal structure (DMET = dimethylethylenedithiodiselenadithiafulvalene, EDTDM = ethylenedithiodimethyltetrathiafulvalene). The organic donor molecules form chain structures, and close S···Br contacts exist between the organic conducting layer and inorganic magnetic layer (Fig. 1.5b). The conductivity behaves metallic down to *ca.* 40 and 15 K for DMET and EDTDM salts, respectively, below which an SDW state appears as nonmetallic ground state. The magnetic anion layer undergoes an antiferromagnetic transition at T_N = 3.7 (DMET) and 3.0 K (EDTDM). Below these temperatures, two successive anomalies appear on the magnetization curves, corresponding to the spin–flop transition of anion *d*-electron spins at $B_{SF\text{-}d}$ and donor π-electron spins at $B_{SF\text{-}\pi}$, respectively. These anomalies also appear on the magnetoresistance traces measured along the interplanar direction. The excellent correlation between the magnetization and the magnetoresistance supports the interplay between the π- and *d*-electron spins in this system.

Since the ground state of the π-electron layer becomes the SDW state, the π-electrons on the organic layer bears partially localized character. The observed magnetization curves are well-reproduced with a model having localized π- and *d*-electron spins with the π–*d* interactions of $J_{\pi d}$ = −0.2 K. The presence of this $J_{\pi d}$, although weak in magnitude, explains the magnetoresistance qualitatively. Below the Néel temperature, the localized spins on the anion layer are antiferromagnetically ordered. As the alternating spin alignment of the anion layer and the donor layer have the same wavelength, the antiferromagnetic coupling of π- and *d*-electron spins via $J_{\pi d} < 0$ stabilizes the SDW state of the donor π-electron layers, which elevates the resistivity of the sample. On applying the external magnetic field above $B_{SF\text{-}d}$, the coupling between the π- and *d*-electron layers are canceled, resulting in a decrease in the resistivity.

The electronic ground state of the π-electron system can be finely tuned by applying an external pressure to $(EDTDM)_2FeBr_4$ salt. At the ambient pressure, the in-plane magnetoresistance shows a small negative value with a minimum around the saturation field. The increase in pressure makes the magnetoresistance more negative, and the largest negative magnetoresistance ($\Delta R(B)/R \sim$ 60%) is observed around 9.2 kbar, which is the phase boundary between nonmetallic and metallic ground states of the π-electron layer (Fig. 1.5c). The remarkable enhancement in the negative magnetoresistance in the vicinity of the phase boundary is considered to be related to the effect of the electronic instability taking place in the critical region. It should be mentioned that these remarkable behavior are observed at temperature (T = 1.8 K) higher than the magnitude of the π–d interaction ($|J_{\pi d}| \sim 0.2$ K), which evidences the effectiveness of the utilization of the instability of the π-electron system.

1.5.2 Radical Ion Salts Containing Polymeric Magnetic Counter Ions

Another trend in creating magnetic molecular conductor is the adaptation of the polymeric magnetic counter-ion system. By using polydentate organic ligands, two- or three-dimensional network of coordination polymer can be constructed. Since organic molecules with several functional groups are used for this purpose, such networks are also called as metal–organic frameworks (MOFs).

In the history of the development of molecular superconductors, the use of coordination polymer counter-ion systems has been accepted as a good guiding principle to obtain higher super-conducting transition temperature. Besides the exceptional case of β'-(BEDT-TTF)$_2$ICl$_2$ under extremely high-pressure condition,[91] organic superconductors with the transition temperature higher than 10 K, such as κ-(BEDT-TTF)$_2$Cu(NCS)$_2$[92] and κ-(BEDT-TTF)$_2$CuN(CN)$_2$Br,[93] have polymeric counter anion system. It is therefore a straight forward idea to use polymeric magnetic transition metal ion in the counterion system of molecular conductors.

The first discovery of superconductivity in organic magnetic conductors was achieved in (BEDT-TTF)$_4$[(H$_3$O)Fe(ox)$_3$]C$_6$H$_5$CN (ox = oxalate).[94,95] The structural feature of this material is the alternate quasi-two-dimensional layers of conducting organic donor molecules and magnetic MOFs. In the anion layers, $Fe(ox)_3^{3-}$ anions arranged in a triangular lattice are bridged with each other by oxonium ions by forming hydrogen bonds to noncoordinating oxygen atoms of the ligands. Inside the cavities of the honeycomb framework, benzonitrile molecules used as a solvent are incorporated (Fig. 1.6a). The transport measurements show the superconducting transition at the onset of $T_C = 8.5$ K. On the other hand, the magnetic interaction between the paramagnetic Fe^{3+} species is negligibly small (Weiss temperature = −0.2 K). The structure and properties of the counter-ion layer can finely be tuned by substituting H_3O^+ cation with NH_4^+ or K^+, or replacing solvent molecules with pyridine or nitrobenzene.[95–97]

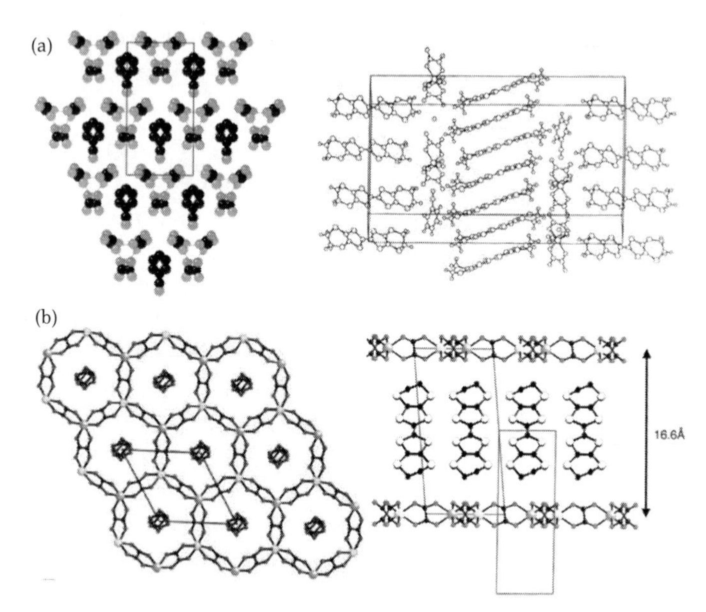

Figure 1.6 Crystal structure of (a) $(BEDT\text{-}TTF)_4[(H_3O)Fe(ox)_3]$[81,94] and (b) $(BEDT\text{-}TTF)_3[MnCr(ox)_3]$,[101] showing the structure of the anion layer and the donor-anion stacking patterns.

The structure of the anion layer in this system is similar to that found in bimetallic tris(oxalate) networks $[M^{II}M^{III}(ox)_3]^-$ (M^{II} = Mn, Fe, Co, Ni, Cu; M^{III} = Fe, Cr) which have ferro-, ferri-, or spin-canted weak ferromagnetic ordered states in relatively high temperatures depending on the selection of two transition metals.[98,99] Based on these findings, the coexistence of metallic conduction and ferromagnetism has been achieved in the $(BEDT\text{-}TTF)_x[MnCr(ox)_3]\cdot CH_2Cl_2$ ($x \approx 3$) system.[100] The structure of this salt consist of an organic donor layer and bimetallic honeycomb layer of $[MnCr(ox)_3]^-$ (Fig. 1.6b). Because of the strong self-assembling tendency in both donor and anion layers, this material is an incommensulate composite crystal and, therefore, nonstoichiometric.[101] The donor layers form stable metallic ground state, and the magnetic anion layer undergoes a long-range magnetic order transition at T_C = 5.5 K, below which a soft ferromagnetic behavior is observed. Since these magnetic features of this material are quite similar to that already reported for other $Mn^{II}Cr^{III}$-oxalato layer magnet, the π–d interaction between the donor and anion layers plays little role in the magnetism of this material. Nevertheless, this is an attractive system because it is possible to modify not only the transition metal of the anion layer but also the donor molecules of the conduction layer from BEDT–TTF to BETS[102,103] or BEDO–TTF,[103,104] etc., to give a large variety of compounds that exhibit metallic conductivity and ferromagnetism. Recently a chiral donor molecule (*S,S,S,S*)-tetramethyl-BEDT-TTF was introduced in the donor layer.[105] Unfortunately no evidence of the magnetochiral anisotropy[106] was observed, presumably because of the incoherency in the interlayer transport.

1.5.3 Introduction of Halogen–Halogen Contacts

In order to increase the magnitude of the π–d interaction, the introduction of specific intermolecular contacts is a useful strategy. For example, halogen–halogen contacts have a covalent character and are evidently stronger than van der Waals contacts; therefore, they are occasionally referred to as "halogen bonds" in analogy to hydrogen bonds.[107–109] Since the halogen–halogen interactions are anisotropic, they have mainly been adopted in molecular materials to control their crystal structures. We can also anticipate that these halogen–halogen contacts with a weak covalent character can mediate exchange interaction between the conduction and magnetic layers.

Radical ion salts $(EDT\text{-}TTFBr_2)_2FeBr_4$ and $(EDO\text{-}TTFBr_2)_2FeX_4$ ($EDT\text{-}TTFBr_2$ = dibromoethylenedithiotetrathiafulvalene, $EDO\text{-}TTFBr_2$ = dibromoethylenedioxotetrathiafulvalene, X = Br, Cl) have the same crystal structure composed of uniform Q1D donor columns and quasi-square lattice of magnetic anions.[110,111] As the halogen–halogen distances within anion layers are longer than the van der Waals distances, the intermolecular anion–anion interactions are regarded to be negligibly small. On the other hand, there are remarkably short halogen–halogen contacts between the bromide substituent of the donor molecules and the halide ligands of the counter anions, as indicated with dashed lines in Fig. 1.7a. Since the donor layers have Q1D characters, these salts show metallic behavior around room temperature and undergo metal–insulator transitions. The ground state of the π-electron system is characterized as SDW state similar to DMET[89] and EDTDM[90] salts, as discussed previously. The most pronounced effect of the halogen substitution in the donor molecule is the presence of magnetic ordered states. (EDO-$TTFBr_2$)$FeCl_4$ undergoes an antiferromagnetic phase transition at $T_N = 4.2$ K, which is higher than the transition temperature of $DMET_2FeCl_4$ (2.8 K). It should be noted that no significant Cl···Cl contacts are observed between the magnetic counter ions in (EDO-$TTFBr_2$)$_2FeCl_4$. The exchange interaction through the donor...anion contact therefore plays an essential role in magnetic ordering. The magnetic properties of (EDO-$TTFBr_2$)$_2FeBr_4$ are somewhat complicated. An antiferromagnetic transition occurs at $T_N = 13.5$ K, as a consequence of multiple exchange paths in the Fe^{3+} spin system via close intermolecular Br···Br contacts. Below the Néel temperature, there is another magnetic phase transition at $T_{C2} = 8.5$ K accompanied with complicated behaviors as shown in Fig. 1.7b, including the possibility of the helical spin system. One of the reasons for the complexity can be ascribed to the axial distortion of the magnetic $FeBr_4$ anions due to the steric effect of the remarkably short Br···Br contacts. This distortion is the origin of single-ion anisotropy when the ligand–metal bond bears covalent character and the Fe d-electron spins delocalize onto the halide ligands. Nevertheless, these high magnetic order temperatures (as molecule-based materials, of course) of EDO-$TTFBr_2$ salts indicate the effectiveness of the halogen–halogen contacts as exchange interaction paths.

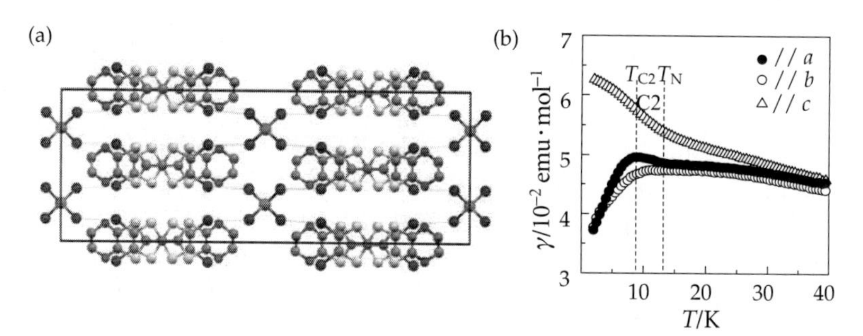

Figure 1.7 (a) Crystal structure and (b) magnetic susceptibility of (EDO-$TTFBr_2$)$_2FeBr_4$.[111]

1.5.4 π–*d* Coordinated System

Although the introduction of intermolecular contacts is an effective way to reinforce the π–d interaction between the conduction and magnetic parts, a more direct approach is to connect these two parts by covalent chemical bonds. This can be achieved by using either a coordination complex or organometallic compounds. In both approaches novel molecular design and synthesis are required, which is in general more difficult for the organometallic case owing to the stability of the compounds. Here, we discuss an example of the coordination complex in which there is distinct interplay between the conduction of the ligands and the magnetism of the transition metal.

The remarkable example of this approach is found in phthalocyanine (Pc) compounds. In the molecular conductor $[Cu_xNi_{1-x}(Pc)](I_3)_{1/3}$,[112–114] the π-electrons in the Pc ligand form one-dimensional conduction bands, which are coordinated to the paramagnetic Cu^{2+} ions with $S = 1/2$ localized moments (Fig 1.8a). Although some interesting phenomena such as negative magnetoresistance have been reported, further systematic studies of this system have not been conducted presumably owing to the low solubility of the starting materials. This difficulty has been somewhat reduced by capping the top and bottom of the metal–Pc unit by capping cyanide ligands. Since these axial ligands are small in comparison to the large Pc ligands, one-dimensional stacking of the metal–Pc unit is preserved. For example, in the crystal of $Ph_4P[Fe(Pc)(CN)_2]_2$, the metal–Pc units stack uniformly to form one-dimensional conduction bands composed of the π-electrons in the Pc ligand.[115] In addition, there are $S = 1/2$ localized moments on the central transition metal atom of this complex unit. In the absence of an external magnetic field, the resistivity of this sample increases with lower temperature[115,116] (Fig. 1.8b). Below around 50 K a large negative magnetoresistance is observed, whose magnitude depends on the direction of the applied field. When the field is applied perpendicular to the conducting chain at 20 K, the resistance at 18 T becomes less than 10% of the zero-field resistance. Since this large negative magnetoresistance is not observed in $Ph_4P[Co(Pc)(CN)_2]_2$ where there is no d-electron spin on the Co atom, this phenomenon is regarded as a consequence of the π–d interaction between the conduction electron on the ligand and the localized moment of the central metal atom. It should be noted that the negative magnetoresistance effect is observed up to 50 K, suggesting that the through-bond π–d interaction is stronger in one-order than in the through-space π–d interaction. Another interesting point is that the emergence of the negative magnetoresistance effect is not accompanied with the metal–insulator transition of the π-electron conduction layer. One possibility is the effect of strong electron correlation in low-dimensional π-electron systems, as theoretically discussed using the extended Hubbard–Kondo model[117,118] that was formerly applied to an organic high-spin polycarbene molecule.[52]

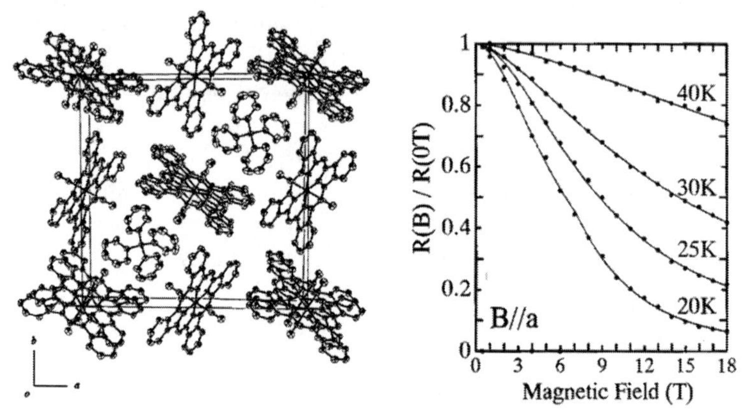

Figure 1.8 (a) Crystal structure[115] and (b) magnetoresistance[116] of $Ph_4P[Fe(Pc)(CN)_2]_2$.

1.6 Spin-Polarized Donor Radical as a Building Block of Organic Magnetic Conductor

This section revisits the electronic structure of molecular crystals of organic donors (Fig. 1.9). While a crystal of a neutral donor is an insulator, a crystal exhibits metallic conductivity when organic donors form a mixed valence state. In contrast, a crystal of a donor radical is basically a Mott insulator because the on-site Coulomb repulsion (U), which is the energy difference between two singly occupied states ($D^{\bullet} + D^{\bullet}$) and when one is doubly occupied and the other is empty (cation) ($D + D^{+}$), is larger than the transfer integral (t). To prepare a conducting crystal of organic radicals, a specific electronic structure must be designed, as depicted in the figure. First, the highest occupied molecular orbital (HOMO) of a donor radical must be energetically higher than the singly occupied molecular orbital (SOMO). Second, donor radicals are required to form a mixed valence state by π-doping, leaving the spin in SOMO intact. Third, two kinds of unpaired electrons in the singly oxidized donor radical, one in SOMO and the other left over in HOMO, must be exchange-coupled ferromagnetically. Then, the crystal is expected to manifest magnetism and conductivity in a cooperative manner.

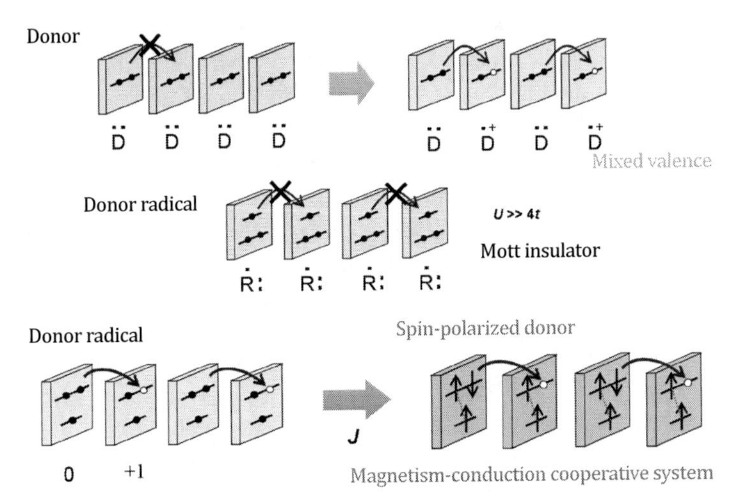

Figure 1.9 Electron transport in organic molecular crystals of donors in a neutral or a mixed valence state. Donor radicals in a neutral, a mixed valence state, or a mixed valence state with the exchange interaction in the singly oxidized state.

The exchange interaction of electrons in molecular orbitals is expressed by the following formula (Eq. 1.6).[119]

$$J_{1,2} = 2K_{1,2} - \{\varepsilon_i(1) - \varepsilon_j(2)\}^2/U \tag{1.6}$$

where $J_{1,2}$ is the effective exchange interaction between electrons (1 and 2) residing in MO_i and MO_j, respectively. The first term, $K_{1,2}$, is the static exchange interaction, and it corresponds to the molecular exchange integral between electrons (1 and 2), which stabilizes the triplet state. The second term is the kinetic exchange interaction, where $\varepsilon_i(1)$ and $\varepsilon_j(2)$ are orbital energies of MO_i and MO_j, and U is the on-site Coulomb repulsion in reference to the electron transfer from MO_i to MO_j. The second term stabilizes the singlet state. If two orbitals are degenerate, $\varepsilon_i(1) = \varepsilon_j(2)$, the second term is zero, which means $J_{1,2} = 2K_{1,2} > 0$. Then the ferromagnetic interaction is operative. Even though two orbitals are not degenerate, if the absolute value of the first term is larger than that of the second, the exchange interaction becomes ferromagnetic: it requires larger U compared with the orbital energy difference in the second term.

To create such an exotic electronic structure, Sugawara proposed a novel donor radical, the electronic structure of which is derived from trimethylenemethane (TMM, Fig. 1.10a).[120–122] The origin of a large ferromagnetic coupling of TMM is rationalized by the molecular orbital theory. TMM has two degenerated nonbonding molecular orbitals (Fig. 1.10b), which means the second term of Eq. 1.6 is null. Since the coefficients of SOMOs are shared by two carbon atoms of TMM, two unpaired electrons have a chance to come close, especially when they have opposite spins. This situation guarantees the large static exchange interaction, $K_{1,2}$. As a result, the triplet state of TMM is significantly more stable than the singlet. The energy difference between the triplet ground state (spin parallel) and the singlet excited state (spin antiparallel) is experimentally determined to be E_{TS} = 0.7 eV (67 kJ/mol).[123] Although Hund's rule is defined as the atomic orbital level, it may also be extended to degenerate or nearly degenerate molecular orbitals, if SOMOs are spatially shared. Thus, the extended Hund's rule can safely predict the spin configuration of TMM.

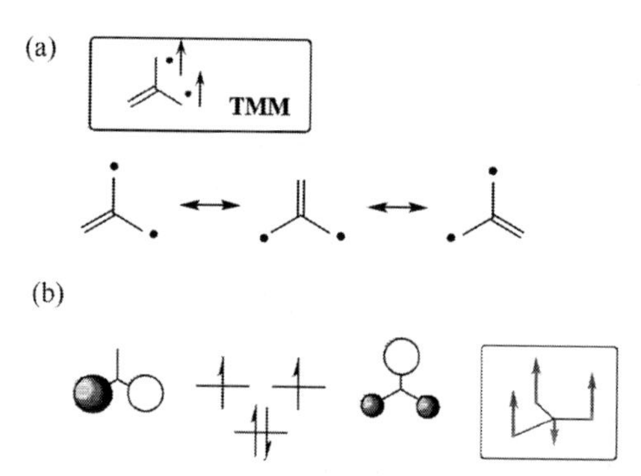

Figure 1.10 (a) Resonance structure of trimethylenemethane (TMM). (b) Degenerate molecular orbitals (SOMO) of TMM and spin-polarization in TMM. Up spin distributes at the peripheral carbons and down spin distributes at the central carbon.

A building block of a conducting magnet (spin-polarized donor in Fig. 1.9) is available by chemical modification of TMM: substitution of an allyl radical part of TMM with a stable π-conjugated radical, e.g., nitronyl nitroxide (NN), and a methylene carbon of TMM with a spin-carrying carbon atom of a singly oxidized donor unit. A precursor of the triplet diradical ($D^{+\bullet}$–R) is supposed to be a donor radical (D–R), in which a donor part and a radical part are connected in a cross-conjugated manner (Fig. 1.11a). Then, single oxidation of the donor radical yields a diradical ($D^{+\bullet}$–R), and the two electron spins, one derived from the radical part and the other from the donor part, must be coupled ferromagnetically. If such donor radicals are stacked in the mixed valence state, only a down-spin (β) in HOMO is able to transfer to the adjacent half-filled HOMO owing to the presence of the ferromagnetic interaction between these two spins in ($D^{+\bullet}$–R) (Fig. 1.11b). When paramagnetic fluctuating localized spins of radical units line up under an applied magnetic field, the scattering of conduction electrons is suppressed. Consequently, this salt is expected to exhibit a large negative magnetoresistance. A donor radical with such an electronic structure is designated as a "spin-polarized" donor radical.

(a)

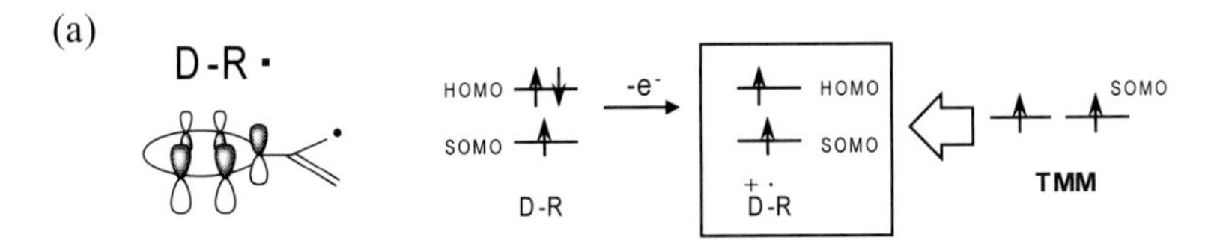

(b)

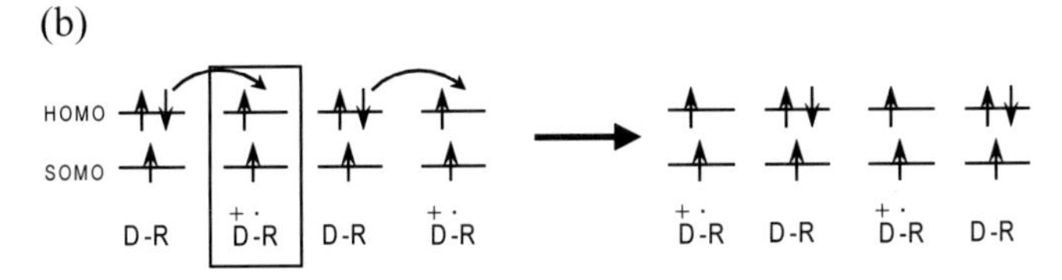

Figure 1.11 (a) Design of spin-polarized donor radical (D–R). Single electron oxidation gives rise to a ground state triplet cation diradical (b) Spin-dependent electron transfer along the alternating arrangement of D–$R^{\bullet}$ and $D^{+\bullet}$–$R^{\bullet}$.

1.7 Conductivity and Magnetism of TTF- and TSF-Based Donor Radicals

1.7.1 Molecular Design of a Building Block of Conducting Magnet

As a building block for such a ferromagnetic conductor, a series of TTF-based donor radicals (TTF–NN, and others) have been designed and synthesized as shown in Fig. 1.12.[124–127] Sugano *et al.* and Sugimoto *et al.* independently prepared a TTF donor connected with a tetramethylpiperidin-*N*-oxyl (TEMPO) radical through an imine linkage. Using this donor radical, they constructed charge transfer (CT) complexes with tetracyanoquinodimethane (TCNQ).[128,129] Ishida *et al.* prepared a pyrene derivative carrying a TEMPO radical.[130] Fujiwara and Kobayashi introduced a proxyl radical into a tetrathiapentalene (TTP) skeleton with a condensed ring structure.[131–133] Although some of the CT complexes and ion-radical salts of these donor radicals have exhibited semiconducting and paramagnetic properties,[134] no direct interaction between conduction electrons and localized spins was detected. This is partly because the exchange interaction between π-donor units and π-radical units through the σ-bond framework is not strong enough to be considered as a direct interaction.

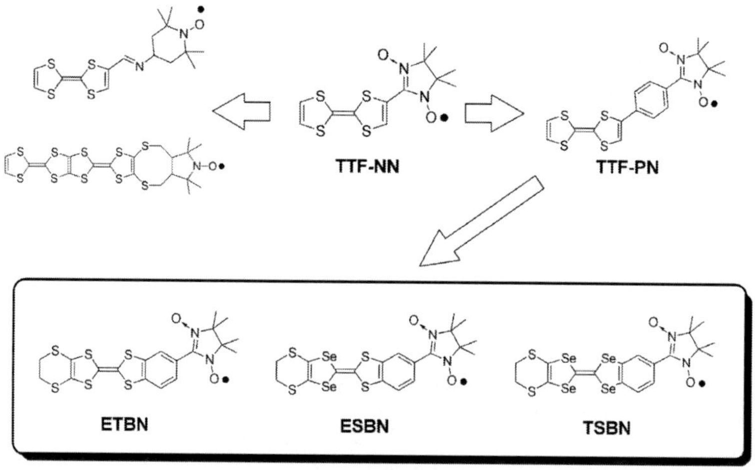

Figure 1.12 TTF-based donor radicals.

According to the discussion in Section 1.6, a TTF-based donor radical, **e**thylenedithio-tetra**t**hiafulvalene **b**enzo-derivative bearing **n**itronyl nitroxide (ETBN) (2-[2-(5,6-Dihydro[1,3]dithio [4,5-b][1,4]dithiin-2-ylidene)-1,3-benzodithiol-5-yl]-4,4,5,5-tetra-methylimidazoline-3-oxide-1-oxyl,[126,127] was synthesized as a spin-polarized donor radical. Replacement of two of the sulfur atoms (van der Waals radius = 1.85 Å) of ETBN with selenium atoms (van der Waals radius = 2.0 Å) afforded **e**thylenedithio-di**s**elenadithiafulvalene **b**enzo-derivative bearing **n**itronyl nitroxide (ESBN) (2-[2-(5,6-dihydro[1,3]diseleno[4,5-b][1,4]dithiin-2-ylidene)-1,3-benzodithiol-5-yl]-4,4,5,5-tetramethylimidazoline-3-oxide-1-oxyl).[135,136] The replacement is expected to cause larger overlap between adjacent donor planes. Further replacement of sulfur atoms with selenium gives rise to a TSF-base donor radical, ethylenedithio-**t**etra**s**elenafulvalene **b**enzo-derivative bearing **n**itronyl nitroxide) (TSBN) (2-[2-(5,6-dihydro[1,3]diseleno [4,5-b][1,4]dithiin-2-ylidene)-1,3-benzodiselenol-5-yl]-4,4,5,5-tetramethyl imidazoline-3-oxide-1-oxyl).[137]

1.7.2 Electronic Structure of ETBN and Physical Properties of Its Radical Ion Salt

It is worthwhile to describe precisely the electronic structure of ETBN. The electronic structures of the neutral and cation radicals of ETBN are shown in Fig. 1.13. One-electron oxidation of ETBN yields a ground state triplet cation diradical as shown in Fig. 1.13a for the following reason. The coefficients of SOMO are localized in the NN group, while those of HOMO are spread over the entire molecule, extending to the NN group. Thus, SOMO and HOMO of ETBN are of a space-sharing nature, which guarantees sufficient exchange interactions between the π spins in SOMO and HOMO. Moreover, the on-site Coulomb repulsion, U, of SOMO is reasonably large because SOMO is spatially constrained. The exchange interaction, J, is estimated as $2J = \sim 200$ K by using B3LYP/6-31G*. Due to the presence of this exchange interaction, the π-electrons in HOMO feel a spin-dependent exchange interaction with the π-localized spin in SOMO. As a result, when the SOMO electron has an up-spin (α), HOMO (β) is pushed up as shown in Fig. 1.13b, according to the spin-unrestricted description, wherein the wave functions of α and β spins are separately expressed. Hence, the ground state of a cation–diradical species becomes a triplet as shown in Fig. 1.13a.[127–138]

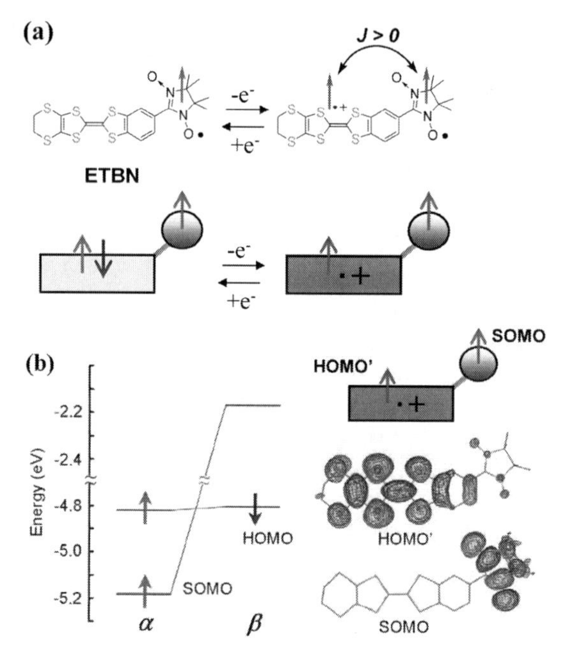

Figure 1.13 (a) Scheme of the generation of ground-state triplet cation-diradical species of ETBN upon one-electron redox process at the donor moiety. (b) Spin-polarized energy diagram of ESBN calculated by a DFT method (B3LYP/6-31G*). The space sharing nature of HOMO' and SOMO is depicted to the right: HOMO' is derived from one-electron oxidation of HOMO.

The benzo-TTF-based donor radical (ETBN) yields a crystalline ion–radical salt, $(ETBN)_2ClO_4(TCE)_{0.5}$ (TCE = 1,1,1-trichloroethane), through galvanostatic electrocrystallization.[135] Although an accurate crystal structure of the salt has not yet been fully elucidated, ESBN forms a columnar stack with donor planes overlapping in an inverted manner to avoid steric repulsion of the NN group: the NN groups alternate on the right and left sides (Fig. 1.14). Donor radicals in adjacent columns are aligned in a head-to-tail manner and there are van der Waals contacts between S···S atoms. To improve the π-overlap of the donor radicals in a column, it is necessary to introduce selenium atoms. Accordingly, a diselena analog of ETBN, ESBN, and a TSF-based donor radical, TSBN, were synthesized. Incidentally, the crystal structures of ESBN and TSBN are considered to be nearly the same on the basis of preliminary diffraction data.

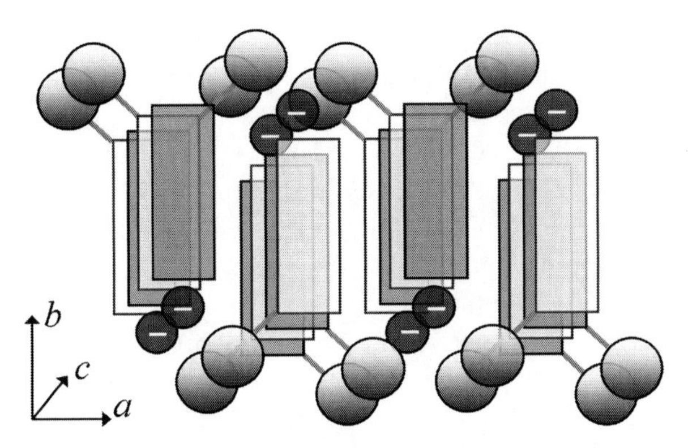

Figure 1.14 Schematic representation of crystal packing of $(ETBN)_2CLO_4(TCE)_{0.5}$.

The temperature dependence of the χT values of the ion–radical salt, $(ETBN)_2ClO_4(TCE)_{0.5}$, showed a paramagnetic behavior with a weak antiferromagnetic interaction of $\theta = -5.2$ K.[125] A Curie constant of 0.73 emu K mol^{-1} for a unit consisting of two donor radicals in the mixed valence state at room temperature indicates that all the radical sites remained intact during the electrocrystallization (Fig. 1.15). This result implies that the generated π-spins become conduction electrons and are coupled antiferromagetically with each other without contributing to the Curie constant. The magnetic properties of $(ESBN)_2ClO_4$ and $(TSBN)_2ClO_4$ were almost the same as those of $(ETBN)_2ClO_4(TCE)_{0.5}$, exhibiting a weak antiferromagnetic interaction of $\theta = -2.5$ and 2.1 K, respectively.

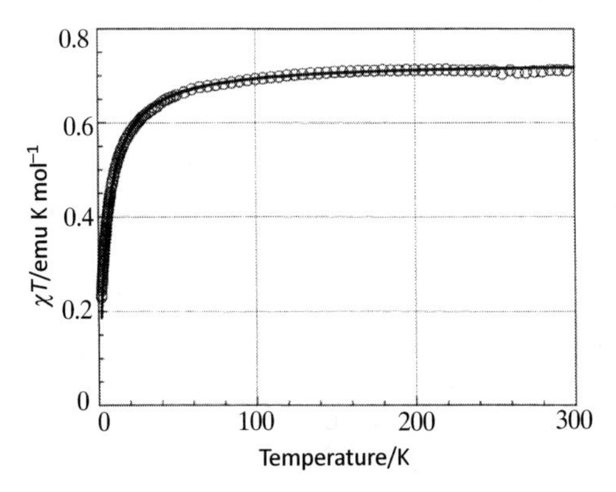

Figure 1.15 Temperature dependence of χT values of $ETBN_2ClO_4$ $(TCE)_{0.5}$.

The salt of $(ETBN)_2ClO_4(TCE)_{0.5}$ exhibits conductivity as well. It is a semiconductor with a room temperature conductivity of 1.5×10^{-4} S/cm. However, the activation energy of the salt (E_a = 0.16 eV) was still large and the temperature dependence could not be measured at temperatures lower than 100 K. Consequently, the direct interaction between conduction electrons and localized spins could not be detected. The low conductivity of $(ETBN)_2ClO_4(TCE)_{0.5}$ may originate from the dimerization and/or the charge localization of the donor column. Since conductivity is expected to increase by substitution of sulfur atoms by selenium, selena-analogs of ETBN, e.g., ESBN and TSBN salts, draw much interest in regards to the coexistence of magnetism and conductivity.

1.8 Magnetic and Conducting Properties of Ion Radical Salts of ESBN and TSBN

1.8.1 Nonlinear Conducting Behavior of $ESBN_2ClO_4$

The TTF-based donor radical, ESBN, yields a 2:1 radical ion salt through electrocrystallization. The crystal structure of $(ESBN)_2ClO_4$ is practically the same as the ETBN salt mentioned in Section 1.7.2.[135] The conductivity of a single crystal of the salt was measured by a two-probe method under constant current (0.5 μA). The activation energy at room temperature was approximately 47 meV, with $\sigma_{RT} = 6.1 \times 10^{-4}$ S cm^{-1}. The replacement of sulfur atoms with selenium in ETBN substantially decreased the activation energy, although the conductivity at room temperature was not remarkably high. The resistance of the salt was measured down to 2 K by using interdigitated electrodes with a gap of 2 μm in the constant voltage mode. The temperature dependence of the resistance could not be fitted by either an Arrhenius model or a variable range hopping model. Matsushita *et al.* claim that the resistance of the salt is proportional to T^{-7} in a wide temperature range from 300 to 50 K, although the origin of this unusual T dependence is not yet clear.[136] However, evident nonlinear I–V characteristics of $(ESBN)_2ClO_4$ were observed that obey a power law of $I \propto V^n$. With lowering temperature, the power n gradually increases from $n = 1.4 \pm 0.03$ at 280 K to $n = 12.0 \pm 0.5$ at 2 K with a bias voltage of $V = 4.5$ V. If 5 V is applied to the crystal across a distance of 2 μm, the applied voltage per molecule is ~1 mV, assuming that the depth of the donor radical is 0.4 nm. Since this applied voltage per molecule does not exceed the activation energy for electron migration, the observed nonlinear *I*–*V* characteristics cannot originate from a Zener breakdown.

Recently it was reported that an ion–radical salt with a 2:1 composition of donor and counterion, $(ET)_2CsZn(SCN)_4$, exhibits similar nonlinear characteristics, obeying the power law of $I \propto V^n$ (n = 8.4 at 0.29 K).[139] For the case of $(ET)_2CsZn(SCN)_4$, the nonlinear characteristics are explained by the electric field–induced unbinding of electron–hole pairs with a potential barrier of $\Delta_{(E)}$ in the background of the charge-ordered state, where the resistance of the salt is expressed as an Arrhenius type by

$$R = R_0 \exp(\Delta_{(E)}/k_B T) \tag{1.7}$$

where $\Delta_{(E)}$ is an effective potential barrier. Since the effective potential barrier of $\Delta_{(E)}$ decreases with the increasing applied electric field, a high power law in the *I*–*V* characteristics emerges. This effect becomes more prominent at lower temperatures.

The difference in the conductivities of $(ETBN)_2ClO_4(TCE)_{0.5}$ and $(ESBN)_2ClO_4$ is discussed on the basis of the above mentioned experimental results. Although the donor to the counter-ion ratio of $(ETBN)_2ClO_4(TCE)_{0.5}$ salt is 2:1, the conductivity of the salt is low (σ_{RT} = 1.5 × 10^{-4} S cm^{-1}) and the activation energy is as large as 160 meV. This suggests that the intermolecular transfer integral (t) between donor radicals is small compared with the Coulomb repulsion, presumably owing to the presence of the bulky groups (geminal dimethyl groups) of the radical site. To raise t, two sulfur atoms of the dithiole ring in ETBN were replaced with selenium. Subsequently, the activation energy of $(ESBN)_2ClO_4$ decreased considerably (E_a = 47 meV). This change is consistent with the interpretation that the charges carried in these ion-radical salts are stuck owing to a Coulomb repulsion that is much larger than t. Although the crystal structures of these salts have not been fully elucidated, there is no sign of strong dimerization of donors in preliminary X-ray crystal analyses. Moreover, the paramagnetic behavior of the spins of the radical sites in $(ESBN)_2ClO_4$ may also exclude the possibility of strong dimerization of donor radicals in these salts. Hence, it may be concluded that the Coulomb repulsion in this salt is not derived from U but from the nearest neighbor Coulomb repulsion (V).

There have been studies related to the melting of the charge-ordered state by an electric field. Matsushita and Sugawara[140] reported on the current-induced transformation from a high-resistance state to a low-resistance state of the ion-radical salt of a TTF-based dimeric donor (CPD) in which two TTF units are linked through trimethylene chains (Fig. 1.16a,c). Although individual donors in the ion–radical salt CPD·Br$(TCE)_2$ exist in the charge disproportionate state (D and D^+) at ambient conditions, the charge disproportionation disappears under application of high voltage and the bond distances of the individual donors become nearly equal. This current-induced structural transformation corresponds to the switching from the class I to the class III mixed valence state (Fig. 1.16b).

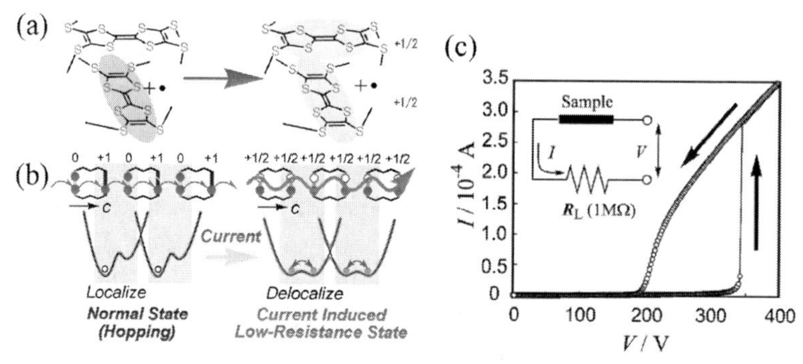

Figure 1.16 (a) Charge disproportionated state of CPD $Br(TCE)_2$ under ambient condition was melted by application of high voltage and the structure becomes almost equivalent. (b) Schematic drawing of the charge-disproportionated state and the current-induced low-resistance state of the partially oxidized donor column in the crystal of CPD $Br(TCE)_2$. Transformation from type I to type III mixed valence state. (c) Current-induced low resistance state with hysteretic behavior.

1.8.2 Variable Range Hopping Mechanism of $TSBN_2ClO_4$

Further substitution of the remaining sulfur atoms of ESBN by selenium increased the conductivity of the ion radical salt of TSBN. The conductivity of the salt of was 6 S/cm at room temperature; it was increased by four orders of magnitude, and the conductivity could be measured by a four-probe method in a constant current mode (1 μA) down to 5 K[141] (Fig. 1.17). The activation energy is only

11 meV, which is smaller than the thermal energy of room temperature. It turned out that the plot in the temperature range of 300–315 K can be fitted by a three-dimensional variable range hopping model.[142] This conducting behavior is a strong proof that the conductivity of the salt is controlled also by the nearest neighbor Coulomb repulsion *V*, but *V* can be overdriven by thermal energy in this salt.

We will notice that the temperature dependence deviates heavily from the extrapolated curve at lower temperatures than 10 K, since the interaction between conduction electrons and localized spins becomes appreciable and the conduction electrons are scattered by the disordered alignment of localized spins.

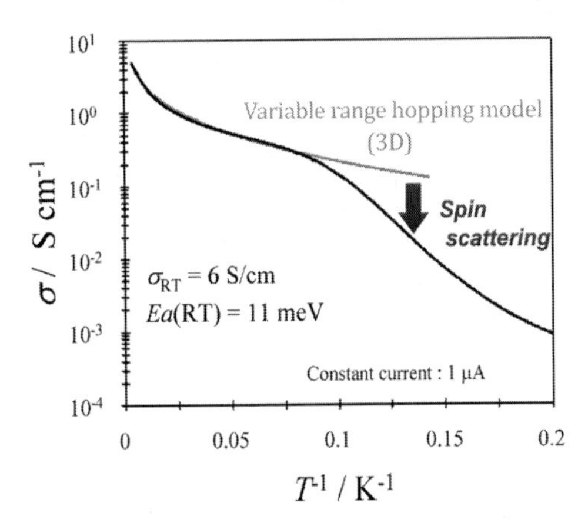

Figure 1.17 Arrhenius plot of the conductivity of $TSBN_2ClO_4$. Dotted line demonstrates the calculated curve of hopping model. The conductivity was decreased by fluctuating radical spins at temperatures lower than 15 K.

1.9 Magnetotransport Properties of Ion Radical Salts of TTF- and TSF-Based Donor Radicals

1.9.1 Magnetoresistance of $ESBN_2ClO_4$ Observed in Nonlinear Conductance Region

The magnetoresistance of $(ESBN)_2ClO_4$ salt was measured using a single crystal mounted on the interdigitated Au electrodes with a constant bias voltage of 7 V by applying magnetic fields in a range of −9 to 9 T: The magnetic field was applied along the parallel direction to the current to suppress the effect of Lorentz force. As shown in Fig. 1.18a, the resistance decreased with increasing magnetic field strength. The negative magnetoresistance, $(R_H - R_0)/R_0$, amounted to −70% under 9 T at 2 K, where R_H is the resistance under the magnetic field and R_0 is the resistance at 0 T. The absolute value of $(R_H - R_0)/R_0$ at 9 T monotonically decreased with increasing temperature and it became almost negligible at 20 K.[136]

These data indicate that the degree of the exchange interaction between organic localized spins and conduction electrons is *ca.* 20 K, which is much smaller than the intramolecular J ($2J \sim 200$ K) of the cation diradical of $ESBN^{+\bullet}$. This difference can be understood by the fact that the hole is delocalized over several molecules in the ion–radical salt. Figure 1.18b shows that the magnetic-field dependence of the resistance of $(ESBN)_2ClO_4$ at 2 K is proportional to the square of the magnetization of an organic radical with $S = 1/2$ against the applied field. The field-dependent profile of magnetoresistance can be well-reproduced by the square of the magnetization. This is good evidence for the interpretation that the observed magnetoresistance is derived from the magnetization of the spins of organic radical units.

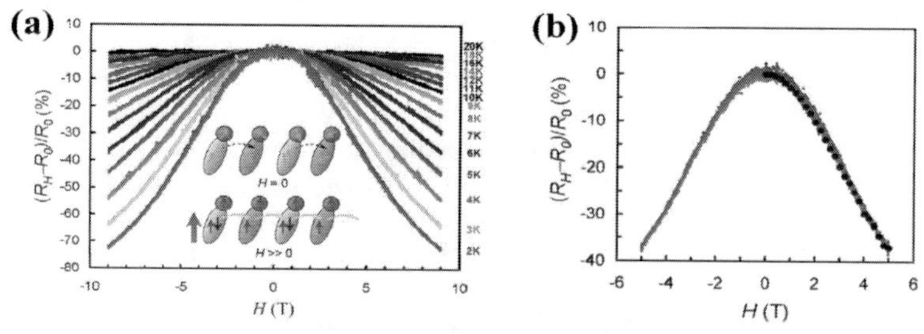

Figure 1.18 (a) Temperature dependence of the ratio of magnetoresistance of $(ESBN)_2ClO_4$ at a bias voltage of 7 V. (b) Magnetic field dependence of magnetoresistance of $(ESBN)_2ClO_4$ at 2 K at a bias voltage of 10 V (red line). The black dots represent a square of the magnetization with a scaled prefactor. See also Color Insert.

Figure 1.19 schematically shows a plausible mechanism of the appearance of the giant negative magnetoresistance in $(ESBN)_2ClO_4$ at temperatures lower than 20 K, which is associated with a nonlinear conducting behavior. Under the low magnetic field (Fig. 1.19a), the mixed valence donor column falls into the charge-ordered state (D and D^+). In such a case, only an electron with a certain spin (e.g., β spin) of D migrates to the neighboring molecule (D^+) with an α localized spin because a π-spin in HOMO of D^+ is an α spin and can accept an electron with a β spin (Fig. 1.11b). Under the high electric field (Fig. 1.19b), the charge-ordered state partially disappears, causing a nonlinear conducting behavior, and then, the coherence length of the conducting electron becomes longer. Finally, when the magnetic field is applied (Fig. 1.19c), most of the localized spins are aligned, and the scattering of the conducting electron is suppressed, causing a large negative magnetoresistance. It has been reported that the thermally activated conduction behavior and the distinct negative magnetoresistance were also found in the ferric–Pc complex as discussed in Section 1.5.4.

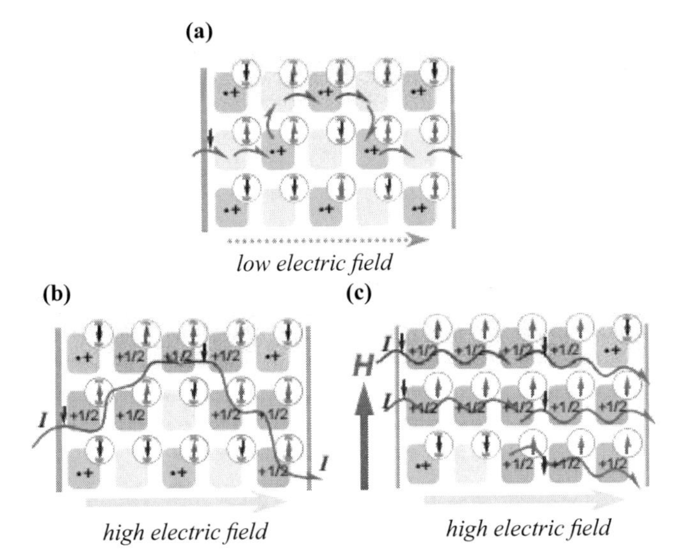

Figure 1.19 Schematic representation of a plausible mechanism of the appearance of giant negative magnetoresistance in $(ESBN)_2ClO_4$ with a nonlinear conducting behavior under (a) low electric field, (b) high electric field, and (c) high electric field with application of a high magnetic field. See also Color Insert.

1.9.2 Magnetoresistance Observed in $(TSBN)_2ClO_4$

Contrasting to the case of ESBN salt, the magnetoresistance effect of TSBN salt, $(TSBN)_2ClO_4$, could be measured using the nearly ohmic current at lower temperatures than 15 K.[143] The switching ratio of the resistivity of $(TSBN)_2ClO_4$ was 50% when measured by on and off external magnetic fields of 5 T at 5 K in a constant current mode of 0.1 μA. It is found that the bell-shaped magnetic field dependence of the magnetoresistance is related to the magnetization of localized spins as in the case of $(TSBN)_2ClO_4$. It suggests that the origin of the negative magnetoresistance is, indeed, due to the alignment of local spins.

1.9.3 Pressure Effect on Magnetotransport of $(TSBN)_2ClO_4$

The external pressure effect on the temperature dependence of the resistivity of the salt was measured in a constant current (5, 10, 15 μA) mode. The room temperature conductivity increased from 14 to 45 to 80 S cm^{-1} by increasing the external pressure from ambient to 7.5 to 15 kbar (Fig. 1.20). As seen in Fig. 1.20, the salt exhibited quasi-metallic conductivity in a temperature range of 290–270 K under 7.5 kbar, and it became more prominent under 15 kbar.[141] The temperature dependence, however, gradually transformed to thermally activated behavior at a temperature lower than about 70 K, presumably owing to the small gap between the conduction band and the valence band owing to the presence of the nearest neighbor Coulomb repulsion (V) in the charge disproportionation regime. In any case, this is the first demonstration that the ion–radical salt of an organic donor radical exhibits quasi-metallic conduction.

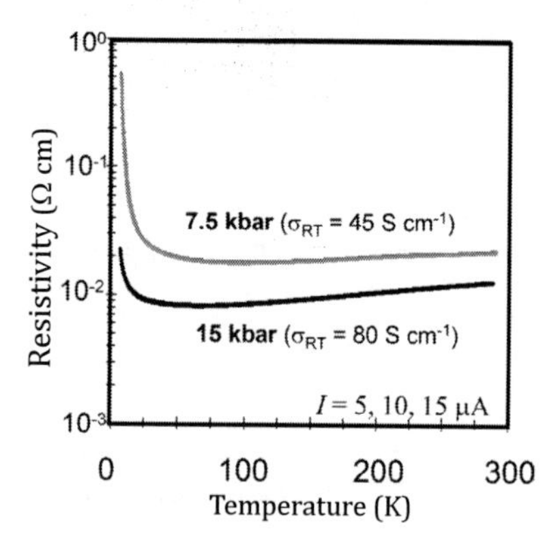

Figure 1.20 Temperature dependence of resistivity of $(TSBN)_2ClO_4$ under 7.5 and 15 kbar.

Magnetic field effect on the temperature dependence of resistivity of the salt under at a pressure of 15 kbar below 30 K is shown in Fig 1.21a. The temperature dependence in this temperature range is of a thermal activation type and the resistance increases sharply at temperatures lower than 15 K. We notice that the increase of the resistivity is almost suppressed under 12 T and the decrease in the resistivity amounts to 1/200 under 12 kbar (ρ_{12T}/ρ_{0T} = 1/200) at 2 K, 12 T.[141] The magnetic field dependence of the ratio of negative magnetoresistance is plotted in Fig 1.21b. It shows a sharp drop and is saturated at higher magnetic fields than 9 T. These results strongly suggest that conductivity of this salt is sharply controlled by the application of the external magnetic field.

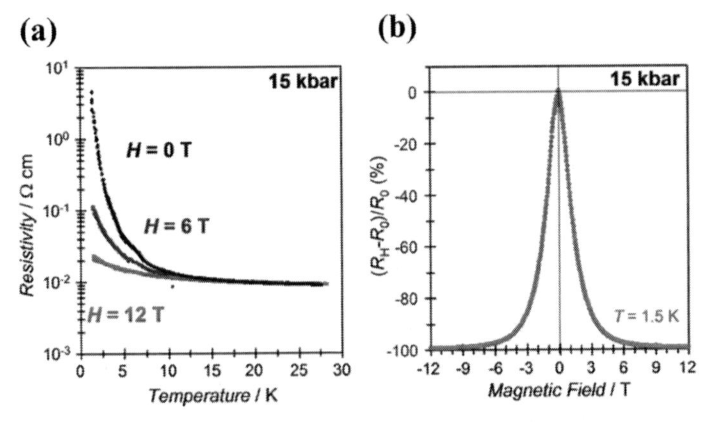

Figure 1.21 (a) Magnetic field effect on the temperature dependence of resistivity of $(TSBN)_2ClO_4$, (b) Negative magnetoresistance of $(TSBN)_2ClO_4$ under pressure. See also Color Insert.

The mechanism of negative magnetoresistance of this ion radical is speculated as follows. This is a 2:1 ion radical salt and the neutral (D) and the cation radical (D^+) are located side-by-side, suffering the nearest neighbor Coulomb repulsion V due to the presence of some disproportionate state. The donor radical is basically a paramagnet with a small antiferromagnetic interaction ($\theta = -2$ K) in this salt, although two types of spins in D^+ are strongly coupled ferromagnetically. Since the electron migration occurs through a double-exchange mechanism, only a β-conduction electron can transfer to D^+ with an α localized spin because HOMO of this molecule has a vacant seat only for β-spin, and *vice versa*. As the magnetic field increases, the probability of finding the α localized spin sites increases. As a result, the passing current through the crystal increases to show the giant negative magnetoresistance. Since the electron transportation between disproportionate sites is considered to occur through a hopping mechanism (Section 1.8.2), the conduction electron sensitively feels the spin regulation at lower temperatures. The presence of a weak antiferromagnetic interaction between localized spins also contributes to the observed giant magnetoresistance.

1.10 Manipulation of Unicomponent Crystal of TTF-Based Donor Radical by Electric and Magnetic Fields

1.10.1 Design of BTBN and Its Self-Assembling Ability

It would be interesting if we could observe the interplay between organic localized spins and positive carriers (holes) injected into a neutral crystal (without counter ions) from an electrode. The transport of charge carriers may be carried out through the space-charge-limited conduction mechanism,[144,145] as in the case of the electron transport mechanism in electroluminescence (EL) devices. When the injected carriers from an electrode are accumulated in an insulator or semiconductor sample, the electric current that flows through the sample becomes proportional to V^2. Such electrical conduction is designated as "space-charge-limited conduction" (SCLC without trap levels). The principle is more precisely explained as follows: the current (I), which flows through a material, is proportional to the product of the number of carriers (n) and the mobility of the carrier (μ). In a regular conducting material, the current obeys Ohm's law. To the contrary, in the above case, the number of carriers also depends on the voltage. Hence the current is proportional to V^2. Since the transport time, τ, which requires to pass through the sample, is expressed as $\tau = d^2/\mu V$, the current density (J_{SCLC}), which corresponds to the current divided by an area of the electrode, is expressed by the following Eq. 1.8.

$$J_{SCLC} = \frac{Q}{\tau} = \frac{Q\mu V}{d^2} = \frac{-\varepsilon\varepsilon_0 \mu V}{d^3} = \propto \mu V^2 \tag{1.8}$$

where Q and ε denote space charge and permittivity, respectively.

On the other hand, a real material contains trap levels; then carriers are populated in these trap levels (SCLC with trap levels); trapped carriers are in a thermal equilibrium with free carriers through electron–phonon coupling in region (a) in the Fig. 1.22 below. Hence the transport time becomes longer compared with the case without traps, but the current increases in proportion to V^2 in this region. The trap in the SCLC is derived from inherent lattice defects and impurities. However, it is to be noted that a neutral donor or an acceptor itself traps the carrier if the structural relaxation to a cation radial or an anion radical, accompanied by the salvation with the surrounding molecules, occurs before delivering the charge to the adjacent molecule.

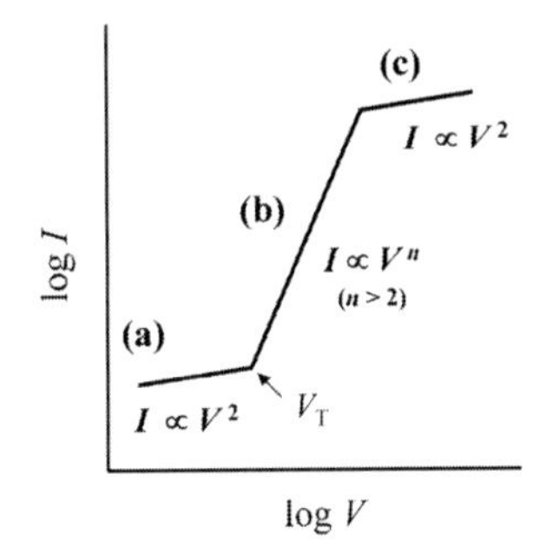

Figure 1.22 I–V characteristic exhibited by a molecular crystal sandwiched by electrodes SCLC region: I is proportional to V^2, (b) TCLC region: I is proportional to $V^{n,}$ (c) trap-filled region: I is proportional to V^2.

When the voltage between the electrodes exceeds a certain threshold value (V_T), carriers are irreversibly trapped because the Fermi level of the sample with the injected electrons is raised and becomes gradually higher than the trap levels. As a result, the current increases abruptly in region (b). This region is called the "trapped charge limited conduction" (TCLC). Eventually all the trap levels are irreversibly occupied and the I–V^2 characteristics is restored (region c).

Although it has been reported that organic thin films, e.g., polythiophene, of neutral compounds without persistent spins exhibited negative magnetoresistance, the mechanism is thoroughly different and it depends on the spin-dependent recombination of excitons under a magnetic field. The theoretical maximum value of the magnetoresistance is 50% for this mechanism.[146,147]

This section describes that a neutral crystal of a novel donor radical (BTBN) of a dibromo-TTF skeleton bearing a NN group exhibited a giant negative magnetoresistance through the SCLC mechanism by means of hole-injection from an electrode. The dibrominated TTF-based donor radical, di**b**romo-substituted tetra**t**hiafluvalene **b**enzo-derivative bearing **n**itronyl nitroxide (BTBN) (2-[2-(4,5-dibromo-[1,3]dithiol-2-ylidene)-1,3-benzodithiol-5-yl]-4,4,5,5-tetramethylimidazoline-3-oxide-1-oxyl, was designed and synthesized so as to increase the intercolumnar interactions in a crystal by introducing two bromine atoms into the dithiole ring (Fig. 1.23a).[148,149] A differential pulse voltammogram of BTBN in benzonitrile in the presence of 0.1 M tetrabutylammonium perchlorate showed redox potentials at 0.82, 0.90, and 1.16 V vs. Ag/AgCl, respectively. The first and third potentials correspond to the redox of the donor unit and the second corresponds to that of the NN group. Hence, the data suggest that BTBN satisfies the requirement of the spin-polarized electronic structure.[136]

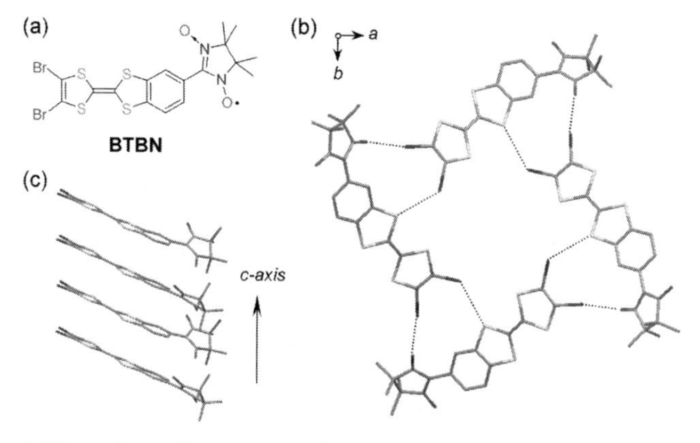

Figure 1.23 Crystal structure of BTBN. (a) Structural formula of BTBN. (b) Molecular arrangement viewed along the *c*-axis. (c) Columnar stack along the *c*-axis.

The crystal packing of a needle crystal of BTBN is characterized as follows (Fig. 1.23b,c). Four BTBN molecules are arranged in a square-like manner, the long molecular axes being orthogonal to each other, and the tetrameric BTBN stacks along the *c*-axis with interplanar distances of about 3.46 Å. The donor planes are overlapped in an inverted manner so as to locate the NN group to the right or left side alternately to avoid steric repulsion. Neighboring stacks are connected with close Br···S contact of approximately 3.6 Å between one of the bromine atoms of the dithiole ring and a sulfur atom of the benzodithiole ring in the adjacent stack, while the other bromine atom has close Br···O contact of around 3.0 Å with an oxygen atom of the O–N unit in the NN group of the neighboring molecule. Close contact between an O–N unit and a halogen atom has been observed in a few crystals of nitroxide derivatives.[150,151] It is also to be noted that the planar stack of the donor units in the neutral crystal is appropriate for the conduction of π-electrons.

Reflecting the black color of the crystal, the CT band of the polycrystalline sample extended over to approximately 1400 cm^{-1} (the band edge is estimated to be about 0.17 eV), suggesting a narrow gap between the valence and conduction bands. The charge-transfer absorption band in the UV-Vis-IR spectrum of the polycrystalline BTBN extended from 10,000 to 1400 cm^{-1}, suggesting the possibility of a self-doped electronic structure, presumably owing to excitation from the valence band to the degenerated SOMO levels.

1.10.2 Conducting Behavior of BTBN Through SCLC Mechanism

The room temperature conductivity, which was measured by the four-probe method, of the neutral single crystal of BTBN was as high as 9×10^{-4} S cm^{-1}. However, since the neutral crystal is a semiconductor of the activation energy of *ca.* 0.28 eV, the degree of resistance of the crystal exceeded the lower limit of the measurement below 240 K.

The I–V characteristics of microcrystalline sample were measured by interdigitated electrodes in a temperature range of 290–292 K and the I–V characteristic was expressed by an $I \propto V^{m+1}$ law as shown in Fig. 1.24a. The order (m) of the V-dependence in the $\log I$–$\log V$ plot at each temperature was determined from the slope of the plot: $m = 2$ at 290 K and $m = 15$ at 2 K, etc.

Provided that the conduction of the BTBN crystal is caused by the SCLC mechanism, the abrupt increase of m is interpreted by the TCLC.[144,145] From the m–T^{-1} plot as shown in Fig. 1.24b, the characteristic temperature, T_C, was determined to be 125 K; T_C is a slope of the m–T^{-1} plot and it represents a temperature characterizing the exponential distribution of traps. At lower temperatures than 10 K, the plot was deviated from the straight line, suggesting that the interaction with localized spins at the radical sites prevailed at lower temperatures. This tendency is rationalized by the magnetoresistance measurement. When the voltages (V) that give the same magnitude of I in the I–V characteristics of BTBN at 2 K are plotted against the gap distance (d) of interdigitated electrodes, e.g., 2, 5, and 8 μm, V turns out to be proportional to d^2. On the basis of the m–T^{-1} and V–d^2 dependence, it was concluded that the nonlinear conductance of BTBN arose from the TCLC mechanism.

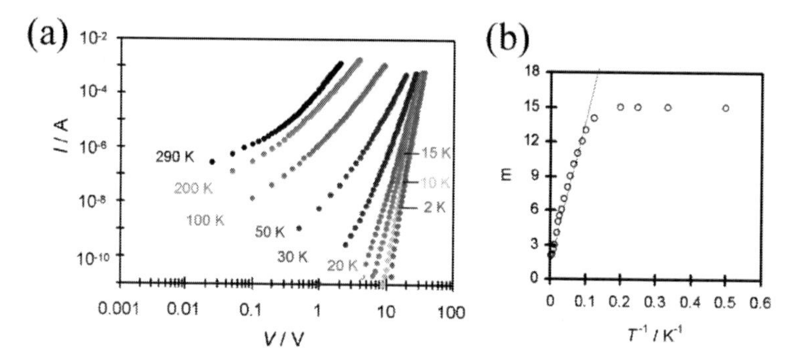

Figure 1.24 (a) Nonlinear I–V characteristics of BTBN in a temperature range of 290–292 K (b) Temperature dependence of m (m–T^{-1} plot).

1.10.3 Magnetoresistance of BTBN

The effect of an external magnetic field on the resistance of microcrystals of BTBN was measured with a bias voltage of 10 V ($E = 5 \times 10^4$ V cm^{-1}) by applying a magnetic field in a range from −5 to 5 T.[149] The resistance of the sample decreased under a magnetic field below 30 K, exhibiting a negative magnetoresistance of $(R_H - R_0)/R_0 = -76\%$ at 2 K under 5 T (Fig. 1.25). The disappearance temperature (around 30 K) of the magnetoresistance corresponds to the degree of interaction between the conduction electrons and the organic localized spins. When the magnetic field dependence of the magneto-resistance of BTBN is compared with that of $(ESBN)_2ClO_4$ (Fig. 1.18),[136] the former plot is appreciably more sensitive to the magnetic field than that of the previously reported one. This is presumably because the conduction electrons in BTBN interact with a cluster of ferromagnetically coupled spins of the one-dimensional chain due to the ferromagnetic interaction.

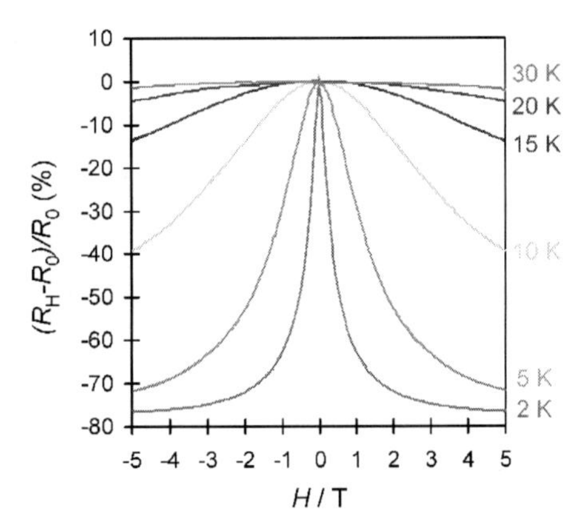

Figure 1.25 Magnetic field dependence of magnetoresistance of BTBN at various temperatures at bias voltage of 10 V.

As mentioned above, the temperature dependence of the magnetic susceptibility of BTBN was well-rationalized by a one-dimensional ferromagnetic chain along the c-axis with $J_{\text{intra}}/k_{\text{B}} = 6.5$ K and a weak antiferromagnetic interchain interaction of $J_{\text{intra}}/k_{\text{B}} = -1.1$ K, where the number (z) of neighboring chains is 2 (Fig. 1.26).[152] Curie constant (χT value) is 0.375 emu K mol^{-1} at room temperature. The ferromagnetic interaction may be derived from the overlap between NN groups along the chain: the NO unit of the NN group is close to the middle carbon of the NN group of the adjacent molecule (Fig. 1.26 right).

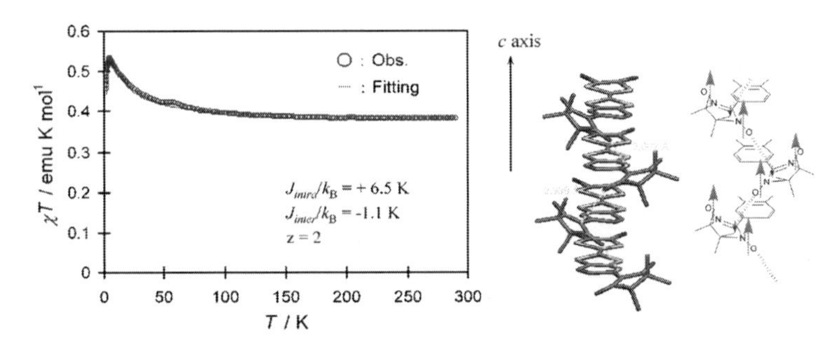

Figure 1.26 Temperature dependence of magnetic susceptibility of BTBN (blue circle). The red line is calculated with the parameters ($J_{\text{intra}}/k_{\text{B}} = 6.5$ K, $J_{\text{inter}}/k_{\text{B}} = -1.1$ K, and $z = 2$). Columnar stacking of the radical units is shown, together with a schematic drawing of magnetic interaction between radical sites. See also Color Insert.

The overall situation of the magnetic field effect on conduction electrons can be explained as follows. When a BTBN crystal is hole-doped by a positive electrode, it exhibits nonlinear conductance caused by TCLC along the columnar donor stack. The current, however, is scattered by the fluctuating paramagnetic spins of the radical sites at higher temperatures. As the external magnetic field increases, the localized spins align in parallel more efficiently than in the salts, as discussed in Section 1.9.1, owing to intracolumnar ferromagnetic coupling. Hence only electrons with the opposite spin can pass through without being scattered, which decreases the resistance of the crystal at temperatures lower than 30 K. A control experiment using crystals of the nonradical acetal precursor of BTBN showed that the nonlinear conductance was attributable to the TCLC, but the magnetoresistance was positive (+1.5% at 2 K under a field of 5 T).

The unicomponent crystal of BTBN became conductive through the injection of holes from the electrodes even at lower temperatures and the conductance of BTBN was controllable by the external magnetic field at temperatures lower than 30 K. This material can be fabricated as a thin film, and exhibit both magnetism and conductivity. It is to be stressed that the conductivity of the neutral molecule-based material can be manipulated by electric and magnetic fields.

1.11 Summary and Future Scope

The coexistence of conductivity and magnetism has been an issue of much concern since the mid-20th century. The *s*–*d* and *s*–*f* interactions in ferromagnetic diluted alloys have been studied extensively. The mechanism of magnetic interaction between localized spins and conduction electrons was discussed by using the Stoner model and the RKKY mechanism. The reason for the presence of a minimum in the temperature dependence of the conductivity was theoretically elucidated by Kondo, and this so-called Kondo effect has been extensively studied by many scientists. Today's hot topic is the construction of spintronic devices.[153] Electronic devices and advancements of the next generation are expected to include spin valves,[154] the detection of spin-polarized currents,[155] exploration of a memory device,[156] and quantum computers.[157]

Accompanied by the rapid progress of materials science in the past two decades, many new materials have emerged, and the stage for the performance of cooperative dynamics of localized spins and conduction electrons has expanded to include metal oxides, graphenes, and metallic nanoparticles. Among molecular materials, transition metal complexes have been actively investigated. Moreover, even organic molecular crystals that exhibit cooperative interplay between localized spins and conduction electrons have been realized.

The appearance temperature of magnetoresistance for organic materials presently seems to be too low for practical purposes. However, it should be recalled that the magnetoresistance discovered in ferromagnetic metals was only −45% at 4.2 K in the first report even though Fe/Cr/Fe nanolayers were used, the Fe layers being coupled antiferromagnetically through a sandwiched Cr layer.[158,159] The appearance temperature of organic materials could be raised by molecular design because this temperature is controlled by the intramolecular ferromagnetic coupling between unpaired electrons and conduction electrons.

Lastly, we mention the merits of molecule-based devices, which in general lie in the ease of fabrication, such as printing a device structure on demand, and the ease of disposal, such as incineration of used devices. Moreover, a more fundamental characteristic is that a molecule can be the ultimate quantum device, which by itself exerts electronic functions by a single electron or a single spin, since it is the smallest entity with an independent geometry and a quantized electronic structure. The spintronic function of spin-polarized molecules will soon be elucidated by means of single molecule conductance measurements using STM[160] or nano-gap electrodes.[161] It is to be stressed that the interplay between the two kinds of electrons discussed in this chapter is crucial for the development of molecule-based spintronics.

References

1. Stoner, E. C. (1938). Collective electron ferromagnetism, *Proc. Roy. Soc. A*, **165**, 372–414.
2. Ruderman, M. A., Kittel, C. (1954). Indirect exchange coupling of nuclear magnetic moments by conduction electrons, *Phys. Rev.*, **96**, 99–102.

3. Kasuya, T. (1956). A theory of metallic ferro- and antiferromagnetism on Zener's model, *Prog. Theor. Phys.*, **16**, 45–57.
4. Yosida, K. (1957). Magnetic properties of Cu-Mn alloys, *Phys. Rev.*, **106**, 893–898.
5. Mattis, D. C. (2006). *The Theory of Magnetism Made Simple*, World Scientific, Singapore.
6. Moriya, T. (1965). Spin polarization in dilute magnetic alloys, *Prog. Theor. Phys.*, **34**, 329–356.
7. Kimura, T., Tomioka, Y., Kuwahara, H., Asamitsu, A., Tamura, M., Tokura, Y. (1996). Interplane tunneling magnetoresistance in a layered manganite crystal, *Science*, **274**, 1698–1701.
8. Kiryukhin, V., Casa, D., Hill, J. P., Keimer, B., Vigliante, A., Tomioka, Y., Tokura, Y. (1997). An X-ray-induced insulator–metal transition in a magnetoresistive manganite, *Nature*, **386**, 813–815.
9. Imada, M., Fujimori, A., Tokura, Y. (1998). Metal-insulator transitions, *Rev. Mod. Phys.*, **70**, 1039–1263.
10. Jonker, G. H., van Santen, J. H. (1950). Ferromagnetic compounds of manganese with perovskite structure, *Physica*, **16**, 337–349.
11. Zener, C. (1951). Interaction between the d-shells in the transition metals. II. Ferromagnetic compounds of manganese with perovskite structure, *Phys. Rev.*, **82**, 403–405.
12. Anderson, P. W., Hasegawa, H. (1955). Considerations on double exchange, *Phys. Rev.*, **100**, 675–681.
13. de Gennes, P. G. (1960). Effects of double exchange in magnetic crystals, *Phys. Rev.*, **118**, 141–154.
14. Bozorth, R. M., Wolff, P. A., Davis, D. D., Compton, V. B., Wernick, J. H. (1961). Ferromagnetism in dilute solutions of cobalt in palladium, *Phys. Rev.*, **122**, 1157–1160.
15. Crangle, J., Scott, W. R. (1965). Dilute ferromagnetic alloys, *J. Appl. Phys.*, **36**, 921–928.
16. Low, G. G., Holden, T. M. (1966). Distribution of the ferromagnetic polarization induced by iron and cobalt atoms in palladium, *Proc. Phys. Soc.*, **89**, 119–127.
17. Miyazaki, A., Ito, Y., Enoki, T. (2010). New classes of nanomagnets created from alkanethiol-coated Pt, Pd nanoparticles and their alloys with Co, *Eur. J. Inorg. Chem.*, **2010**, 4279–4287.
18. Ito, Y., Miyazaki, A., Fukui, K., Valiyaveettil, S., Yokoyama, T., Enoki, T. (2008). Pd nanoparticle embedded with only one Co atom behaves as a single-particle magnet, *J. Phys. Soc. Jpn.*, **77**, 103701.

19. Ito, Y., Miyazaki, A., Valiyaveettil, S., Enoki, T. (2010). Magnetic properties of Fe–Pd alloy nanoparticles, *J. Phys. Chem. C*, **114**, 11699–11702.

20. Brust, M., Walker, M., Bethel, D., Schiffrin, D. J., Whyman, R. J. (1994). Synthesis of thiol-derivatised gold nanoparticles in a two-phase liquid–liquid system, *J. Chem. Soc. Chem. Commun.*, **1994**, 801–802.

21. Brust, M., Fink, J., Bethel, D., Schiffrin, D. J., Kiely, C. (1995). Synthesis and reactions of functionalized gold nanoparticles, *J. Chem. Soc. Chem. Commun.*, **1995**, 1655–1656.

22. Kubo, R. (1962). Electronic properties of metallic fine particles. I, *J. Phys. Soc. Jpn.*, **17**, 975–986.

23. Respaud, M., Broto, J. M., Rakoto, H., Fert, A. R., Thomas, L., Barbara, B., Verelst, M., Snoeck, E., Lecante, P., Mosset, A., Osuna, J., Ould Ely, T., Amiens, C., Chaudret, B. (1998). Surface effects on the magnetic properties of ultrafine cobalt particles, *Phys. Rev. B.*, **57**, 2925–2935.

24. Ishida, K., Nishizawa, T., (1991). The Co–Pd [Cobalt-Palladium] system, *J. Phase Equilib.*, **12**, 83.

25. Massalski, T. B., (1990). *Binary Alloy Phase Diagrams,* ASM International.

26. Lu, A.-H., Salabas, E. L., Schüth, F. (2007). Magnetic nanoparticles: synthesis, protection, functionalization, and application, *Angew. Chem. Int. Ed.*, **2007**(46), 1222–1244.

27. Waldmann, O. (2005). Magnetic molecular wheels and grids — the need for novel concepts in "zero-dimensional" magnetism, *Coord. Chem. Rev.*, **249**, 2550–2566.

28. Batlle, X., Labarta, A. (2002). Finite-size effects in fine particles: magnetic and transport properties, *J. Phys. D: Appl. Phys.*, **35**, R15–R42.

29. Kroto, H. W., Heath, J. R., Obrien, S. C., Curl, R. F., Smalley, R. E. (1985). C_{60}: Buckminsterfullerene, *Nature,* **318**, 162–163.

30. Iijima, S., Ichihashi, T. (1993). Single-shell carbon nanotubes of 1-nm diameter, *Nature*, **363**, 603–605.

31. Novoselov, K. S., Geim, A. K., Morozov, S. V., Jiang, D., Zhang, Y., Dubonos, S. V., Grigorieva, I. V., Firsov, A. A. (2004). Electric field effect in atomically thin carbon films, *Science*, **306**, 666–669.

32. Geim, A. K., Novoselov, K. S. (2007). The rise of graphene, *Nat. Mater.*, **6**, 183–191.

33. Novoselov, K. S., Geim, A. K., Morozov, S. V., Jiang, D., Katsnelson, M. I., Grigorieva, I. V., Dubonos, S. V., Firsov, A. A. (2005). Two-dimensional gas of massless Dirac fermions in graphene, *Nature,* **438**, 197–200.

34. Zhang, Y., Tan, Y.-W., Stormer, H. L., Kim, P. (2005). Experimental observation of the quantum Hall effect and Berry's phase in graphene, *Nature,* **438**, 201–204.

35. Nair, R., Blake, P., Grigorenko, A., Novoselov, K., Booth, T., Stauber, T., Peres, N., Geim, A. (2008). Fine structure constant defines visual transparency of graphene, *Science*, **320**, 1308.

36. Enoki, T., Takai, K. (2009). The edge state of nanographene and the magnetism of the edge-state spins, *Solid State Commun.*, **149**, 1144–1150.

37. Enoki, T., Takai, K. Osipov, V., Baidakova, M., Vul', A. (2009). Nanographene and nanodiamond; new members in the nanocarbon family, *Chem. Asian J.*, **4**, 796–804.

38. Wallace, P. R. (1947). 1946 Annual Meeting of the American Physical Society, The band theory of graphite, *Phys. Rev.*, **77**, 476.

39. McClure, J. W. (1957). Band structure of graphite and de Haas — van Alphen effect, *Phys. Rev.*, **108**, 612–618.

40. Slonczewski, J. C., Weiss, P. R. (1958). Band structure of graphite, *Phys. Rev.*, **109**, 272–279.

41. Stein, S. E., Brown, R. L. (1987). π-Electron properties of large condensed polyaromatic hydrocarbons, *J. Am. Chem. Soc.*, **109**, 3721 -3729.

42. Yoshizawa, K., Okahara, K., Sato, T., Tanaka, K., Yamabe T. (1994). Molecular orbital study of pyrolytic carbons based on small cluster models, *Carbon*, **32**, 1517–1522.

43. Fujita, M., Wakabayashi, K., Nakada, K., Kusakabe K. (1996). Peculiar localized state at zigzag graphite edge, *J. Phys. Soc. Jpn.*, **65**, 1920–1923.

44. Nakada, K., Fujita, M., Dresselhaus, G., Dresselhaus, M. (1996). Edge state in graphene ribbons: Nanometer size effect and edge shape dependence, *Phys. Rev. B*, **54**, 17954–17961.

45. Lieb, E. H. (1989). Two theorems on the Hubbard model, *Phys. Rev. Lett.*, **62**, 1201–1204.

46. Longuet-Higgins, H. C. (1950). Some studies in molecular orbital theory I. Resonance structures and molecular orbitals in unsaturated hydrocarbons, *J. Chem. Phys.*, **18**, 265–274.

47. Goto, K., Kubo, T., Yamamoto, K., Nakasuji, K., Sato, K., Shiomi, D., Takui, T., Kobayashi, T., Yakushi, K., Ouyang, J. (1999). A stable neutral hydrocarbon radical: synthesis, crystal structure, and physical

properties of 2,5,8-tri-*tert*-butylphenalenyl, *J. Am. Chem. Soc.*, **121**, 1619 –1620.

48. Inoue, J., Fukui, K., Shiomi, D., Morita, Y., Yamamoto, K., Nakasuji, K., Takui, T., Yamaguchi, K. (2001), The first detection of a Clar's hydrocarbon, 2,6,10-tri-*tert*-butyltriangulene: A ground-state triplet of non-Kekulé polynuclear benzenoid hydrocarbon, *J. Am. Chem. Soc.*, **123**, 12702–12703.
49. Morita, Y., Suzuki, S., Fukui, K., Nakazawa, S., Kitagawa, H., Kishida, H., Okamoto, H., Naito, A., Sekine, A., Ohashi, Y., Shiro, M., Sasaki, K., Shiomi, D., Sato, K., Takui T., Nakasuji, K. (2008). Thermochromism in an organic crystal based on the coexistence of σ- and π-dimers, *Nature Mater.*, **7**, 48–51.
50. Sugawara, T., Bandow, S., Kimura, K., Iwamura, H., Itoh, K. (1984). Magnetic behavior of nonet tetracarbene, *m*-phenylenebis((diphenyl methylene-3-yl)methylene), *J. Am. Chem. Soc.*, **106**, 6449–6450.
51. Sugawara, T., Bandow, S., Kimura, K., Iwamura, H., Itoh, K. (1986). Magnetic behavior of nonet tetracarbene as a model for one-dimensional organic ferromagnets, *J. Am. Chem. Soc.*, **108**, 368.
52. Nasu, K. (1986). Periodic Kondo-Hubbard model for a quasi-one-dimensional organic ferromagnet *m*-polydiphenylcarbene: Cooperation between electron correlation and topological structure, *Phys. Rev. B*, **33**, 330–338.
53. Niimi, Y., Matsui, T., Kambara, H., Tagami, K., Tsukada, M., Fukuyama, H. (2005). Scanning tunneling microscopy and spectroscopy studies of graphite edges, *Appl. Surf. Sci.*, **241**, 43–48.
54. Kobayashi, Y., Fukui, K., Enoki, T., Kusakabe, K., Kaburagi, Y. (2005). Observatoin of zigzag and armchair edges of graphite using scanning tunneling microscopy and spectroscopy, *Phys. Rev. B*, **71**, 193406.
55. Kobayashi, Y., Fukui, K., Enoki, T., Kusakabe, K. (2006). Edge state on hydrogen-terminated graphite edges investigaed by scanning tunneling microscopy, *Phys. Rev. B*, **73**, 193406.
56. Wakabayashi, K., Sigrist, M., Fujita, M. (1998). Spin wave mode of edge-localized magnetic states in nanographite zigzag ribbons, *J. Phys. Soc. Jpn.*, **67,** 2089–2093.
57. Shibayama, Y., Sato, H., Enoki, T., Bi, X. X., Dresselhaus, M. S., Endo, M. (2000). Novel electronic properties of a nano-graphite disordered network and their iodine doping effects, *J. Phys. Soc. Jpn.*, **69**, 754–767.

58. Shibayama, Y., Sato, H., Enoki, T., Endo, M. (2000). Disordered magnetism at the metal-insulator threshold in mano-graphite-based carbon materials, *Phys. Rev. Lett.*, **84**, 1744–1747.

59. Hao, S., Takai, K., Kang F., Enoki, T. (2008). Electronic and magnetic properties of acid-adsorbed nanoporous activated carbon fibers, *Carbon*, **46**, 110–116.

60. H. Sato, Kawatsu, N., Enoki, T., Endo, M., Kobori, R., Maruyama, S., Kaneko K. (2006). Physisorption-induced change in the magnetism of microporous carbon, *Carbon*, **45**, 214–217.

61. Takahara, K., Takai, K., Enoki, T., Sugihara, K. (2007). Effect of oxygen adsorption on the magnetoresistance of a disordered nanographite network, *Phys. Rev. B*, **76**, 035442.

62. Takai, K., Sato, H., Enoki, T., Yoshida, N., Okino, F., Touhara, H., Endo M. (2001). Effect of fluorination on nano-sized π-electron systems, *J. Phys. Soc. Jpn.*, **70**, 175–185.

63. Takai, K., Kumagai, H., Sato, H., Enoki, T. (2006). Bromine-adsorption-induced change in the electronic and magnetic properties of nanographite network systems, *Phys. Rev. B*, **73,** 035435.

64. Tajima, N., Sugawara, S., Tamura, M., Kato, R., Nishio, Y., Kajita, K. (2006). Electronic phases in an organic conductor α-(BEDT-TTF)$_2$I$_3$: ultra narrow gap semiconductor, superconductor, metal, and charge-ordered insulator, *J. Phys. Soc. Jpn.*, **75**, 051010.

65. Tajima, N., Sugawara, S., Tamura, M., Kato, R., Nishio, Y., Kajita, K. (2007). Transport properties of massless Dirac fermions in an organic conductor α-(BEDT-TTF)$_2$I$_3$ under pressure, *Europhys. Lett.*, **80**, 47002.

66. Kobayashi, A., Katayama, S., Noguchi, K., Suzumura, Y. (2004). Superconductivity in charge ordered organic conductor — α-(ET)$_2$I$_3$ Salt, *J. Phys. Soc. Jpn.*, **73**, 3135.

67. Katayama, S., Kobayashi, A., Suzumura, Y. (2006). Pressure-induced zero-gap semiconducting state in organic conductor α-(BEDT-TTF)$_2$I$_3$ salt, *J. Phys. Soc. Jpn.*, **75**, 054705.

68. Kino, H., Miyazaki, T. (2006). First-principles study of electronic structure in α-(BEDT-TTF)$_2$I$_3$ at ambient pressure and with uniaxial strain, *J. Phys. Soc. Jpn.*, **75**, 034704.

69. Tajima, N., Sugawara, S., Kato, R., Nishio, Y., Kajita, K. (2009). Effect of the zero-mode Landau level on interlayer magnetoresistance in multilayer massless Dirac Fermion systems, *Phys. Rev. Lett.*, **102**, 176403.

70. Tajima, N., Sato, M., Sugawara, S., Kato, R., Nishio, Y., Kajita, K. (2010). Spin and valley splittings in multilayered massless Dirac fermion system, *Phys. Rev. B*, **82**, 121420.

71. Kagoshima, S., Kanoda, K., Mori, T. (eds.) (2006). Organic conductors, special topics section in *J. Phys. Soc. Jpn.*, **75**.

72. Batail, P. (ed.) (2004). Molecular conductors, special issue in *Chem. Rev.*, **104**.

73. Ishiguro, T., Yamaji, K., Saito, G. (1998). *Organic Superconductors*, Springer-Verlag, Heidelberg, Germany.

74. Grüner, G. (1994). *Density Waves in Solids*, Frontiers in Physics, Addison-Wesley, Reading, Massachusetts.

75. Kino, H., Fukuyama, H. (1996). Phase diagram of two-dimensional organic conductors: $(BEDT\text{-}TTF)_2X$, *J. Phys. Soc. Jpn.*, **65**, 2158–2169.

76. Takahashi, T., Nogami, Y., Yakushi, K. (2006). Charge ordering in organic conductors, *J. Phys. Soc. Jpn.*, **75**, 051008.

77. Yamada, J., Sugimoto, T. (2004). *TTF Chemistry: Fundamentals and Applications of Tetrathiafulvalene*, Springer, Berlin, Germany.

78. Shieber, M. M. (1967). *Experimental Magentochemistry — Nonmetallic Magnetic Materials*, North-Holland, Amsterdam.

79. Kosower, E. M., Martin, R. L., Meloche, V. W. (1957). Copper(II) bromide complexes, *J. Am. Chem. Soc.*, **79**, 1509–1510.

80. Shannon, R. D. (1976). Revised effective ionic radii and systematic studies of interatomic distances in halides and chalcogenides, *Acta Cryst. A*, **32**, 751–767.

81. Coronado, E., Day, P. (2004). Magnetic molecular conductors, *Chem. Rev.*, **104**, 5419–5448, and references cited therein.

82. Day, P., Kurmoo, M., Mallah, T., Marsden, I. R., Friend, R. H., Pratt, F. L., Hayes, W., Chasseau, D., Gaultier, J., Bravic, G., Ducasse, L. (1992). Structure and properties of tris[bis(ethylenedithio)tetrathiafulval enium]tetrachlorocopper(II) hydrate, $(BEDT\text{-}TTF)_3CuCl_4{\cdot}H_2O$: First evidence for coexistence of localized and conduction electrons in a metallic charge-transfer salt, *J. Am. Chem. Soc.*, **114**, 10722–10729.

83. Kobayashi, H., Cui, H. (2004). Organic metals and superconductors based on BETS (BETS = bis(ethylenedithio)tetraselenafulvalene), *Chem. Rev.*, **104**, 5265–5288, and references cited therein.

84. Uji, S., Shinagawa, H., Terashima, T., Yakabe, T., Terai, Y., Tokumoto, M., Kobayashi, A., Tanaka, H., Kobayashi, H. (2001). Magnetic-field-induced

superconductivity in a two-dimensional organic conductor, *Nature*, **410**, 908–910.

85. Balicas, L., Brooks, J. S., Storr, K., Uji, S., Tokumoto, M., Tanaka, H., Kobayashi, H., Kobayashi, A., Barzykin, V., Gor'kov, L. P. (2001). Superconductivity in an organic insulator at very high magnetic fields, *Phys. Rev, Lett.*, **87**, 067002.

86. Jaccarino, V., Peter, M. (1962). Ultra-high-field superconductivity, *Phys. Rev. Lett.*, **9**, 290–292.

87. Meul, H. W., Rossel, C., Decroux, M., Fischer, Ø., Remenyi, G., Briggs, A. (1984). Observation of magnetic-field-induced superconductivity, *Phys. Rev. Lett.*, **53**, 497–500.

88. Mori, T., Katsuhara, M. (2002). Estimation of πd-interactions in organic conductors including magnetic anions, *J. Phys. Soc. Jpn.*, **71**, 826–844.

89. Enomoto, K., Yamaura, J., Miyazaki, A., Enoki, T. (2003). Electronic and magnetic properties of organic conductors $(DMET)_2MBr_4$ (M = Fe, Ga), *Bull. Chem. Soc. Jpn.*, **76**, 945–953.

90. Okabe, K., Yamaura, J.-I., Miyazaki, A., Enoki, T. (2005). Electronic and magnetic properties of π-d interaction system $(EDTDM)_2FeBr_4$, *J. Phys. Soc. Jpn.*, **74**, 1508–1520.

91. Taniguchi, H., Miyashita, M., Uchiyama, K., Satoh, K., Mori, N., Okamoto, H., Miyagawa, K., Kanoda, K., Hedo, M., Uwatoko, Y. (2003). Superconductivity at 14.2 K in layered organics under extreme pressure, *J. Phys. Soc. Jpn.*, **72**, 468–471.

92. Urayama, H., Yamochi, H., Saito, G., Nozawa, K., Sugano, T., Kinoshita, M., Sato, S., Oshima, K., Kawamoto, A., Tanaka, J. (1988). A new ambient pressure organic superconductor based on BEDT-TTF with T_C higher than 10 K (T_C = 10.4 K), *Chem. Lett.*, **17**, 55–58.

93. Kini, A. M., Geiser, U., Wang, H. H., Carlson, K. D., Williams, J. M., Kwok, W. K., Vandervoort, K. G., Thompson, J. E., Stupka, D. L. (1900). A new ambient-pressure organic superconductor, κ-$(ET)_2Cu[N(CN)_2]Br$, with the highest transition temperature yet observed (inductive onset T_C = 11.6 K, resistive onset = 12.5 K), *Inorg. Chem.*, **29**, 2555–2557.

94. Graham, A. W., Kurmoo, M., Day, P. (1995). β''-$(BEDT\text{-}TTF)_4[(H_2O)Fe(C_2O_4)_3]\cdot PhCN$: The first molecular superconductor containing paramagnetic metal ions, *J. Chem. Soc., Chem. Commun.*, 2061–2062.

95. Kurmoo, M., Graham, A. W., Day, P., Coles, S. J., Hursthouse, M. B., Caulfield, J. M., Singleton, J., Ducasse, L., Guionneau, P. (1995). Superconducting and semiconducting magnetic charge transfer salts: $(BEDT\text{-}TTF)_4AFe(C_2O_4)_3\cdot C_6H_5CN$ (A = H_2O, K, NH_4), *J. Am. Chem. Soc.*, **117**, 12209–12217.

96. Turner, S. S., Day, P., Malik, K. M. A., Hursthouse, M. B., Tent, S. J., MacLean, E. J., Martin, L. L. (1999). Effect of included solvent molecules on the physical properties of the paramagnetic charge transfer salts β''-(BEDT-TTF)$_4$[(H$_3$O)Fe(C$_2$O$_4$)$_3$]·Solvent (BEDT-TTF = Bis(ethylenedi thio)tetrathiafulvalene), *Inorg. Chem.*, **38**, 3543–3549.

97. Rashid, S., Turner, S. S., Day, P., Howard, J. A. K., Guinonneau, P., McInnes, E. J. L., Mabbs, F. E., Clark, R. J. H., Firth, S., Biggs, T. (2001). New superconducting charge-transfer salts (BEDT-TTF)$_4$[A·M(C$_2$O$_4$)$_3$]·C$_6$ H$_5$NO$_2$ (A = H$_3$O or NH$_4$, M = Cr or Fe, BEDT-TTF = bis(ethylenedithio) tetrathiafulvalene), *J. Mater. Chem.*, **11**, 2095–2102.

98. Tamaki, H., Zhong, Z. J., Matsumoto, N., Kida, S., Koikawa, M., Achiwa, N., Hashimoto, Y., Ôkawa, H. (1992). Design of metal-complex magnets. Syntheses and magnetic properties of mixed-metal assemblies {NBu$_4$[MCr(ox)$_3$]}$_x$ (NBu$_4^+$ = tetra(*n*-butyl)ammonium ion; ox$_2^-$ = oxalate ion; M = Mn^{2+}, Fe^{2+}, Co^{2+}, Ni^{2+}, Cu^{2+}, Zn^{2+}), *J. Am. Chem. Soc.*, **114**, 6974–6979.

99. Mathionière, C., Nuttal, C. J., Carling, S. G., Day, P. (1996). Ferrimagnetic mixed-valency and mixed-metal tris(oxalato)iron(III) compounds: Synthesis, structure, and magnetism, *Inorg. Chem.*, **35**, 1201–1206.

100. Coronado, E., Galán-Mascarós, J. R., Gómez-García, C. J., Laukhin, V. (2000). Coexistence of ferromagnetism and metallic conductivity in a molecule-based layered compound, *Nature*, **408**, 447–449.

101. Galán-Mascarós, J. R., Coronado, E. (2008). Molecule-based ferromagnetic conductors: Strategy and design, *C. R. Chimie*, **11**, 1110–1116.

102. Alberola, A., Coronado, E., Galán-Mascarós, J. R., Giménez-Saiz, C., Gómez-García, C. (2003). A molecular metal ferromagnet from the organic donor bis(ethylenedithio)tetraselenafulvalene and bimetallic oxalate complexes, *J. Am. Chem. Soc.*, **125**, 10774–10775.

103. Alberola, A., Coronado, E., Galán-Mascarós, J. R., Giménez-Saiz, C., Gómez-García, C. J., Martínez-Ferrero, E., Murcia-Martínez, A. (2003). Multifunctionality in hybrid molecular materials: Design of ferromagnetic molecular metals, *Synth. Met.*, **135–136**, 687–689.

104. Yamochi, H., Kawasaki, T., Nagata, Y., Maesato, M., Saito, G. (2002). BEDO-TTF complexes with magnetic counter ions, *Mol. Cryst. Liq. Cryst.*, **376**, 113–120.

105. Galán-Mascarós, J. R., Coronado, E., Goddard, P. A., Singleton, J., Coldea, A. I., Wallis, J. D., Coles, S. J., Alberola, A. (2010). A chiral ferromagnetic molecular metal, *J. Am. Chem. Soc.*, **132**, 9271–9273.

106. Rikken, G. L. J. A., Folling, J., Wyder, P. (2001). Electrical magnetochiral anisotropy, *Phys. Rev. Lett.*, **87**, 236602.

107. Desiraju, G. R., Parthasarathy, R. (1989). The nature of halogen···halogen interactions: are short halogen contacts due to specific attractive forces or due to close packing of nonspherical atoms?, *J. Am. Chem. Soc.*, **111**, 8725–8726.

108. Legon, A. C. (1999). Prereactive complexes of dihalogens XY with Lewis bases B in the gas phase: A systematic case for the halogen analogue B···XY of the hydrogen bond B···HX, *Angew. Chem. Int. Ed. Engl.*, **38**, 2686–2714.

109. Ouvrard, C., Le Questel, J. Y., Berthelot, M., Laurence, C. (2003) Halogen-bond geometry: a crystallographic database investigation of dihalogen complexes, *Acta Cryst. B*, **59**, 512–526.

110. Nishijo, J., Miyazaki, A., Enoki, T., Watanabe, R., Kuwatani, Y., Iyoda, M. (2005). d-Electron-induced negative magnetoresistance of a π–d interaction system based on a brominated-TTF donor, *Inorg. Chem.*, **44**, 2493–2506.

111. Miyazaki, A., Yamazaki, H., Aimatsu, M., Enoki, T., Watanabe, R., Ogura, E., Kuwatani, Y., Iyoda, M. (2007). Crystal structure and physical properties of conducting molecular antiferromagnets with a halogen-substituted donor: $(EDO\text{-}TTFBr_2)_2FeX_4$ (X = Cl, Br), *Inorg. Chem.*, **46**, 3353–3366.

112. Ogawa, M. Y., Martinsen, J., Palmer, S. M., Stanton, J. L., Tanaka, J., Greene, R. L., Hoffman, B. M., Ibers, J. A. (1987). The (phthalocyaninato)copper iodide complex Cu(pc)I: a molecular metal with a one-dimensional array of local moments embedded in a "Fermi sea" of charge carriers, *J. Am. Chem. Soc.*, **109**, 1115–1121.

113. Ogawa, M. Y., Hoffman, B.M., Lee, S., Yudkowsky, M., Halperin, W.P. (1986). Transition of local moments coupled to itinerant electrons in the quasi one-dimensional conductor copper phthalocyanine iodide, *Phys. Rev. Lett.*, **57**, 1177–1180.

114. Quirrion, G., Poirier, M., Liou, K. K., Hoffman, B. M. (1991). Strong carrier scattering by a Cu^{2+} local moment array in one-dimensional molecular conductors Cu_xNi_{1-x}(phthalocyaninato)I, *Phys. Rev. B*, **43**, 860–864.

115. Matsuda, M., Naito, T., Inabe, T., Hanasaki, N., Tajima, H., Otsuka, T., Awaga, K., Narymbetov, B., Kobayashi, H. (2000). A one-dimensional macrocyclic π-ligand conductor carrying a magnetic center. Structure and electrical, optical and magnetic properties of $TPP[Fe(Pc)(CN)_2]_2$ {TPP = tetraphenylphosphonium and $[Fe(Pc)(CN)_2]$ = dicyano(phthalo cyaninato)iron(III)}, *J. Mater. Chem.*, **10**, 631–636.

116. Hanasaki, N., Tajima, H., Matsuda, M., Naito, Inabe, T. (2000). Giant negative magnetoresistance in quasi-one-dimensional conductor TPP[Fe(Pc)$(CN)_2]_2$: Interplay between local moments and one-dimensional conduction electrons, *Phys. Rev. B*, **62**, 5839–5842.

117. Hotta, C., Ogata, M., Fukuyama, H. (2005). Interaction of the ground state of quarter-filled one-dimensional strongly correlated electronic system with localized spins, *Phys. Rev. Lett.*, **95**, 216402.

118. Hotta, C. (2010). Interplay of strongly correlated electrons and localized Ising moments in one dimension, *Phys. Rev. B*, **81**, 245104.

119. Hay, P. J., Thibeault, J. C., Hoffmann, R. (1975). Orbital interactions in metal dimer complexes, *J. Am. Chem. Soc.*, **97**, 4884–4899.

120. Dowd, P. (1972). Trimethylenemethane, *Acc. Chem. Res.*, **5**, 242–248.

121. Borden, W. T., Iwamura, H., Berson, J. A. (1994). Violations of Hund's rule in Non-Kekule hydrocarbons: Theoretical prediction and experimental verification, *Acc. Chem. Res.*, **27**, 109–116.

122. Borden, W. T. (1999) In: Magnetic Properties of Organic Materials, Lahti, P. M. (ed.), Marcel Dekker, Inc., New York, Chapter 5.

123. Wenthold, P. G., Hu, J., Squires, R. R., Lineberger, W. C. (1996). Photoelectron spectroscopy of the trimethylene-methane negative ion. The singlet–triplet splitting of trimethylenemethane, *J. Am. Chem. Soc.*, **118**, 475–476.

124. Kumai, R., Matsushita, M. M., Izuoka, A., Sugawara, T. (1994). Intramolecular Exchange Interaction in a Novel Cross-Conjugated Spin System Composed of pi-Ion Radical and Nitronyl Nitroxide, *J. Am. Chem. Soc.*, **116**, 4523–4524.

125. Nakazaki, J., Matsushita, M. M., Izuoka, A., Sugawara, T. (1999). Novel spin-polarized TTF donors affording ground state triplet cation diradicals, *Tetrahedron Lett.*, **40**, 5027–5030.

126. Ishikawa, Y., Miyamoto, T., Yoshida, A., Kawada, Y., Nakazaki, J., Izuoka, A., Sugawara, T. (1999). New synthesis of 2-[1,3-dithiol-2-ylidene]-5,6-dihydro-1,3-dithiolo[4,5-b][1,4]dithiins with formyl group on fused benzene, [1,4]dithiin, or thiophene ring, *Tetrahedron Lett.*, **40**, 8819–8822.

127. Nakazaki, J., Ishikawa, Y., Izuoka, A., Sugawara, T., Kawada, Y. (2000). Preparation of isolable ion-radical salt derived from TTF-based spin-polarized donor, *Chem. Phys. Lett.*, **319**, 385–390.

128. Sugano, T., Fukasawa, T., Kinoshita, M. (1991). Magnetic interactions among unpaired electrons in charge-transfer complexes of organic donors having a neutral radical, *Synth. Met.*, **41**, 3281–3284.

129. Sugimoto, T., Yamaga, S., Nakai, M., Tsuji, M., Nakatsuji, H.,d Hoshino, N. (1993). Different magnetic properties of charge-transfer complexes and cation radical salts of Tetrathiafulvalene derivatives substituted with Imino Pyrolidine- and Piperidine-1-oxyls, *Chem. Lett.*, 1817–1820.

130. Ishida, T., Tomioka, K., Nogami, T., Yamaguchi, K., Mori, W., Shirota, Y. (1993). Intermolecular ferromagnetic interaction of 4-(1-Pyrenylmethyleneamino)-2,2,6,6-Tetra Methylpiperidin-1-Oxyl, *Mol. Cryst. Liq. Cryst.*, **232**, 99–102.

131. Fujiwara, H., Fujiwara, E., Kobayashi, H., (2002). Novel π-electron donors for magnetic conductors containing a PROXYL radical, *Chem. Lett.*, 1048–1049.

132. Fujiwara, H., Lee, H.-J., Cui, H.-B., Kobayashi, H., Fujiwara, E., Kobayashi, A. (2004). Synthesis, structure, and physical properties of a new organic conductor based on a π-extended donor containing a stable 2,2,5,5-tetramethyl-1-pyrrolidinyloxy radical, *Adv. Mater.*, **16**, 1765–1769.

133. E. Fujiwara, S. Aonuma, H. Fujiwara, T. Sugimoto, Y. Misaki, (2008). New π-electron donors with a 2,2,5,5-Tetramethylpyrrolin-1-yloxyl radical designed for magnetic molecular conductors, *Chem. Lett.*, **37**, 84–85.

134. Mukai, K., Semba, N., Hatanaka, T., Minakuchi, H., Ohara, K., Taniguchi, T., Misaki, Y., Hosokoshi, Y., Inoue, K., Azuma, N. (2004). Molecular paramagnetic semiconductor: Crystal structures and magnetic and conducting properties of the Ni(dmit)2 salts of 6-Oxoverdazyl radical cations (dmit = 1,3-Dithiol-2-thione-4,5-dithiolate), *Inorg. Chem.*, **43**, 566–576.

135. Matsushita, M. M., Kawakami, H., Kawada, Y., Sugawara, T. (2007). Negative magneto-resistance observed on an ion-radical salt of a TTF-based spin-polarized donor, *Chem. Lett.*, **36**, 110–111.

136. Matsushita, M. M., Kawakami, H., Sugawara, T., Ogata, M. (2008). Molecule-based system with coexisting conductivity and magnetism and without magnetic inorganic ions, *Phys. Rev. B*, **77**, 195208.

137. Komatsu, H., Mogi, R., Matsushita, M. M., Miyagi, T., Kawada, Y., Sugawara, T. (2009). Synthesis and properties of TSF-based spin-polarized donor, *Polyhedron*, **28**, 1996–2000.

138. Hiraoka, S., Okamoto, T., Kozaki, M., Shiomi, D., Sato, K., Takui, T., Okada, K. (2004). A stable radical-substituted radical cation with strongly ferromagnetic interaction: Nitronyl nitroxide-substituted 5,10-Diphenyl-5,10-dihydrophenazine radical cation, *J. Am. Chem. Soc.*, **126**, 58–59.

139. Takahide, Y., Konoike, T., Enomoto, K., Nishimura, M., Terashima, T., Uji, S., Yamamoto, H. M. (2006). Current–voltage characteristics of charge-ordered organic crystals, *Phys. Rev. Lett.*, **96**, 136602.

140. Matsushita, M. M., Sugawara, T. (2005). Current-induced low-resistance state and its crystal structure of a TTF-based dimeric donor salt, *J. Am. Chem., Soc.* **127**, 12450.

141. Sugawara, T., Matsushita, M. M., Komatsu, H., Mogi, R., Suzuki, K., Kondo, R. (2010). Interplay between *Conduction Electron and Localized Spins in Organic Spin System, International Conference on Science and Technology of Synthetic Metals*, Kyoto, Abstract, 6E-1, p. 164.

142. Mott, N. F. (1969). *Philos. Mag.*, **19**, 835.

143. McKenzie, R. H., Merino, J., Marston, J. B., Sushkov, O. P. (2001). Charge ordering and antiferromagnetic exchange in layered molecular crystals of the θ type, *Phys. Rev. B*, **64**, 085109.

144. Kao, K. C., Hwang, W. (1981). *Electrical Transport in Solids*, Pergamon Press, Oxford.

145. Burrows, P. E., Shen, Z., Bulovic, V., McMarty, D. M., Forrese, S. R., Cronin, J. A., Thompson, M. E. (1996). Relationship between electroluminescence and current transport in organic heterojunction light-emitting devices, *J. Appl. Phys.* **79**, 7991–8006.

146. Prigodin, V. N., Bergeson, J. D., Lincoln, D. M., Epstein, A. (2006). Anomalous room temperature magnetoresistance in organic semiconductors, *J. Synth. Met.*, **156**, 757–761.

147. Bergeson, J. D., Prigodin, V. N., Lincoln, D. M., Epstein, A. (2008). Inversion of magnetoresistance in organic semiconductors, *J. Phys. Rev. Lett.*, **100**, 067201.

148. Shirahata, T., Kibune, M., Maesato, M., Kawashima, T., Saito, G., Imakubo, T. (2006). New organic conductors based on dibromo- and diiodo-TSeFs with magnetic and non-magnetic MX_4 counter anions (M = Fe, Ga; X = Cl, Br), *J. Mater. Chem.*, **16**, 3381–3390.

149. Komatsu, H., Matsushita, M. M., Yamamura, S., Sugawara, Y., Suzuki, K., Sugawara, T. (2010). Influence of magnetic field upon the conductance of a unicomponent crystal of a tetrathiafulvalene-based Nitronyl Nitroxide, *J. Am. Chem. Soc.*, **132**, 4528–4529.

150. Boubekeur, K., Syssa-Magalé, J-L., Palvadeau, P., Schöllhorn, B. (2006). Self-assembly of nitroxide radicals via halogen bonding — directional NO···I interactions, *Tetrahedron Lett.*, **47**, 1249–1252.

151. Fourmigué, M. (2008). Halogen bonding in conducting or magnetic molecular materials, *Struct. Bonding*, **126**, 181–207.

152. de Panthou, F. L. Luneau, D., Laugier, J., Rey, P. (1993). Crystal structures and magnetic properties of a nitronyl nitroxide and of its imino analog. Crystal packing and spin distribution dependence of ferromagnetic intermolecular interactions, *J. Am. Chem. Soc.,* **115**, 9095–9100.

153. Gary A. P. (2005). Magnetoelectronics, *Science,* **282**, 1660–1663.

154. Sanvito, S. (2007). Injecting and controlling spins in organic materials, *J. Mater. Chem.,* **17**, 4455–4459.

155. Fiederling, R., Keim, M., Reuscher, G., Ossau, W., Schmidt, G., Waag, A., and Molenkamp, W. (1999). Injection and detection of a spin-polarized current in a light-emitting diode, *Nature,* **402**, 787–790.

156. Ono, K., Austing, D. G., Tokura, Y., Tarucha, S. (2002). Current rectification by Pauli exclusion in a weakly coupled double quantum dot system, *Science,* **297**, 1313–1317.

157. Hanson, R., Awschalom, D. D. (2008). Coherent manipulation of single spins in semiconductors, *Nature,* **453**, 1043–1049.

158. Baibich, M. N., Broto, J. M., Fert, A., F. Nguyen van Dau, Petroff, F., Eitenne, P., Creuzet, G., Friedrich, A., Chazelas, J. (1988). Giant magnetoresistance of (001)Fe/(001)Cr magnetic superlattices, *Phys. Rev. Lett.,* **61**, 2472–2475.

159. Binasch, G., Grünberg, P., Saurenbach, F., Zinn, W. (1989). Enhanced magnetoresistance in layered magnetic structures with antiferromagnetic interlayer exchange, *Phys. Rev. B,* **39**, 4828–4830.

160. Bumm, L. A., Arnold J. J., Cygan, M. T., Dunbar, T. D., Burgin, T. P., Jones, L., Allara, D. L., Tour. J. M., Weiss, P. S., (1996). Are single molecular wires conducting ?, *Science,* **271**, 1705–1707.

161. Reed, M. A., Zhou, C., Muller, C. J., *et al.* (1997). Conductance of a molecular junction, *Science,* **278**, 252–254.

Chapter 2

Multifunctionalities of Single-Molecule Magnets with Electrical Conductivities

Kazuya Kubo, Hiroki Hiraga, Hitoshi Miyasaka, and Masahiro Yamashita

Institute of Electronic Science, Hokkaido University, N20, W10, Kita-ku, Sapporo, Hokkaido, 001-0020, Japan

kkubo@es.hokudai.ac.jp

Abstract

A new strategy for the preparation of hybridized materials with coexisting magnetism and conductivity was developed using single-molecule magnets (SMMs) and molecular conductors. The new strategy has various advantages including easy modification of their molecular and electronic features. The hybridized materials, $[\{Mn^{II}_2Mn^{III}_2(hmp)_6(MeCN)_2\}\{Pt(mnt)_2\}_x][Pt(mnt)_2]_2 \cdot n$MeCN ($x$ = 2 or 4, n = 0, or 2) and $[Mn_2(5\text{-MeOsaltmen})_2(solvent)_x][Ni(dmit)_2]_y \cdot n$ solvent (x = 0 or 2, y = 2, or 7, n = 0 or 4, solvent = acetone or MeCN), were prepared from cationic SMMs and acceptor types of molecular conductors by diffusion or electrochemical crystallization methods. In this review, the strategy for preparation of hybrid materials based on SMMs and molecular conductors and physical properties of the salts are discussed.

Multifunctional Molecular Materials
Edited by Lahcène Ouahab

ISBN 978-981-4364-29-4 (Hardcover), 978-981-4364-30-0 (eBook)
www.panstanford.com

Abbreviations

BEDT-TTF	bis(ethylenedithio)tetrathiafulvalene
BETS	bis(ethylenedithio)tetraselenafulvalene
Hhmp	2-hydroxymetylpyridine
$salen^{2-}$	*N,N'*-ethylene-bis(salicylideneimminate) (2-)
$saltmen^{2-}$	*N,N'*-(1,1,2,2-tetramethylethylene)bis (salicylideneiminate) (2-)
$5\text{-}Mesaltmen^{2-}$	*N,N'*-(1,1,2,2-tetramethylethylene)bis(5-methylsalicylideneiminate) (2-)
$5\text{-}MeOsaltmene^{2-}$	*N,N'*-(1,1,2,2-tetramethylethylene)bis(5-methoxysalicylideneiminate) (2-)
$dmit^{2-}$	1,3-dithiole-2-thiol-4,5-dithiolate(2-)
HOMO	highest occupied molecular orbital
LUMO	lowest unoccupied molecular orbital
mnt^{2-}	maleonitriledithiolate(2-)
MeCN	acetonitrile
ox^{2-}	oxalate(2-)
ZFS	zero-field splitting
QTM	quantum-tunneling magnetization

2.1 General Introduction of Conducting Single-Molecule Magnet

2.1.1 General Feature of Single-Molecule Magnets

In the early 1990s, the storage of information at the molecular level became potentially feasible owing to the discovery of multinuclear transition metal complexes acting as single-molecule magnets (SMMs).[1–2] To date, several SMM families based on complexes of Mn,[3] Fe,[4] Ni,[5] V,[6] Co,[7] or mixed transition metals[8] have been reported. These systems exhibit slow relaxation of their magnetization induced by the combined effect of a high-spin ground state S_T and uniaxial anisotropy D. These two ingredients create an energy barrier $\Delta = |D|S_T^2$ between spin-up and spin-down states, which leads to a hysteresis phenomena with reversal of the magnetization, similar to that observed in bulk magnets (Fig. 2.1).

For example, mixed-valence Mn^{III}_2/Mn^{II}_2 tetramer $[Mn^{III}_2 Mn^{II}_2 (hmp)_6 (MeCN)_4 (H_2O)_2] (ClO_4)_4 \cdot 2CH_3CN$ (Fig. 2.2) possesses $S_T = 9$ spin ground state and exhibits SMM behavior.[9] Dimeric manganese(III) tetradentate Schiff base complexes such as $[Mn^{III}(saltmen)(ReO_4)]_2$ are also SMMs with $S_T = 4$ spin ground state.[10] When a magnetic field is applied to saturate the magnetization of these complexes and then removed, the magnetization decays with a relaxation time τ that follows the Arrhenius law with an activation energy equal to Δ. At lower temperatures, τ may reach saturation as the thermally activated relaxation pathway becomes slower than quantum tunneling through the energy barrier. Hence, SMM complexes appear to be unique systems for studying fundamental phenomena, such as quantum spin tunneling and quantum phase interference, which may be used for future applications in molecular electronics.[11]

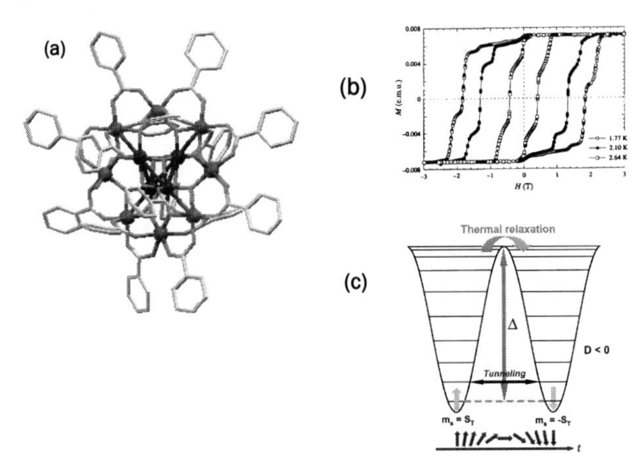

Figure 2.1 (a) Crystal structure of Mn_{12} SMM and (b) its magnetization and (c) energy barrier.

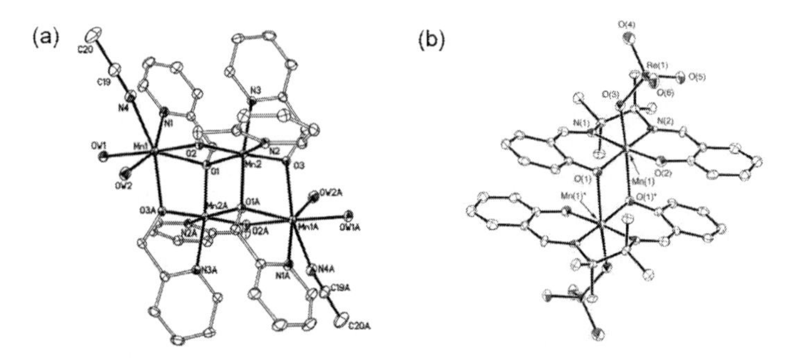

Figure 2.2 Molecular structures of (a) cationic parts for $[Mn^{III}{}_2Mn^{II}{}_2(hmp)_6(CH_3CN)_4(H_2O)_2](ClO_4)_4 \cdot 2CH_3CN$ and (b) $[Mn^{III}(saltmen)(ReO_4)]_2$.

2.1.2 General Feature of Molecular Conductors Based on Metal Ditholene Complex System

In general, component molecules for molecular conductors (Fig. 2.3) belong to the π-conjugated system such as BEDT-TTF and are divided into two categories, electron donor and electron acceptor. The component molecule is an insulator in itself. In order to obtain the metallic state, the formation of at least one partially filled energy band is required. A straightforward access to the molecular metal can be achieved by arranging open-shell molecules (radicals) so as to enable intermolecular electron transfer. In most cases, cation radicals or anion radicals generated from the donor or the acceptor have been used for the formation of metallic molecular crystals, and the conduction band originates from HOMO of the donor or LUMO of the acceptor.

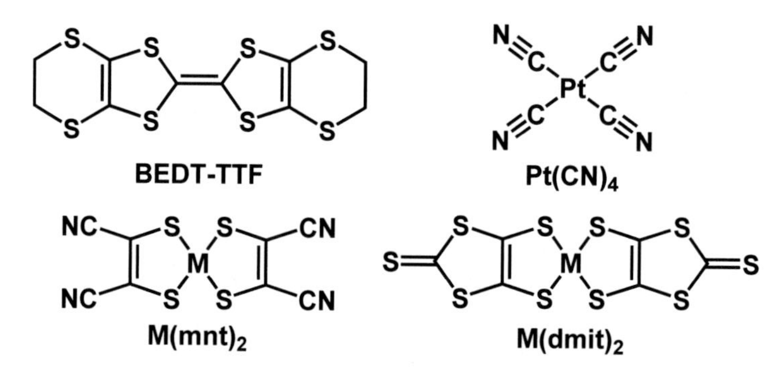

Figure 2.3 Examples of component molecules for molecular conductors.

In the early stage of the development of molecular conductors based on metal complexes, partially oxidized tetracyanoplatinate salts (e.g., KCP; $K_2[Pt(CN)_4]Br_{0.30}{\cdot}3H_2O$) and related materials were intensively studied.[12] In this system, the square-planar platinum complexes are stacked to form a linear Pt-atom chain. The conduction band originates from the overlap of $5dz^2$ orbitals of the central platinum atom and exhibits the one-dimensional (1-D) character.

On the other hand, metal dithiolene complexes possess a delocalized electron system as a planar central core $M(C_2S_2)_2$. The conduction band is formed by the ligand π orbitals or mixed-metal ligand orbitals where the sulfur atoms play an important role.[13] Depending on the choice of substituent groups attached to the central core, metal dithiolene complexes behave as both the donor and the acceptor. Development of molecular conductors based on the dithiolene complexes was triggered by the discovery of the metallic behavior in an anion radical salt $(H_3O)_{0.33}Li_{0.8}[Pt(mnt)_2]\cdot 1.67H_2O$.[14] Among metal dithiolene complexes, the metal-dmit complexes $M(dmit)_2$ (M = Ni and Pd; Fig 2.2) have been the most studied system. In the $M(dmit)_2$ molecule, HOMO has b_{1u} symmetry, while LUMO has b_{2g} symmetry. The metal *d* orbital can mix into the LUMO, but cannot contribute to the HOMO due to the symmetry, which destabilizes the HOMO and leads to a small energy splitting between HOMO and LUMO. The side-by-side intermolecular interaction, which leads to the formation of the 2-D electronic structure, is strong for the HOMO and weak for the LUMO. This is because some of the overlapping integrals for the intermolecular S···S pairs are canceled out due to the b_{2g} symmetry of the LUMO. Although the $M(dmit)_2$ molecule belongs to the acceptor, the nature of the conduction band in their anion radical salts strongly depends on the central metal. In general, the conduction band of the Pd system originates from the HOMO, while the conduction band of the Ni system originates from the LUMO. This unusual feature of the Pd system, HOMO–LUMO band inversion, is due to the strong dimerization and the small energy splitting between HOMO and LUMO.[15] In a series of anion radical salts with closed-shell cations (cation)$[Pd(dmit)_2]_2$ (cation = $Et_x Me_{4-x}Z^+$; Z = N, P, As, Sb and x = 0, 1, 2), the dimer units $[Pd(dmit)_2]_2^-$ form a strongly correlated 2-D system with a quasi triangular lattice and exotic properties derived from frustration and strong correlation are reported.[13] In these Pd salts, the choice of the counter cation tunes the degrees of frustration and correlation which are associated with the molecular arrangement. On the other hand, in the Ni salts, the choice of the counter cation provides a variety of molecular arrangements.[13]

2.1.3 General Feature of Hybrid Materials Based on Molecular Nano-sized Magnets

The design of multifunctional materials is one of the most challenging themes in the field of molecular materials science. One of the most attractive targets is magnetic/conducting bifunctional materials.[16] One way to design such materials is to hybridize *in situ* individual functioning parts, that is, magnetic and conducting frames, using a molecular self-assembly. Three representative hybrid materials have been synthesized so far: paramagnet/superconductor,[17] antiferromagnet/superconductor,[18] and ferromagnet/metal.[19] With emphasis on the third material, it is worth noting that Coronado *et al.* were the first to synthesize a hybrid material of coexisting ferromagnetic layers of $[Mn^{II}Cr^{III}(ox)]^-$ and metallic layers of (BEDT-TTF) molecules.[19]

These compounds as shown above are hybridized materials based on molecular conductors and classical magnets. However, material scientists have been most interested in the correlation between relaxing local spins and conducting itinerant electrons in such superparamagnetic/conducting hybrid materials in order to develop quantum spintronics system. As shown in section 1.1., SMMs show slow relaxation of the magnetization and quantum phenomena. From recent results of studies on SMM-based supramolecular oligmers[20] and network compounds,[21–26] it can be anticipated that inter-SMM interaction via conducting electrons, albeit small, has a mutual influence on both SMMs.

Scheme 2.1.1

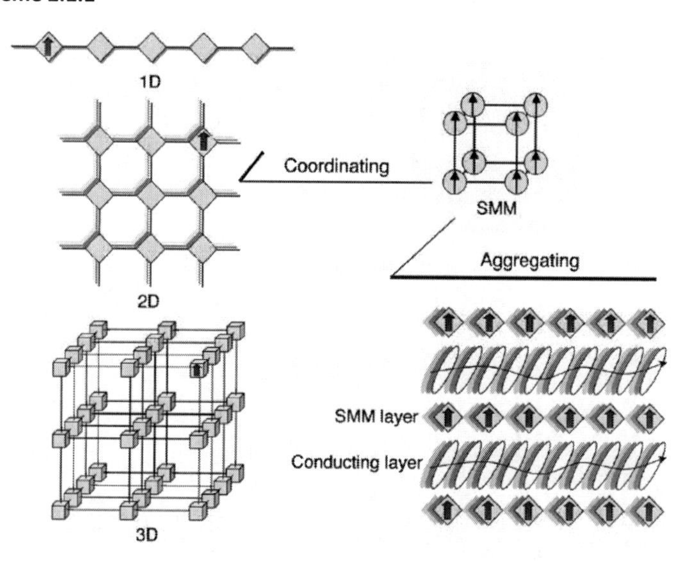

Miyasaka *et al.* suggested that SMMs are available as a magnetic building block for further molecule assemblies.[27] In developing their work on crystal engineering based on SMM or anisotropic molecule building blocks, their group has designed new classes of magnetic materials including single-chain magnets (SCMs)[28] and multidimensional SMM networks (they call them nano-dots networks),[29] which function as coordinating polymers of SMMs (Scheme 2.1.1). Recently, they have revealed the first example of hybrid materials aggregating both SMM layers and molecular conducting layers (Scheme 2.1.1).[30] This compound was designed in a Coulomb set combining a cationic SMM and an anionic molecular conductor for which we have chosen a double-cuboidal mixed-valent $[Mn_4]^{4+}$ cluster, $[Mn^{II}_2Mn^{III}_2(hmp)_6(MeCN)_2(H_2O)_4](ClO_4)_4$,[9] and $(NBu_4)[Pt(mnt)_2]$,[31] respectively. Electrochemical oxidation in a solution containing both starting building blocks successfully yielded a hybrid compound containing six noninteger-valent $[Pt(mnt)_2]^{n-}$ molecules per a unit without decomposition of basic $[Mn_4]^{4+}$ SMM core, which finally behaved as an SMM/semiconductor, while the diffusion reaction of respective starting materials led simply to a counter-ion exchanging assembly that has four integer-valent $[Pt(mnt)_2]^-$ molecules vs. a $[Mn_4]^{4+}$ SMM, which was an SMM/insulator. In order to improve their conducting properties, they have also prepared another type of hybrid materials based on Mn-salen dimer type SMMs with $[M(dmit)_2]^{n-}$ (M = Ni, Au).[32,33] In this review, we exhibit these new concepts for design of the nano-dot networks.

2.2 Tetra-Nuclear Manganese Cluster SMMs with Metal–Dithiolene Complex System

2.2.1 Introduction

Although there are a large number of hybridized materials based on magnetic ions with molecular conductors as described in the previous section,[19] there are only a few applications of them constructed by SMMs with molecular conductors. Pioneering work in this field was started in the 2000s by Miyasaka *et al.* They suggested that the SMM clusters are available as a magnetic building block for the molecular assemblies coexisting SMM and conducting behaviors.[27]

As our first approach to obtain such hybridized materials, we chose a Mn^{II}_2/Mn^{III}_2 tetranuclear SMM, $[Mn^{II}_2Mn^{III}_2(hmp)_6(MeCN)_2$

$(H_2O)_4](ClO_4)_4 \cdot 2MeCN$ (hereinafter, the SMM unit is abbreviated as $[Mn_4]^{4+}$),[9] and $(NBu_4)[Pt^{III}(mnt)_2]$[31] as starting materials. The $[Pt^{III}(mnt)_2]^{n-}$ unit has the potential to form a molecular aggregation having conducting property in assembling materials with cations.[14,34–37] The choice of such ionic starting materials ensured the coexistence of two distinct units, that is, an SMM unit and a conducting unit, by the mutual exchange of counter ions. Then, we performed two synthetic reactions: (i) the diffusion reaction of respective starting materials, and (ii) electrochemical oxidation in a solution containing both starting materials. These reactions, as expected, yielded two novel compounds based on $[Mn_4]^{4+}$ SMM units and $[Pt(mnt)_2]^{n-}$ molecules, $[\{Mn^{II}_2Mn^{III}_2(hmp)_6(MeCN)_2\}\{Pt(mnt)_2\}_2][Pt(mnt)_2]_2 \cdot 2MeCN$ (**1**) and $[\{Mn^{II}_2Mn^{III}_2(hmp)_6(MeCN)_2\}\{Pt(mnt)_2\}_4][Pt(mnt)_2]_2$ (**2**), respectively. While **1** is an SMM/insulator, **2** is an SMM/semiconductor possessing a frame composed of $[Pt(mnt)_2]^{n-}$ units possessing noninteger average valence. Although these materials are not the initially intended materials that exhibit SMM characteristics and high conductivity, 2 is the first example of a hybridized material that shows both SMM behavior and electronic conductivity. In addition, the present synthetic strategies and the products obtained give us a foresight into the design of materials with coexisting superparamagnetic/conducting properties in the same temperature range. In this section, the structures and physical properties of **1** and **2** are exhibited.

2.2.2 Structures of $[\{Mn^{II}_2Mn^{III}_2(hmp)_6(MeCN)_2\}\{Pt(mnt)_2\}_2][Pt(mnt)_2]_2 \cdot 2MeCN$ and $[\{Mn^{II}_2Mn^{III}_2(hmp)_6(MeCN)_2\}\{Pt(mnt)_2\}_4][Pt(mnt)_2]_2$

Complex **1** crystallized in the triclinic $P\bar{1}$ (#2) space group with $Z = 1$. ORTEP drawings of the asymmetric units of **1** are shown in Fig. 2.4. Material **1** is composed of one $[Mn^{II}_2Mn^{III}_2(hmp)_6(MeCN)_2]^{4+}$ double-cuboidal cluster unit and four $[Pt(mnt)_2]^-$ units, where two independent $[Pt(mnt)_2]^-$ units are present: coordinated ones, **A**; and uncoordinated ones, **B** (Fig. 2.4). Despite such differences as coordinated and uncoordinated units, there are no significant differences in the intra-$[Pt(mnt)_2]$ bond lengths.[30] The core structure of the $[Mn_4]^{4+}$ unit is very similar to that of the original $[Mn_4]^{4+}$ complex,[9] where the outer manganese ions

(Mn(1) and Mn(1*)) are divalent and the inner manganese ions (Mn(2) and Mn(2*)) are trivalent, as demonstrated by bond valence sum calculations. Each hexa-coordinated Mn^{III} ion has a distorted octahedral geometry revealing a Jahn–Teller elongation axis of [N(1*)–Mn(2)–O(2*)] with Mn(2)–O(2*) = 2.218(3) Å, Mn(2)–N(1*) = 2.206(4) Å, and N(1*)–Mn(2)–O(2*) = 160.15(15)°. In a $[Mn_4]^{4+}$ unit, the Jahn–Teller axes at Mn(1) and Mn(1*) are parallel to each other. The outer Mn^{II} ions are hepta-coordinated with one nitrogen atom and three oxygen atoms from the chelating hmp-ligands, one acetonitrile molecule and two $[Pt(mnt)_2]$-molecules (**A**) (Mn(1)–N(8*)**A**) 2.384(5) Å and (Mn(1)–N(5)**A**) 2.375(4) Å. Complex **2** crystallized in the triclinic $P\bar{1}$ (#2) space group with $Z = 1$. An ORTEP drawing of the asymmetrical unit of **1** is depicted in Fig. 2.5. Hybrid material **2** is composed of one $[Mn^{II}_2Mn^{III}_2(hmp)_6(MeCN)_2]^{4+}$ double-cuboidal unit and six $[Pt(mnt)_2]^{n-}$ units, where three independent $[Pt(mnt)_2]^{n-}$ units are present: coordinated ones, **A** and **B**; and uncoordinated one, **C** (Fig. 2.5). Similar to **1**, variations of respective bond lengths of the $[Pt(mnt)_2]^{n-}$ units are not characteristic among **A**, **B**, and **C** units and even between **1** and **2**. The $[Mn^{II}_2Mn^{III}_2(hmp)_6(MeCN)_2]^{4+}$ unit is basically the same as that of **1**. The inner hexa-coordinated Mn^{III} ions have a Jahn–Teller elongation axis of [N(2)–Mn(2)–O(1*)] with Mn(2)–O(1*) = 2.262(6) Å, Mn(2)–N(2) = 2.207(7) Å, and N(2)–Mn(2)–O(1*) = 159.8(2)°. The outer Mn^{II} ion is hepta-coordinated with two $[Pt(mnt)_2]^{n-}$ units (**A** and **B**) with $Mn(1)–N(4)_A$ = 2.409(7) Å and $Mn(1)–N(11)_B$ = 2.321(5) Å, forming a discrete unit, $\{[Pt(mnt)_2]_2–[Mn_4]–[Pt(mnt)_2]_2\}$.

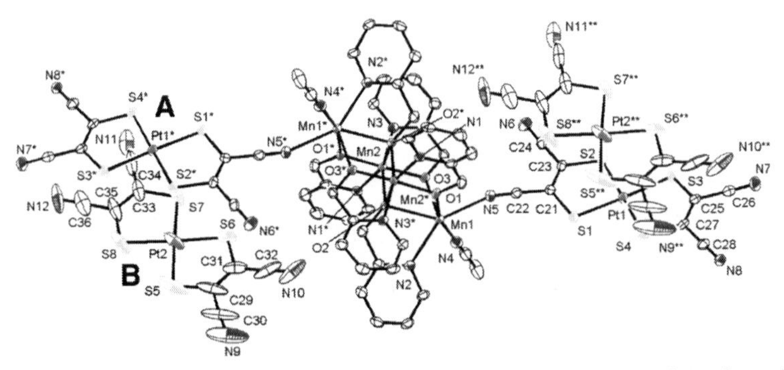

Figure 2.4 ORTEP view of **1** with atomic numbering scheme for selected atoms (50% probability thermal level), where symmetry operations (*): $1 - x, 2 - y, 1 - z$; (**): $-1 + x, -1 + y, -1 + z$.

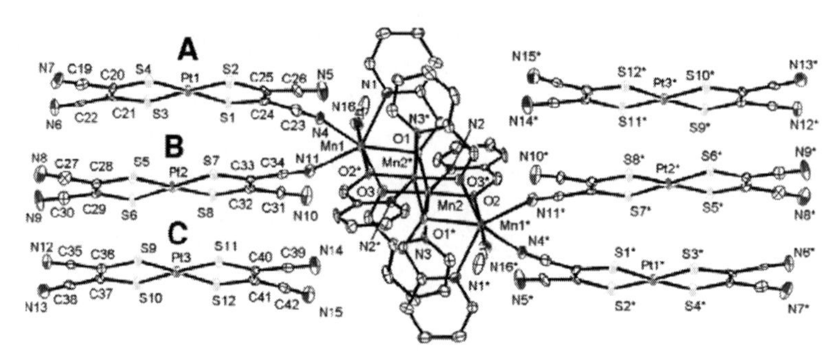

Figure 2.5 ORTEP view of **2** with atomic numbering scheme for selected atoms (50% probability thermal level), where symmetry operation (*): $2 - x, 1 - y, 2 - z$.

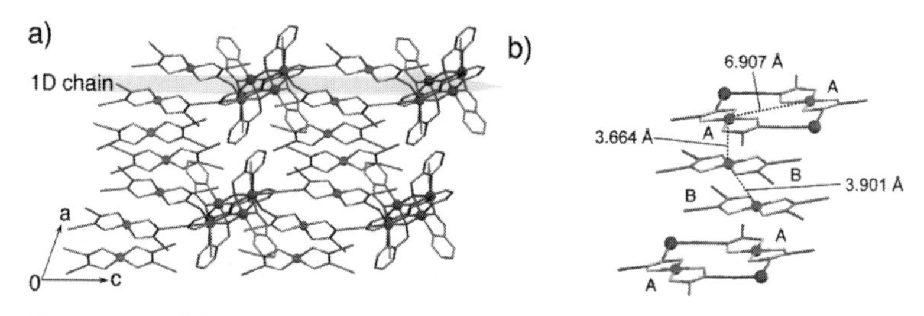

Figure 2.6 (a) Packing view of **1** and (b) arrangement of part of $[Pt(mnt)_2]^{n-}$ moieties.

Packing view of **1** is shown in Fig. 2.6. The bonding of $[Mn_4]^{4+}$ with two $[Pt(mnt)_2]^-$ molecules (**A**) forms a 1-D chain of {–$[Mn_4]$–$(\mathbf{A})_2$–} running along the (011) direction. The uncoordinated $[Pt(mnt)_2]^-$ units (**B**) are located between chains, forming a stair-like broken column of [**A**···**B**···**B**···**A**] arrangement, as shown in Fig. 2.6b, in which the Pt···Pt distances of **A**···**B**, **B**···**B**, and intrachain **A**···**A** are 3.664, 3.901, and 6.907 Å, respectively. Note that the platinum ions of **A** and **B** are assigned to Pt^{3+} (integer state) from the charge balance, the structure, and the IR spectra, all of the $[Pt(mnt)_2]^-$ units being local paramagnetic species with $S = ½$.[30] The packing view of **2** is shown in Fig. 2.7. The uncoordinated $[Pt(mnt)_2]^{n-}$ (**C**) and coordinated **A** and **B** are mutually stacked along the *a*-axis to form a segregated 1-D double-column possessing an [···**A**···**B**···**C**···] repeat unit (Fig. 2.7b) in which the Pt···Pt distances of **A**···**B**, **B**···**C**, and **C**···**A** are 3.577, 3.464, and 4.363 Å, respectively. Considering the charge balance, the structures of **A**, **B**, and **C**, and the IR spectra, the Pt(1), Pt(2), and Pt(3) ions have an average charge of +3.33 (noninteger average state), that is, forming $[Pt(mnt)_2]^{0.66-}$. Note that in both **1** and **2**, the easy axis of the $[Mn_4]^{4+}$ unit is aligned parallel to the same direction in their crystal packing.

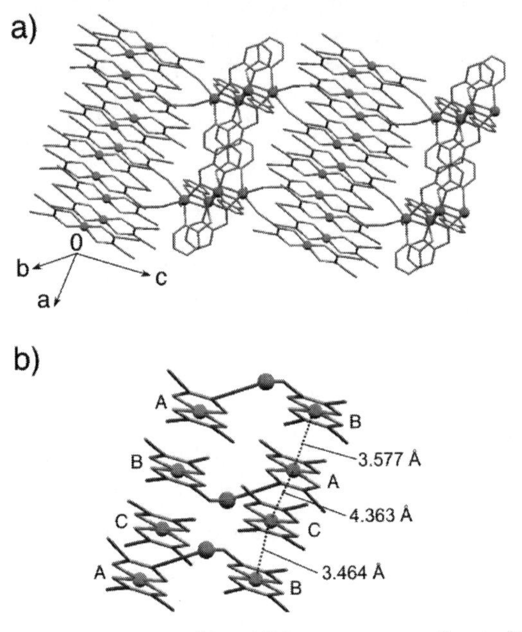

Figure 2.7 (a) Packing view of **2** and (b) arrangement of part of $[Pt(mnt)_2]^{n-}$ moieties.

2.2.3 Physical Properties of $[\{Mn^{II}_2Mn^{III}_2(hmp)_6(MeCN)_2\}\{Pt(mnt)_2\}_2][Pt(mnt)_2]_2\cdot 2MeCN$ and $[\{Mn^{II}_2Mn^{III}_2(hmp)_6(MeCN)_2\}\{Pt(mnt)_2\}_4][Pt(mnt)_2]_2$

2.2.3.1 Electrical conductivity

The electronic conductivity of the single crystals of **1** and **2** was measured by the two- and four-probe methods in the temperature range of 5–300 K. Material **1** is an insulator ($\rho > 10^7$ Ω cm at room temperature). In contrast, **2** shows transport property when the probes are attached in the direction of the $[Pt(mnt)2]^{0.66-}$ column (a-axis direction). The conductivity at room temperature is σ = 0.22 S cm^{-1}, which decreases gradually with the decreasing temperature, and **2** is no longer conducting at approximately

110 K (Fig. 2.8), as indicated by its semiconducting behavior with an activation energy of 136 meV, between a valence band and a conducting band. This insulating behavior with the decreasing temperature would produce magnetically interacting spins on the $[Pt(mnt)_2]^{n-}$ moieties. Indeed, interactions between localized spins in the $[Pt(mnt)_2]^{n-}$ moieties were detected by measuring high-field and high-frequency EPR (HF-EPR) spectra even in the semiconductor **2**.[30]

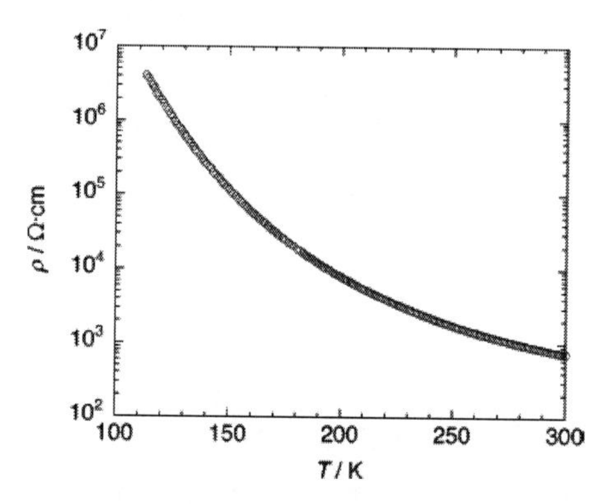

Figure 2.8 Temperature dependence of resistivity for a single crystal of **2** using four-probe attachment along the $[Pt(mnt)_2]^{n-}$ column.

2.2.3.2 Magnetic properties

The temperature dependence of magnetic susceptibilities for **1** and **2** were measured on a polycrystalline sample at 0.1 T in the temperature range of 1.81–300 K. Figure 2.9 displays χT vs. T plots of **1** and **2**. The χT products of **1** and **2** increase from 17.04 and 16.44 $cm^3 \cdot K \cdot mol^{-1}$ at 300 K to 44.10 $cm^3 \cdot K \cdot mol^{-1}$ at 2.2 K and 38.80 $cm^3 \cdot K \cdot mol^{-1}$ at 2.7 K, respectively, and then decrease slightly to 43 and 37.75 $cm^3 \cdot K \cdot mol^{-1}$ at 1.81 K, respectively. That the χT value of **1** at 300 K is large compared to that of only a Mn_4 cluster (15–16 $cm^3 \cdot K \cdot mol^{-1}$) is due to the contribution of four $[Pt(mnt)_2]^-$ units with $S = 1/2$, whereas that of **2** is the same degree as that for an isolated $[Mn_4]$ cluster, in which the contribution from the six $[Pt(mnt)_2]^{n-}$ units should be negligible.[38] It should be noted that the entire plot displays a typical intracluster ferromagnetic behavior seen in related $[Mn_4]$ clusters.[3,9,25,26] The slight decrease at low temperatures is probably due to the zero-field splitting (ZFS) effect and/or due to the interunit antiferromagnetic interaction between $[Mn_4]^{4+}$ and $[Pt(mnt)_2]^{n-}$ units through coordination bonds and/or between respective units through space, although the contribution of the interunit interaction is minimal. Therefore, to evaluate intracluster couplings between Mn^{III} ions (J_{bb}) and Mn^{III} and Mn^{II} ions (J_{wb}), the χT behavior at temperatures above 15 K was simulated based on a Heisenberg–Van Vleck model [$H_1 = -2J_{bb}(S_{Mn2}S_{Mn2*}) - 2J_{wb}(S_{Mn1} + S_{Mn1*})-(S_{Mn2} + S_{Mn2*})$], which is generally employed for related compounds, where $\mathbf{S}_{Mn1}$ and $\mathbf{S}_{Mn1*}$ are the $S = 5/2$ operator for Mn^{II} ions and $\mathbf{S}_{Mn2}$ and $\mathbf{S}_{Mn2*}$ are the $S = 2$ operator for Mn^{III} ions.[3,9,25,26] The contribution of four noninteracting $S = 1/2$ spins from the $[Pt(mnt)_2]^-$ units in **1** were taken into account ($g = 2$). With temperature independence of paramagnetism (TIP) at 6×10^{-4} $cm^3 \cdot mol^{-1}$ common to both compounds,[30] the following parameter sets gave respective adequate fits as shown by the red lines in Fig. 2.9: $g = 1.99$, $J_{wb}/k_B = +0.62(2)$ K, and $J_{bb}/k_B = +4.33(6)$ K for **1**, and $g = 2.02$, $J_{wb}/k_B = +0.58(2)$ K, and $J_{bb}/k_B = +10.03(10)$ K for **2**.[30] These parameter sets are consistent with those of related $[Mn_4]$ SMM clusters [3,9,25,26] and induce an $S_T = 9$ ground state.

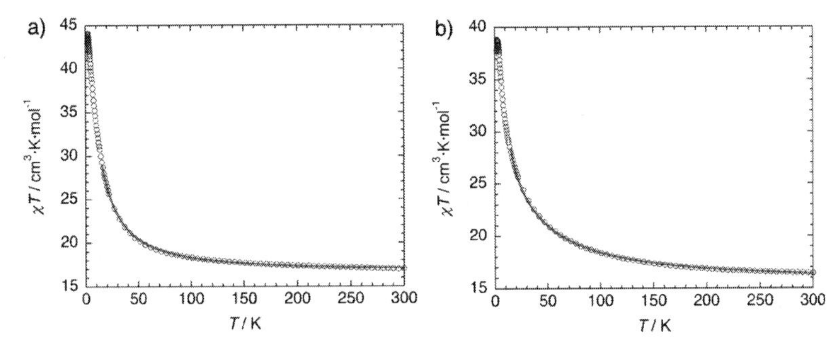

Figure 2.9 Temperature dependence of χT for (a) **1** and (b) **2** measured at 1 kOe. Solid lines represent the best fit obtained with a tetranuclear model, for which the contribution of four noninteracting paramagnetic species with $S = 1/2$ as $g = 2$ is taken into account for **1**. See also Color Insert.

Examination of the magnetic behavior at low temperatures allows us to know the SMM character of the $[Mn_4]^{4+}$ unit and possible interunit interactions. The temperature dependence of ac susceptibilities measured at several ac frequencies (3 Oe ac field and zero dc field) exhibited frequency dependence of both in-phase (χ') and out-of-phase (χ'') components below 3 K (Fig. 2.10 for **1** and

Fig. 2.11 for **2**). Within the frequency range of 1–1488 Hz, quantifiable distinct peaks of χ' were observed only in **2** at temperatures above 1.8 K, inducing $\tau_0 = 3.8 \times 10^{-8}$ s and $\Delta_{eff}/k_B = 18.7$ K from the Arrhenius relation $[\tau(T) = \tau_0 \exp(\Delta_{eff}/k_B T)]$ (inset of Fig. 2.11). Distinct peaks were detectable in both **1** and **2** as the dc field was increased and the characteristic dc field dependence could be observed, as shown in Fig. 2.12, revealing a thermal magnetization relaxation process in which quantum tunneling of the magnetization (QTM) is suppressed.[39] Figure 2.13 shows the field dependence of relaxation time estimated from the data of Fig. 2.12. The Δ_{eff} value was changed to 20.8 K at 1700 Oe for **1** and 21.9 K at 600 Oe for **2**, where these values were obtained by measuring temperature dependence of ac susceptibility (e.g., Fig. 2.14 for **1**). These behaviors prove that the $[Mn_4]^{4+}$ unit in **1** and **2** has SMM character. It should be noted that we could observe in **1** a growing small anomaly of χ'' in the low-frequency region in addition to the frequency-/field-dependent main signals in the high-frequency region (Fig. 2.14). This anomaly might be due to interactions between the $[Mn_4]^{4+}$ units via the coordinating $[Pt(mnt)_2]^-$ units, although it should be very weak, but it is not certain as this effect does not affect the essentials of the $[Mn_4]^{4+}$ SMM confirmed by other magnetic data. Figure 2.11 shows the field dependence of the magnetization measured on field-oriented single crystals of **1** and **2** at 470 mK with a 2 Oe/s sweep rate. For both compounds, a hysteresis loop typical of SMM was observed with two distinct steps (H_1 and H_2), although the coercivity of **1** was smaller than that of **2** (170 Oe for **1** and 900 Oe for **2**). The central step (H_1) of the hysteresis associated with the ground state QTM shifted to ±200 Oe for **1**, whereas it was observed at zero field for **2**. Neglecting ambiguity of the crystal orientation, this shift in **1** confirms the presence of weak interunit antiferromagnetic (zJ') already suggested by the anomaly of low-frequency ac susceptibility under dc fields, where zJ' is defined in a Hamiltonian $H = H_1 + 2zJ' \langle S_T \rangle$, using mean-field approximation with the number (z) of interacting neighbor units. Nevertheless, the estimated zJ' value is approximately −0.001 K and is negligible.[20] Thus, the $[Mn_4]$ SMM magnetism is almost isolated in the magnetic systems of **1** and **2**, even if the paramagnetic $[Pt(mnt)_2]^{n-}$ units are linked to the $[Mn_4]$ SMM unit. The second step (H_2) was observed at *ca.* 3000 Oe for both compounds; however, it remains uncertain if the second anomaly is due to QTM from $m_s = +9$ to $m_s = -8$.

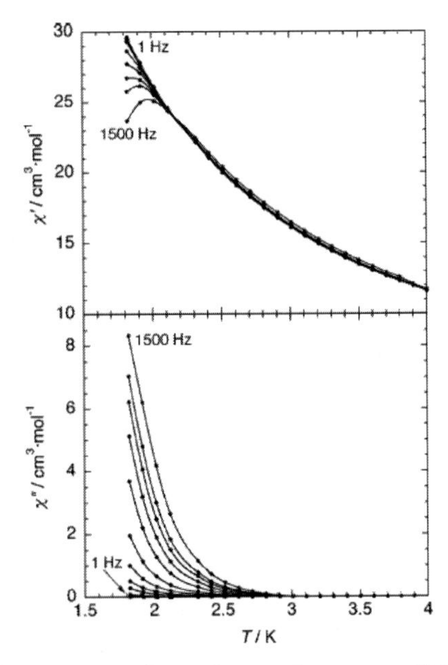

Figure 2.10 Temperature dependence of ac susceptibilities of **1** measured at several frequencies in the range of 1–1500 Hz.

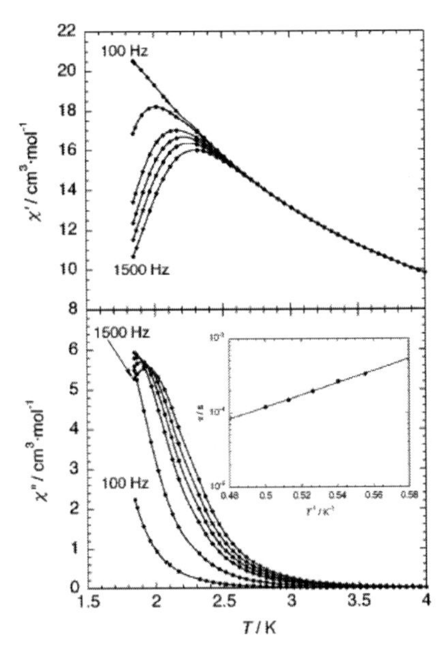

Figure 2.11 Temperature dependence of ac susceptibilities of **2** measured at several frequencies in the range of 100–1500 Hz.

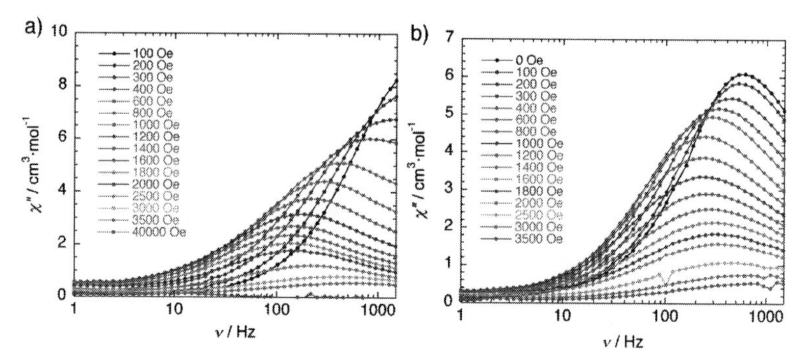

Figure 2.12 Frequency dependence of χ'' for (a) **1** and (b) **2** at 1.8 K as a function of the dc field. See also Color Insert.

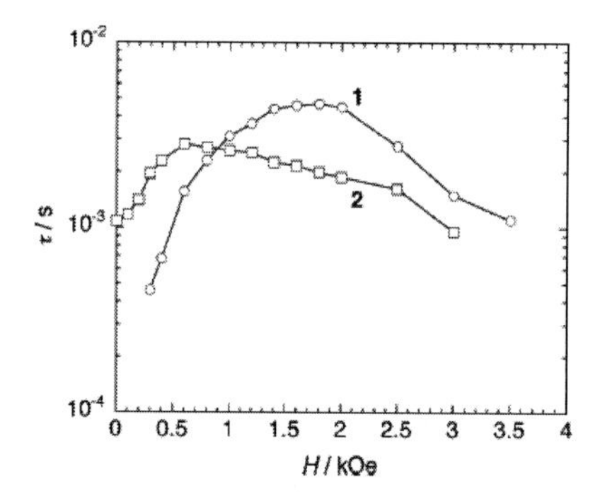

Figure 2.13 Field dependence of relaxation time (τ) estimated from the peak top of χ'' in Fig. 2.12.

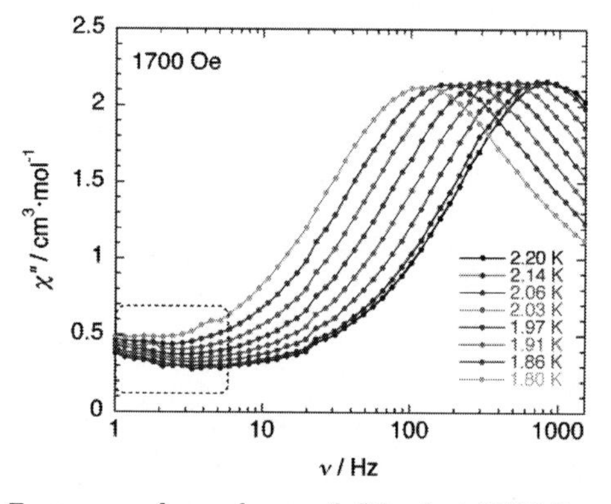

Figure 2.14 Frequency dependence of χ'' for **1** at 1700 Oe as a function of temperature. An anomaly could be seen in the low-frequency region (dotted square). See also Color Insert.

2.2.4 Conclusion

Two unique materials **1** and **2** based on the $[Mn_4(hmp)_6(MeCN)_2]^{4+}$ SMM unit and the electron-conductible $[Pt(mnt)_2]^{n-}$ unit were selectively synthesized by two methods, the material diffusion method and electrochemical oxidation, respectively. In both compounds, the $[Mn_4(hmp)_6(MeCN)_2]^{4+}$ unit was preserved, whereas the $[Pt(mnt)_2]^{n-}$ unit was present in 1:4 and 1:6 ratios of $[Mn_4]^{4+}$/$[Pt(mnt)_2]^{n-}$ for **1** and **2**, respectively. Material **1** has an SMM/insulating character, whereas **2** exhibits an SMM/semiconducting character. It is interesting to note that such an example as **2** in which noninteger oxidized $[Pt(mnt)_2]^{n-}$ subunits allowed conductivity is rare; only lithium salts of $[Pt(mnt)_2]^{n-}$ have been reported so far.[14,36] However, the conductivity was present only in the temperature region, where the SMM unit behaved as a general paramagnet. In addition, both characteristic components of SMM and conducting aggregations acted independently in the entire temperature range. Nevertheless, this work is the first to open doors to the design of superparamagnetic/*highly* conductive hybrid materials. The modification of respective starting materials and the variation of their combination would make it possible to yield intriguing materials composed of sets of SMM and conductible molecules.

2.3 Manganese-Salen Type Out-of-Plane Dimer SMMs with Metal–Dithiolene Complex: Integer Oxidation State of the Metal–Dithiolene Moiety

2.3.1 Introduction

An interesting strategy toward the tuning and modification of SMM characteristics is to design multifunctional molecules,[16,19,40] in which other properties such as electrontransfer phenomenon, electrical conductivity, dielectric property, spin-crossover behavior, and so forth would be targets in synergy with the SMM behavior. However, such SMM-based hybrid systems have never been reported so far because of the difficulty in associating the desired functional building units in such a way that they interact. Hence, it is becoming of great importance for this field of research to find useful approaches for the molecular design of these multifunctional systems. The first example, as shown in previous section, $[\{Mn^{II}_2Mn^{III}_2(hmp)_6(MeCN)_2\}\{Pt(mnt)_2\}_4][Pt(mnt)_2]_2$, has been synthesized electrochemically in a solution containing functional building blocks, $[Mn^{II}_2Mn^{III}_2(hmp)_6(MeCN)_2(H_2O)_4](ClO_4)_4$ and $(NBu_4)[Pt(mnt)_2]$.[30] This material behaves as an SMM and also possesses semiconductor properties, thanks to the network of mixed-valent $[Pt(mnt)_2]^{n-}$ stacks. In addition, four $[Pt(mnt)_2]^{n-}$ molecules coordinate with the SMM moiety to form a unique $[\{Mn^{II}_2Mn^{III}_2(hmp)_6(MeCN)_2\}\{Pt(mnt)_2\}_4]^{m-}$ SMM unit. No significant electronic interaction between the SMM unit and the conducting column was observed. Following these new synthetic strategies toward SMM-based multifunctional materials with a synergy of electrical and magnetic properties, Hiraga *et al.* present in this paper a new class of conducting and magnetic materials: $[Mn^{III}(5\text{-}Rsaltmen)\{M^{III}(dmit)_2\}]_2$ (R = MeO, M^{III} = Ni, **3a**; M^{III} = Au, **3b**; R = Me, M^{III} = Ni, **4a**; M^{III} = Au, **4b**). These compounds have a simple Mn(III) saltmen out-of-plane dinuclear core of formula (Scheme 2.3.1) and $[Mn^{III}(5\text{-}Rsaltmen)(X)]_2$, where X– is a coordinating counter anion corresponding to $[M^{III}(dmit)_2]^-$ in the present compounds. Since the coordinating $[M^{III}(dmit)_2]^-$ moiety is a monoanion, these compounds consist of a single-component neutral unit having a linear tetranuclear $[M^{III}–Mn^{III}–Mn^{III}–M^{III}]$ skeleton. Intermolecular contacts of the $[Ni^{III}(dmit)_2]^-$ (S = 1/2) moieties in **3a** and **4a** induce a strong spin dimerization that makes them magnetically silent even at room temperature. Nevertheless, **3a** and **4a** exhibit semiconducting properties, while **3b** and **4b** behave as electrical insulators, as expected in the presence of diamagnetic $[Au^{III}(dmit)_2]^-$ anions. Interestingly, the magnetic properties of **3a** and **4a** are completely different despite their similar structures; **3a**, **3b**, and **4b** behave as SMMs because of the good magnetic isolation provided by the $[M^{III}(dmit)_2]^-$ column, acting only as a magnetic buffer layer, while **4a** exhibits an antiferromagnetic 3-D order induced by a significant intermolecular Mn···Mn exchange through the singlet $[Ni^{III}(dmit)_2]_2^{2-}$ dimer that makes the semiconductor network. The present series of materials is a unique case in which a single-component molecule displays various electrical and magnetic properties such as semiconductor/SMM (**3a**), insulator/SMM (**3b** and **4b**), and semiconductor/antiferromagnet (**4a**).[41] In this section, crystal structures, electrical conductivity, and magnetic properties of these compounds are exhibited.

Scheme 2.3.1

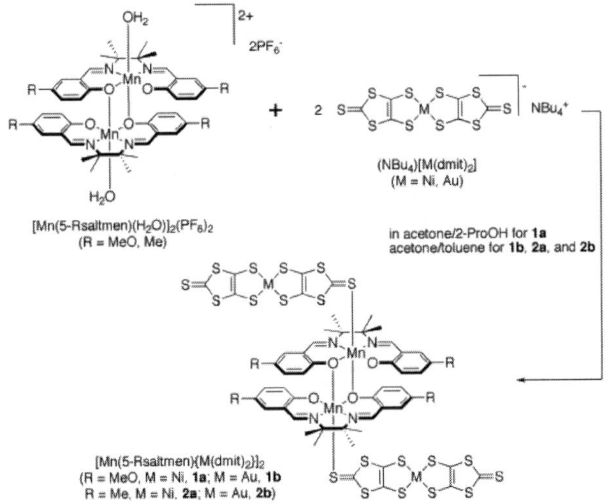

$[Mn(5\text{-Rsaltmen})(H_2O)]_2(PF_6)_2$
(R = MeO, Me)

$(NBu_4)[M(dmit)_2]$
(M = Ni, Au)

in acetone/2-ProOH for **1a**
acetone/toluene for **1b**, **2a**, and **2b**

$[Mn(5\text{-Rsaltmen})\{M(dmit)_2\}]_2$
(R = MeO, M = Ni, **1a**; M = Au, **1b**
R = Me, M = Ni, **2a**; M = Au, **2b**)

2.3.2 Structure of $[Mn(5\text{-Rsaltmen})\{M^{III}(dmit)_2\}]_2$ (R = Me, MeO; M = Ni, Au)

All compounds crystallize in the triclinic $P\bar{1}$ (no. 2) space group with the inversion center located at the midpoint of the Mn···Mn vector ($Z = 1$). The structures of the sets **3a**/**3b** and **4a**/**4b** are, respectively, isomorphous with a very similar structural building unit. ORTEP drawings of the asymmetrical units of **3a** and **4a** are depicted in Fig. 2.15, respectively. All compounds have an out-of-plane $[Mn^{III}_2(5\text{-Rsaltmen})_2]^{2+}$ dinuclear core capped in apical positions by one of 2-thioketone sulfur atoms of $[M(dmit)_2]^-$ anion, forming a linear-type assembly of $[(dmit)\text{–}M\text{–}(dmit)\text{–}Mn\text{–}(OPh)_2\text{–}Mn\text{–}(dmit)\text{–}M\text{-}(dmit)]$, where $-(OPh)_2^-$ is the biphenolate bridge of the out-of-plane dimer. The hexa-coordinated Mn^{III} ion has an apically distorted octahedral geometry with a Jahn–Teller elongation axis of [S(8)–Mn(1)–OPh(1a)] with Mn(1)–S(8) = 2.9936(12) Å for **3a**, 2.9978 (12) Å for **3b**, 2.7172(10) Å for **4a**, and 2.7257(12) Å for **4b**; Mn(1)–O(1a) = 2.389(2) Å for **3a**, 2.395(3) Å for **3b**, 2.557(2) Å for **4a**, and 2.501(3) Å for **4b**; and S(8)–Mn(1)–O(1a) = 163.21(7)° for **3a**, 163.39(8)° for **3b**, 163.96(5)° for **4a**, and 163.68(6)° for **4b**. On the basis of this coordination sphere geometry, the Mn ions (Mn(1) and Mn(1a)) are clearly trivalent. The Jahn–Teller axes at Mn(1) and Mn(1a) are parallel to each other because of the existence of an inversion center located at the midpoint of the out-of-plane dimer (see Scheme 2.3.1). The most important structural difference between the sets **3a**/**3b** and **4a**/**4b** is found in the dihedral angle made by the coordinating dmit plane defined by S(7)–S(8)–S(9)–C(27) and the Mn(saltmen) plane defined by O(1)–O(2)–N(2)–N(1)–Mn(1): 88.87(6)° and 88.46(11)° for **3a** and **3b**, respectively, and 14.84(15)° and 14.23(11)° for **4a** and **4b**, respectively. The bond angle of Mn(1)–S(8)–C(27) is 126.97(15)° and 127.78(15)° for **3a** and **3b**, respectively, and 103.01(11)° and 101.98(12)° for **4a** and **4b**, respectively, which also reveals a large difference between the structures of the sets **3a**, **3b**, **4a**, and **4b**.

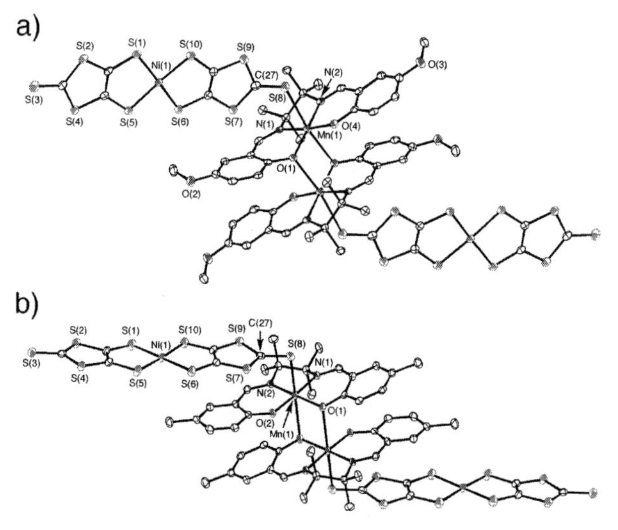

Figure 2.15 ORTEP views of (a) **3a** and (b) **4a** with atomic numbering schemes for selected atoms (50% probability thermal level).

The present single-component materials made of [Mn(5-Rsaltmen){M(dmit)$_2$}]$_2$ complexes display strong intermolecular interactions formed by π–π and S···S contacts from both dinuclear Mn core and [M(dmit)$_2$]$^-$ moieties. A true segregation is observed with two kinds of layers (Fig. 2.16 shows the packing features of **3a** and **4a**, those of **3b** and **4b** being essentially identical to them, respectively). The [Mn$_2$(5-Rsaltmen)$_2$] moieties arranged in the a-axis direction make weak C···C, C···O, and O···O contacts in the phenyl ring and 5-R substituent of the 5-Rsaltmen ligand (C···O ≈ 3.1 Å for **3a** and **3b** and C···C ≈ 3.4 Å for **4a** and **4b**). The most important contacts are seen between [M(dmit)$_2$]$^-$ moieties that stack and form stair-like zigzag columns along the *a*-axis direction. These columns are composed of face-to-face π–π and S···S interactions (M···M distance: 4.129 Å for **3a**, 4.164 Å for **3b**, 4.256 Å for **4a**, 4.238 Å for **4b**) and side-by-side S···S interactions (M···M distance: 6.433 Å for **3a**, 6.422 Å for **3b**, 7.384 Å for **4a**, 7.512 Å for **4b**) between adjacent $[M(dmit)_2]^-$ molecules. In both interacting modes, the shortest one is found in the range of 3.5–3.7 Å for **3a** and **3b** and 3.6–3.7 Å for **4a** and **4b**.[33]

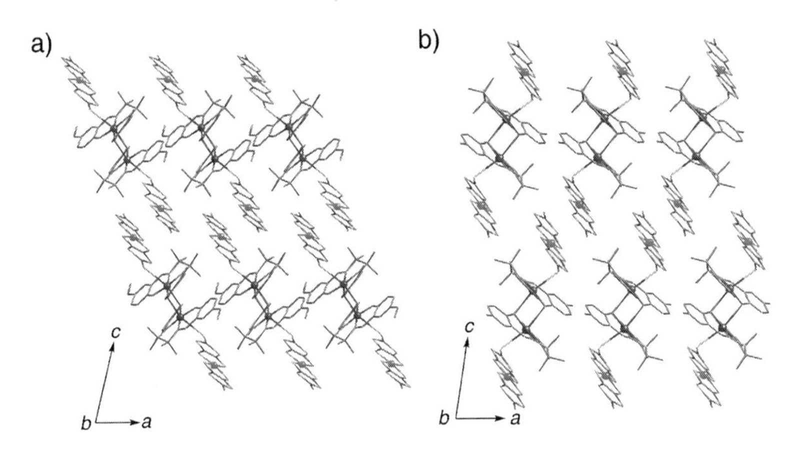

Figure 2.16 Packing arrangements of (a) **3a** and (b) **4a** projected along the *b*-axis.

2.3.3 Physical Properties of $[Mn(5\text{-}Rsaltmen)\{M^{III}(dmit)_2\}]_2$ (R = Me, MeO; M = Ni, Au)

2.3.3.1 Electrical conductivity

While for **3b** and **4b,** the conductivity, σ, was at less than 10^{-6} S cm^{-1} (insulator) at room temperature, **1a** and **2a** yielded the measurable conductivity as $\sigma = 7 \times 10^{-4}$ S cm^{-1} and 1×10^{-4} S cm^{-1} at 300 K (using the two-probe method), respectively. Therefore, the temperature dependence of σ was measured for **3a** and **4a** between 300 and 50 K (using a two-probe method). These compounds show a measurable transport property when the probes are attached in the direction of the $[M(dmit)_2]^-$ stair-like column (the *a*-axis direction). The conductivity of **3a** and **4a** in this direction decreases gradually with decreasing temperature and becomes too small to be measured by our technique at approximately 150 and 220 K, respectively. This temperature dependence of **3a** and **4a** reveals their semiconducting behavior with activation energies of 0.182 and 0.292 eV, respectively.[33] Despite the integer valence of the $[Ni(dmit)_2]^-$ moiety, the observed conductivities are surprisingly large.

2.3.3.2 DC magnetic property

Temperature dependence of χT measured at 1 KOe for **3a, 3b, 4a, 4b** are shown in Fig. 2.17. The susceptibility obeys the Curie–Weiss law through the whole temperature range for **3a, 3b,** and **4b** and above 20 K for **4a** with $C = 5.9$ cm^3·K·mol^{-1}, $\theta = +5$ K for **3a**; $C = 5.9$ cm^3·K·mol^{-1}, $\theta = +5$ K for **3b**; $C = 6.32$ cm^3·K·mol^{-1}, $\theta = -16$ K for **4a**; and $C = 5.8$ cm^3·K·mol^{-1}, $\theta = +5$ K for **4b**. The Curie constants are very similar to each other independent of the M^{III} metal ion (Ni^{III}, $S = 1/2$, or Au^{III}, $S = 0$) and very close to the expected spin-only value at 6 cm^3·mol^{-1} for two isolated $S = 2$ Mn(III) spins ($g = 2$).[33] The χT products of **1a** slightly decrease from 6.09 cm^3·K·mol^{-1} at 300 K to 6.03 cm^3·K·mol^{-1} at 185 K and then increase to 7.67 cm^3·K·mol^{-1} at 1.41 K followed by a decrease to 6.16 cm^3·K·mol^{-1} at 0.48 K. Although the slight decrease at high temperatures might be attributed to the contribution of the excited triplet spins of dimerized $[Ni^{III}(dmit)_2]^-$, the χT value at room temperature is consistent with

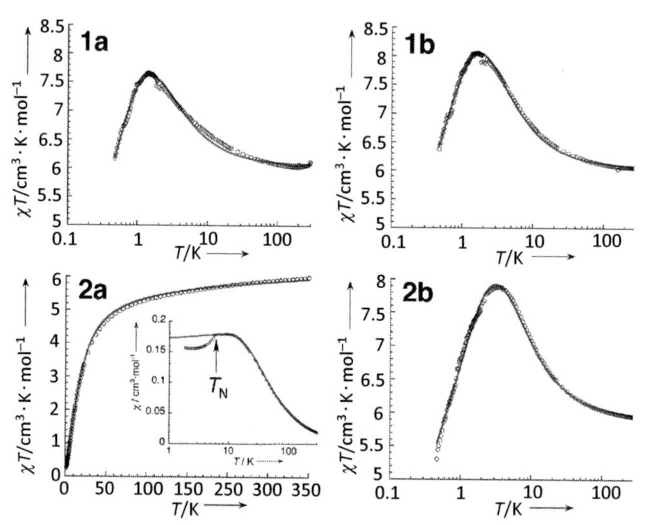

Figure 2.17 Temperature dependence of χT measured at 1 kOe for **3a, 3b, 4a,** and **4b**. Solid red lines represent the best fits obtained with the models described in the text. See also Color Insert.

the value expected from the spin-only value of two $S = 2$ excluding the diamagnetic pairs of $S = 1/2$ $[Ni^{III}(dmit)_2]^-$ units. As a confirmation, the global χT behaviors of **3a** and **3b** with $[Au^{III}(dmit)_2]^-$ ($S = 0$) are essentially identical: the χT products of **3b** slightly increase from 6.06 at 300 K to 8.06 $cm^3 \cdot K \cdot mol^{-1}$ at 1.6 K and then decrease to 6.36 $cm^3 \cdot K \cdot mol^{-1}$ at 0.47 K. Therefore, the magnetic behavior of **3a** as well as of **3b** should be only due to the contribution of Mn(III) dimer. To evaluate the intradimer magnetic coupling between Mn^{III} ions via a biphenolate bridge, the χT vs. T plots of **3a** and **3b** were simulated by a Heisenberg $S = 2$ dimer model, taking into account the magnetic anisotropy arisen from each Mn(III) ion (ZFS, D_{Mn}) and intermolecular interactions (zJ') introduced in the frame of the mean-field approximation ($H = -2J_{Mn\text{-}Mn}S_{Mn1}S_{Mn2} + 2D_{Mn}S_{Mn,z}{}^2$, where $S_{Mn,z}$ is the z component of the S_{Mni} spin vectors).[10,39,42] The best simulations for **3a** and **3b** are given as red lines in Fig. 2.17, and the obtained parameter sets are $g = 2.01$, $J_{Mn\text{-}Mn}/k_B = +0.23$ K, $D_{Mn}/k_B = -0.71$ K, and $zJ'/k_B = -0.03$ K for **3a** and $g = 2.01$, $J_{Mn\text{-}Mn}/k_B = +0.31$ K, $D_{Mn}/k_B = -0.95$ K, and $zJ'/k_B = -0.02$ K for **3b**. These parameters are very similar to each other, but $J_{Mn\text{-}Mn}$ and D_{Mn} are relatively smaller than those of related compounds previously reported.[10,39,42–45] On the other hand, the temperature dependence of χT for **4a** is completely different from those of **3a** and **3b**. The χT product monotonically decreases from 5.68 $cm^3 \cdot K \cdot mol^{-1}$ at 300 K to 0.27 $cm^3 \cdot K \cdot mol^{-1}$ at 1.8 K. In addition, χ shows a maximum at 6.4 K followed by a decrease, suggesting the occurrence of a long-range antiferromagnetic order at this temperature (T_N; inset of Fig. 2.17). As in **3a**, the χT value at 300 K suggests that the magnetic contribution of $[Ni^{III}(dmit)_2]^-$ units can be negligible, but it is not easy to judge if the main contribution for the antiferromagnetic coupling is coming from the Mn···Mn interaction via the biphenolate bridge or via the singlet $[Ni(dmit)_2]^-$ dimer. However, the magnetic behavior of **4b** is typical for ferromagnetically coupled Mn(III) dimers: the χT products slightly increase from 5.94 $cm^3 \cdot K \cdot mol^{-1}$ at 300 K to 7.90 $cm^3 \cdot K \cdot mol^{-1}$ at 3.43 K followed by a decrease to 5.30 $cm^3 \cdot K \cdot mol^{-1}$ at 0.46 K, essentially identical to those of **3a** and **3b**. The best set of parameters obtained fitting the experimental χT vs. T data with the previously anisotropic Heisenberg $S = 2$ dimer model is $g = 1.99$, $J_{Mn\text{-}Mn}/k_B = +0.57$ K, $D_{Mn}/k_B = -1.91$K, and $zJ'/k_B = -0.03$ K for **4b**. Considering the isostructural packing of **4a** and **4b** and magnetic properties of **4b**, the dominant antiferromagnetic coupling in **4a** should be due to an exchange pathway through a singlet $[Ni^{III}(dmit)_2]^-$ pair (i.e., $\{Mn^{III}–[Ni^{III}(dmit)_2]\cdots[Ni^{III}(dmit)_2]–Mn^{III}\}$ path; see Fig. 2.16). Indeed, on the basis of the crystal packing, chains of $\{Mn^{III}–[Ni^{III}(dmit)_2]\cdots[Ni^{III}(dmit)_2]–Mn^{III}\}$ units are formed as shown in Fig. 2.16, neglecting the much weaker inter-$[Ni^{III}(dmit)_2]_2$ dimer exchange.[33] Therefore, to evaluate the effective exchange between Mn(III) ions via the singlet $[Ni^{III}(dmit)_2]_2^{2-}$ dimer, J_{eff}, the magnetic properties (χ vs. T and χT vs. T data) of **4a** were fitted to a Heisenberg chain model with classical $S = 2$ spins and two alternated magnetic interactions: $J_{Mn\text{-}Mn}$ and J_{eff} (using the following Hamiltonian: $H = -2J_{eff}\Sigma S_{2i}S_{2i+1} - 2J_{Mn\text{-}Mn}\Sigma S_{2i+1}S_{2i+2}$). The analytical expression of the susceptibility derived from this model was reported by Cortés *et al.*,[46] who applied the Fisher approach.[47] It is worth noting that the effect of the magnetic anisotropy was neglected, as the data have been fitted only above 7 K in the paramagnetic phase. Using the full set of parameters (g, $J_{Mn\text{-}Mn}$, and J_{eff}) free leads to excellent fits of both χ vs. T and χT vs. T data, but multiple solutions were found as an indication of overparameterization of the model. Therefore, according to the isostructural packing of **4a** and **4b**, $J_{Mn\text{-}Mn}/k_B$ has been fixed at +0.57 K, found for **4b**. In this condition, the fits of both χ vs. T and χT vs. T data stay excellent as shown in Fig. 2.17, with $g = 1.97$ and $J_{eff}/k_B = -2.85$ K. Despite a long distance between Mn^{III} spins, the exchange through the singlet $[Ni^{III}(dmit)_2]\cdots[Ni^{III}(dmit)_2]$ pair, J_{eff} is surprisingly large. The origin of such a large exchange should be associated with the singlet–triplet excitation of the dimeric $[Ni^{III}(dmit)_2]^-$ moiety and might be similar to what is seen in a superexchange mechanism.

2.3.3.3 Magnetostructural correlation

Despite a similar structural motif of **3a** and **4a**, their magnetic properties are completely different. Considering the similarity between **3a** and **4a**, this result can be easily attributed to different Mn(III)$\cdots$Mn(III) interactions through $[Ni^{III}(dmit)_2]^-$ units that are effective in **4a** or not in **3a**. Indeed, to discuss this result, two important overlaps are relevant: (i) the one between $[Ni^{III}(dmit)_2]^-$ moieties and also (ii) those between the $[Ni^{III}(dmit)_2]^-$ group and the coordinating Mn(III) ion, especially the overlap between the HOMO on the terminal S (2-thioketone S) of $[Ni^{III}(dmit)_2]^-$ and the d_z^2 orbital of Mn(III) ion (note that when Mn(III) ion has $^5B_{1g}$ ground state with elongated Jahn–Teller distortion, four unpaired electrons are assigned as $d_{xy}^1\, d_{yz}^1\, d_{xz}^1\, d_{x2}^1$, where the Jahn–Teller axis is parallel to the z-axis[45]). According to molecular orbital calculations, the HOMO on the terminal S is made from one of the p orbitals,[41] which align perpendicular to the dmit plane, so the overlap with the d_{z2} orbital is strongly dependent on the angles of Mn–S–C and the dihedral angle between the dmit plane and the equatorial plane around the Mn ion. As mentioned in the structural section, huge geometrical differences are indeed observed on the **3a** and **4a** complex core. In particular, the small dihedral angle in **4a** (14.84°) should favor a strong σ-type overlap, which generally contributes to antiferromagnetic coupling (Fig. 2.18).[48] On the other hand, the large dihedral angle in **1a** (88.87°) should lead to a nonbonding situation (Fig. 2.18) and thus very weak Mn···$[Ni^{III}(dmit)_2]^-$ interactions. These structural differences are well in line with the magnetic behavior reported for **3a** and **4a** (Fig. 2.17), especially considering that the overlaps between the $[Ni^{III}(dmit)_2]^-$ moieties described above are quite similar and thus should not dramatically influence the observed properties.[33]

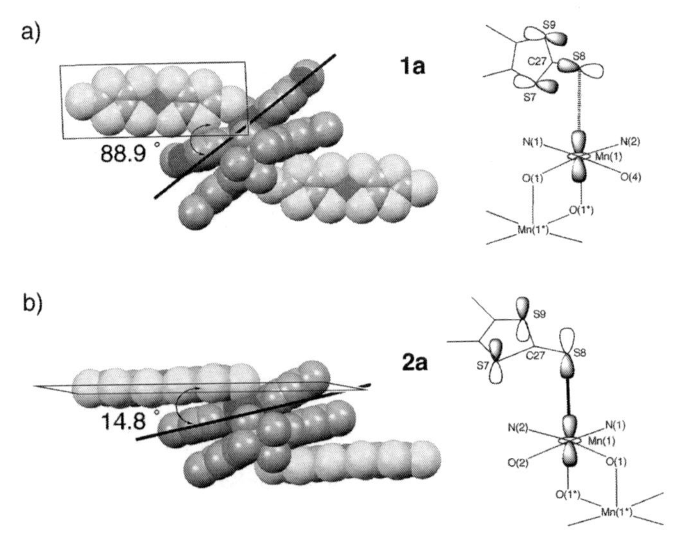

Figure 2.18 View of coordination modes between the $[Ni(dmit)_2]^-$ moiety and the Mn(III) metal ion in (a) **3a** and (b) **4a** (left), where the given angles are the dihedral angles between least-squares planes defined by S(7)–S(8)–S(9)–C(27) (for dmit) and O(1)–O(2)–N(2)–N(1)–Mn(1) (for Mn(saltmen)) and on the right, schematic orbital configurations of the d_{z2} orbital of Mn ion and a π orbital of S(8) in (a) **3a** and (b) **4a**.

2.3.3.4 AC magnetic property

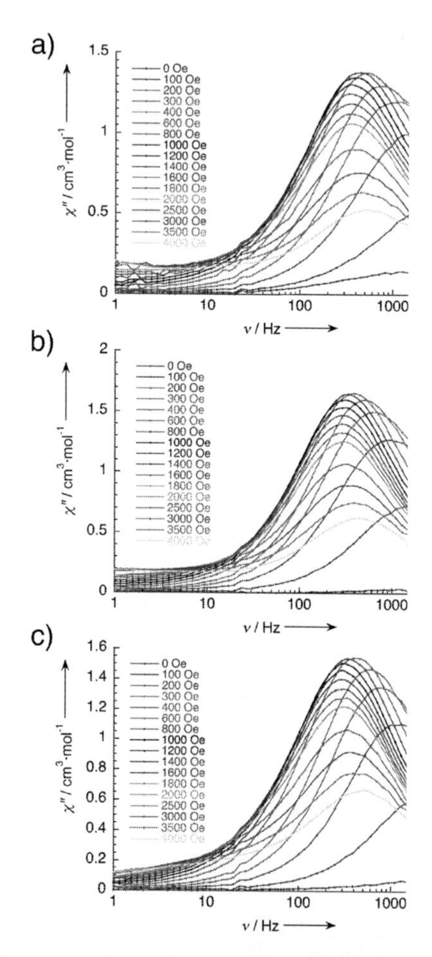

Figure 2.19 Frequency dependence of the out-of-phase component of ac susceptibility for (a) **3a**, (b) **3b**, and (c) **4b** at several applied dc fields at 1.8 K. See also Color Insert.

Compounds **3a**, **3b**, and **4b**, which are essentially ferromagnetic $[Mn^{III}_2(saltmen)_2]^{2+}$ dimers with an $S_T = 4$ ground state, are candidates to behave as SMMs as shown previously.[10,39,44] The temperature dependence of ac susceptibilities (10–1500 Hz) of these compounds were performed at 1.8 K under several dc fields (Fig. 2.19), at which the QTM pathway of relaxation is expected to be at least partially suppressed by Zeeman effects that remove the degeneracy of the $\pm m_S$ state levels.[39,43,49,50] With increasing dc field, clear χ'' peaks are observed for **3a**, **3b**, and **4b** (Fig. 2.19), and the characteristic dc field dependence of the relaxation time can be plotted as shown in Fig. 2.20. The maximum value was observed around 1500 Oe for **3a** and **3b** and 1200 Oe for **4b**, and τ_0 and Δ_{eff} at these fields were obtained as 8.2×10^{-7} s and 11.4 K for **3a**, 1.1×10^{-6}s and 12.5 K for **3b**, and 2.9×10^{-7} s and 13.6 K for **4b** by measuring temperature dependence of ac susceptibility and making Arrhenius plots as shown in Fig. 2.21. These behaviors prove that **3a**, **3b**, and **4b** exhibit typical SMM behavior.

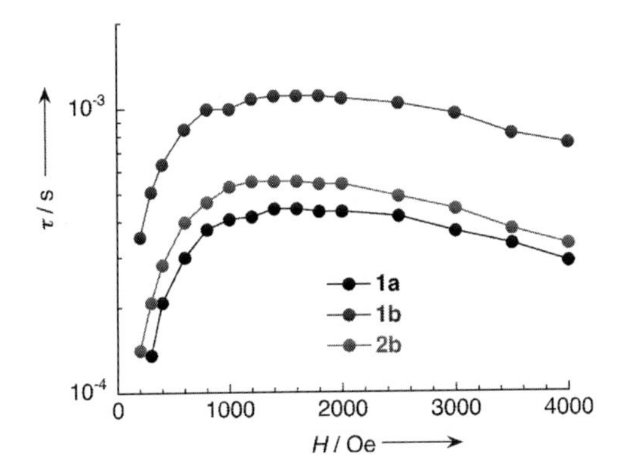

Figure 2.20 Field dependence of relaxation time (τ) at 1.8 K estimated from maximum of χ'' as shown in Fig. 2.19. See also Color Insert.

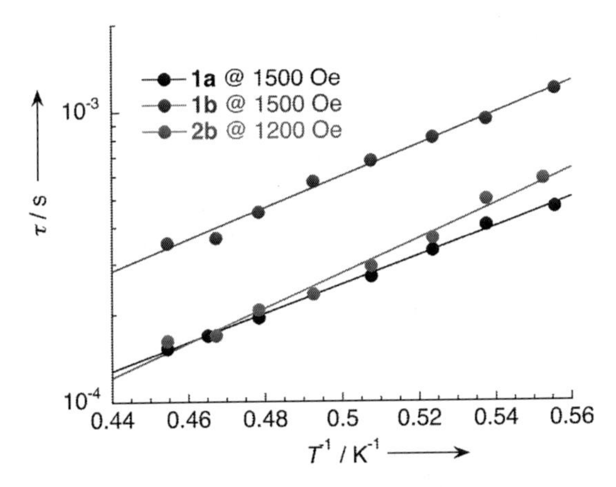

Figure 2.21 Arrhenius plots of the relaxation times measured at 1500 Oe for **3a** and **3b** and 1200 Oe for **4b**. See also Color Insert.

2.3.4 Conclusion

The aerobic reaction of $[Mn_2(5\text{-Rsaltmen})_2(H_2O)_2](PF_6)_2$ (R = MeO, Me) with $(NBu_4)[M(dmit)_2]$ (M = Ni, Au) in a 1:2 molar ratio (Mn: M = 1:1) in acetone/2-propanol or toluene yielded stoichiometric neutral assemblies with a formula of $[Mn(5\text{-Rsaltmen})\{M(dmit)_2\}]_2$ (R = MeO, M = Ni, **3a**; M = Au, **3b**; R = Me, M = Ni, **4a**; M = Au, **4b**) by a simple ion exchange removing $(NBu_4)PF_6$. These compounds are Mn(III) saltmen out-of-plane dimers decorated by $[M(dmit)_2]^-$ in apical positions, forming a linear type tetranuclear complex, $[M\text{–}(dmit)\text{–}Mn\text{–}(OPh)_2\text{–}Mn\text{–}(dmit)\text{–}M]$. The Mn ions are trivalent from the structural characteristics, and the $[M(dmit)_2]^-$ unit is consequently monovalent. Despite such a simple assembly composed of a single component, that is, single neutral complex, **3a** and **4a**, the Ni analogues (M = Ni) behave as semiconductors with $\sigma_{rt} = 7 \times 10^{-4}$ S·cm^{-1} and E_a = 182 meV for **3a** and $\sigma_{rt} = 1 \times 10^{-4}$ S·cm^{-1} and E_a = 292 meV for **4a** (σ_{rt} is the conductivity at room temperature and E_a is activation energy between the valence band and the conducting band), while **3b** and **4b** with M = Au are insulators. This conductivity results from the 1-D columnar arrangement of $[Ni(dmit)_2]^-$ moieties with π–π and S···S contacts along the a-axis direction. Nevertheless, the obtained electrical conductivity values of about 10^{-4} S·cm^{-1} stay weak because of the absence of mixed valence and the presence of a strong dimerization along the $[Ni(dmit)_2]^-$ columns. Indeed, Hückel calculations confirm the strong intermolecular dimerization in $[Ni(dmit)_2]^-$ columns leading to singlet $[Ni(dmit)_2]^-$ pairs. Thus, the $[Ni(dmit)_2]^-$ spins stay silent below 300 K, and the magnetic properties of **3a** are essentially identical to those of **3b** and **3b**. These compounds exhibit SMM properties with slow magnetization relaxation at low temperatures and fast zero-field QTM. On the other hand, the magnetic properties of **4a** are dominated by the inter-complex Mn···Mn antiferromagnetic interactions (J_{eff}/k_B = −2.85 K) via the singlet $[Ni(dmit)_2]^-$ dimer that stabilize a long-range antiferromagnetic order at T_N = 6.4 K. The relatively strong antiferromagnetic interactions via this singlet pair (despite a long Mn–Mn distance) are probably associated with the singlet-triplet excitation of the dimeric $[Ni^{III}(dmit)_2]^-$ moiety and might be similar to what is seen in a superexchange mechanism. The magnitude of this intermolecular Mn···Mn interaction is indeed mainly dependent on the Mn/$[Ni(dmit)_2]^-$ orbital orientations and the strength of the σ-type orbital overlap on the Mn-S bond.

2.4 Manganese–Salen Type Out-of-Plane Dimer with Metal–Dithiolene Complex: Noninteger Oxidation State of the Metal–Dithiolene Moiety

2.4.1 Introduction

Based on the success of $[\{Mn^{II}_2Mn^{III}_2(hmp)_6(MeCN)_2\}\{Pt(mnt)_2\}_2][Pt(mnt)_2]_2 \cdot 2MeCN$ (**1**) and $[\{Mn^{II}_2Mn^{III}_2(hmp)_6(MeCN)_2\}\{Pt(mnt)_2\}_4][Pt(mnt)_2]_2$ (**2**) (see section 2), Hiraga *et al.* have then tried a combination of the $[Mn_4]^{4+}$ SMM and $(NBu_4)[Ni(dmit)_2]$ in the same electrochemical oxidation method, because the use of $[Ni(dmit)_2]^{n-}$ promised to likely provide better electronic conductivity than the $[Pt(mnt)_2]^{n-}$ one due to its smaller HOMO–LUMO gap and preferring 2-D aggregation for electronic interaction.[13] Nevertheless, this reaction derived decomposition of the SMM core. These results indicate that the stability, i.e., structural stability and redox stability, of used SMM clusters not only versus the reaction medium but also versus used conducting molecules are very important factors to choose the SMM building block. The SMM family of Mn(III) salen-type out-of-plane dimers is one of the smallest nuclear SMMs[10,44] and is relatively redox stable at general condition.[45] Hence, this SMM family is a good candidate of the SMM building block for this type of reactions. In this vein, the reaction of a Mn(III) out-of-plane dimer, $[Mn^{III}{}_2(5\text{-MeOsaltmen})_2(H_2O)_2](PF_6)_2$, with $(NBu_4)[Ni(dmit)_2]$ was carried out in a set-up for electronic oxidation with different reaction media using acetone and MeCN, and two assembled compounds, $[Mn_2(5\text{-MeOsaltmen})_2(acetone)_2][Ni(dmit)_2]_7 \cdot 4acetone$ (**5**) and $[Mn_2(5\text{-MeOsaltmen})_2(MeCN)_2][Ni(dmit)_2]_7 \cdot 4MeCN$ (**6**), were finally obtained dependent on the used solvent. These compounds have a similar hybrid structure aggregating $[Mn_2]^{2+}$ layers and $[Ni(dmit)_2]^{n-}$ layers. As the third example of multifunctional materials exhibiting SMM property and electronic conductivity, the details of these compounds are described in this section.

2.4.2 Structures of $[Mn(5\text{-}MeOsaltmen)(solvent)]_2[Ni(dmit)_2]_7 \cdot 4(solvent)$ (solvent = acetone, MeCN)

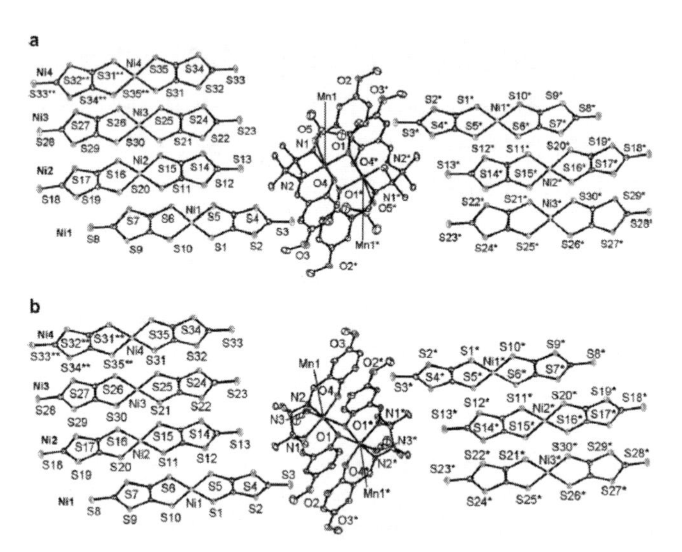

Figure 2.22 ORTEP drawings of (a) **5** and (b) **6** with 50% probability thermal level (symmetry operations *: $-x$, $1-y$, $1-z$; **: $2-x$, $1-y$, $-z$ for **5** and *: $1-x$, $1-y$, $2-z$; **: $3-x$, $1-y$, $1-z$ for **6**). Hydrogen atoms and crystallization solvents were omitted for clarity.

Both compounds crystallize in the same triclinic $P\bar{1}$ (#2) space group with $Z = 1$. ORTEP drawings of the asymmetrical unit of **5** and **6** are depicted in Fig. 2.22. Compounds **5** and **6** are not isomorphous but have similar structures composed of one Mn^{III} out-of-plane dimer, seven $[Ni(dmit)_2]^{n-}$, and four crystallization solvents that occupy void space in packing, in which the crystallographically independent moiety is half of this set, one $[Mn^{III}(5\text{-}MeOsaltmen)(S)]^+$ (S = acetone, **5**; MeCN, **6**), three complete $[Ni(dmit)_2]^{n-}$ moieties (Ni1, Ni2, Ni3), one-half $[Ni(dmit)_2]^{n-}$ moiety (Ni4), and two crystallization solvent molecules (S), with inversion centers on the midpoint of Mn···Mn vector in the dimer and the Ni4 of $[Ni(dmit)_2]^{n-}$. The out-of-plane dimer $[Mn^{III}_2(5\text{-}MeOsaltmen)_2(S)_2]^{2+}$ has a typical form of this type of compounds with shorter equatorial bonds [average bond distances: $[Mn\text{–}N]_{av}$ = 1.969 and 1.975 Å and $[Mn\text{–}O]_{av}$ = 1.885 and 1.883 Å for **5** and **6**] and elongated out-of-plane axis of [O(5)–Mn(1)–O(1*)] for **5** [Mn(1)–O(5) = 2.281(3) Å, Mn(1)–O(1*) = 2.404(3) Å, and O(5)–Mn(1)–O(1*) = 171.88(12)°] and [N(3)–Mn(1)–O(1*)] for **6** [Mn(1)–N(3) = 2.3215(18) Å, Mn(1)–O(1*) = 2.4258(7) Å, and N(3)–Mn(1)–O(1*) = 167.97(4)°], where the elongation of an axis is due to the Jahn–Teller distortion generally observed in hexa-coordinated Mn(III) geometry (symmetry operation (*): – *x*, 1 – *y*, 1 – *z* for **5**; 1 – *x*, 1 – *y*, 2 – *z* for **6**). The coordination geometry around Ni ion in $[Ni(dmit)_2]^{n-}$ is square planner and the bond distances of Ni–S are in the range of 2.140–2.175 Å in **5** and **6**, from which the distinct trend to define the oxidation state of the respective molecules cannot be found. Considering such a structural feature of $[Ni(dmit)_2]^{n-}$ units, in addition to IR data and the charge balance of components, the $[Ni(dmit)_2]^{n-}$ units possess a noninteger average oxidized state with *ca.* –0.29 for each without strong charge localization on specific one.[32]

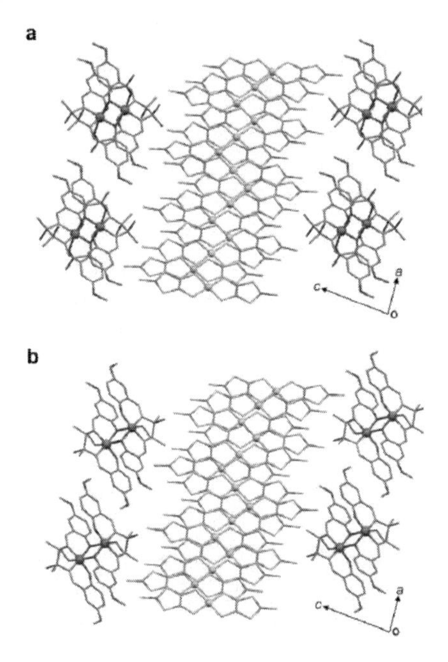

Figure 2.23 Packing diagrams of (a) **5** and (b) **6** projected along the *c**-axis. The crystallization solvents, acetone for **5** and MeCN for **6**, were omitted for clarity.

Packing views of **5** and **6** are depicted in Figs. 2.23. They are very similar to each other and are constructed in hybrid frames of aggregates consisting of $[Mn^{III}_2]^{2+}$ SMM layers and $[Ni(dmit)_2]^{n-}$ layers stacking along the c^* direction. Each component layer is spreading on the ab plane. In the $[Mn^{III}_2]^{2+}$ SMM layer, the arrangement along the a-axis direction forms weak π–π contacts of the 5-MeOsaltmen phenyl rings with its neighbors [the shortest contact is 3.471 Å for **5** and 3.247 Å for **6**]. Meanwhile, there is no significant contact in the b direction arrangement. In the $[Ni(dmit)_2]^{n-}$ aggregating layer, the $[Ni(dmit)_2]^{n-}$ units connected through a 2-D network of S···S contacts. This layer is arranged in a β-type mode.[32]

2.4.2 Physical Properties of $[Mn(5\text{-MeOsaltmen})(solvent)]_2[Ni(dmit)_2]_7 \cdot 4(solvent)$ (solvent = acetone, MeCN)

2.4.2.1 Electrical conductivity

The electrical conductivity of single crystals of **5** and **6** was measured by the four-probe method in the temperature range of 300–50 K. Complex **5** and **6** shows transport property when the probes are attached in the direction of the $[Ni(dmit)_2]^{n-}$ layers (a-axis direction), while the transport property of the layer-stacking direction (c^* direction) is very low like ρ_{rt} < 10^{-3} S·cm^{-1}. The conductivity at room temperature of **5** and **6** is ρ = 1.6 S·cm^{-1} and 2.8 S·cm^{-1}, respectively, which decreases gradually with decreasing temperature and then, is hardly detectable at approximately 80 K (Fig. 2.24), indicating their semiconducting behavior.[32]

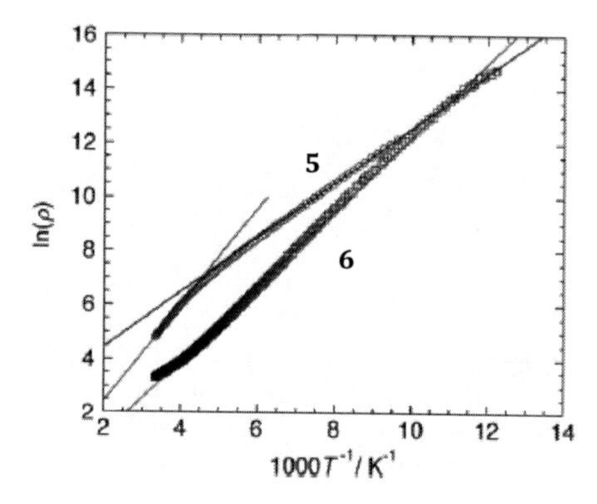

Figure 2.24 Ln(ρ) vs. 1/T plots for a single crystal of **5** and **6** using a four-probe attachment along the c^*-axis, where the solid lines represent least-squares linear fits.

2.4.2.2 Magnetic property

The temperature dependence of dc magnetic susceptibility was measured on polycrystalline samples of **5** and **6** at 0.1 T in the temperature range of 1.8–300 K. It should be first noted that the entire feature of dc magnetic properties of **5** and **6** is very similar to that of isolated $[Mn_2]^{2+}$ SMM. The susceptibility obeys the Curie–Weiss law with C = 5.9 $cm^3 \cdot K \cdot mol^{-1}$, θ = +5.5 K for **5** and C = 5.3 $cm^3 \cdot K \cdot mol^{-1}$, θ = +4.5 K for **6**. Despite that the positive Weiss constant (θ) indicating a ferromagnetic contribution among magnetic centers being dominant, the Curie constant (C) estimated from the high temperature behavior is close to or smaller than the spin-only value of 6.02 $cm^3 \cdot mol^{-1}$ (g = 2) only for isolated two S = 2 spins of Mn(III) ions, indicating that the magnetic contribution from the set of seven $[Ni(dmit)_2]^{0.29-}$ molecules is negligible at least in the high temperature region. This is because of their semiconductor character possessing diluted thermally excited carriers, and such a conclusion has been similarly accepted in the $[Mn_4]^{4+}/[Pt(mnt)_2]^{n-}$ hybrid compound[30] and other related compounds reported previously.[38,51] Figure 2.25 displays χT–T plots of **5** and **6**. The χT products of **5** and **6** increase from 5.95 $cm^3 \cdot K \cdot mol^{-1}$ for **5**, 5.35 $cm^3 \cdot K \cdot mol^{-1}$ for **6** at 300 K to 9.01 $cm^3 \cdot K \cdot mol^{-1}$ at 8.4 K for **5**, 7.72 $cm^3 \cdot K \cdot mol^{-1}$ at 8.3 K for **6**, and then decrease to 7.93 $cm^3 \cdot K \cdot mol^{-1}$ at 1.8 K for **5**, 6.55 $cm^3 \cdot K \cdot mol^{-1}$ at 1.8 K for **6**.

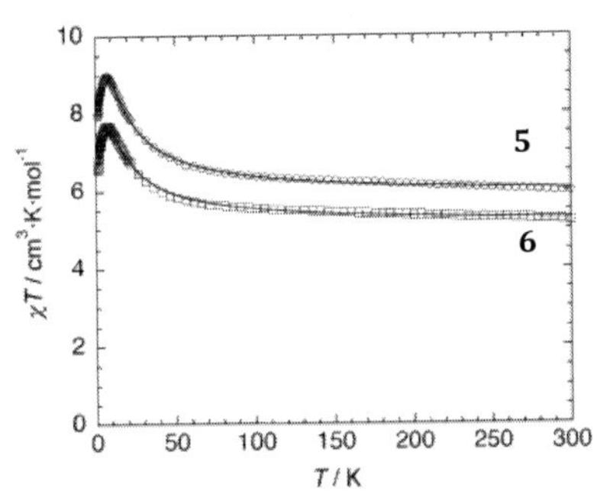

Figure 2.25 Temperature dependence of χT of **5** and **6** measured at 1 kOe. The solid lines represent best fit curves obtained with a dimer model of S = 2 (see text).

Taking into account the above mentioned hypothesis on magnetic contributors, the increase of χT is purely due to the ferromagnetic coupling between Mn(III) ions via the bis-phenolate bridge in the $[Mn_2]^{2+}$ dimer, and the slight decrease at low temperatures is due to the ZFS attributed to Mn(III) ions and/or intermolecular interactions. Therefore, the χT behavior of **5** and **6** were simulated using a Heisenberg dimer model of $S = 2$ taking into account ZFS of Mn(III) ion (D_{Mn}): $H = -2JS_{Mn1}S_{Mn2} + 2D_{Mn}S_{Mn,z}{}^2$, where $S_{Mn,z}$ is the z component of the $S_{Mn,i}$ spin vectors. The intermolecular interaction (zJ') was introduced in the frame of the mean-field approximation to simulate adequately the experimental data. The best-fitting curves for **5** and **6** are shown as solid lines in Fig. 2.25 and the parameters are $g = 1.98$, $J/k_B = 2.14$ K, $D_{Mn}/k_B = -4.41$ K, $zJ'/k_B = 0.03$ K for **5** and $g = 1.85$, $J/k_B = 2$ K, $D_{Mn}/k_B = -4.21$ K, $zJ'/k_B = 0.004$ K for **6**. The exchange coupling constants and ZFS parameters are typical values found in related [Mn_2] dimers reported previously, which induce an $S_T = 4$ ground state.[10,42,44] Only the g value of 2 seems to be a little small compared with others.

The temperature dependence of ac susceptibilities was measured at 3 Oe ac field and zero dc field changing ac frequencies from 1 Hz to 1500 Hz. As shown in Fig. 2.26, both in-phase (χ') and out-of-phase (χ'') components of **5** and **6** are frequency dependent on temperatures below 4 K, suggesting SMM behavior, but quantifiable distinct peaks of χ'' were not found above 1.8 K even at 1500 Hz frequency.[32]

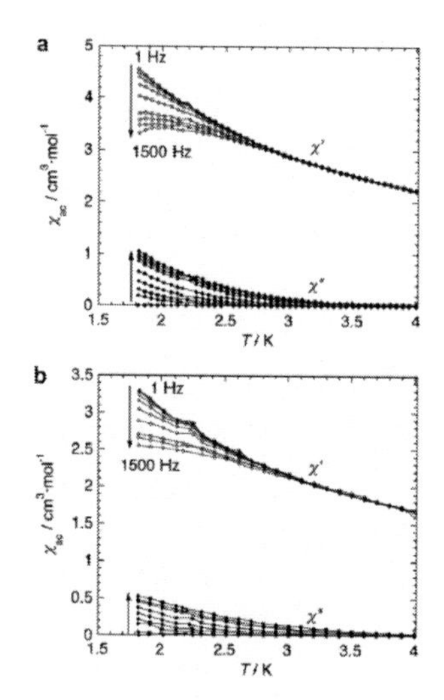

Figure 2.26 Temperature and frequency dependence of the real (χ') and imaginary (χ'') parts of the ac susceptibility for (a) **5** and (b) **6** measured at 3 Oe ac field and zero dc field (the solid lines are only guide for the eyes).

2.4.3 Conclusion

The third example of structurally hybridized compounds aggregating SMM layers and molecular conducting layers was shown in this article. The choice of a structurally and electrically stable cationic $[Mn_2]^{2+}$ SMM, $[Mn^{III}_2(5\text{-MeOsaltmen})_2(S)_2]^{2+}$ (S = acetone or MeCN), made it possible to yield such assemblies with conductive $[Ni(dmit)_2]^{n-}$ molecules, following a failure of the combination of $[Mn_4]^{4+}$ SMM and $[Ni(dmit)_2]^{n-}$ in previous works. Compounds **5** and **6** exhibit an SMM/semiconductor character, in which both characteristic components, however, act independently in the whole temperature range.

2.5 General Conclusion

As shown in Section 2.1, the design of multifunctional materials is one of the most challenging themes in the field of molecular material sciences. One of the most attractive targets is magnetic/conducting bifunctional materials.[16] One way to design such materials is to hybridize *in situ* individual functioning parts, that is, magnetic and conducting frames, using a molecular self-assembly. Three representative hybrid materials have been synthesized so far: paramagnet/superconductor,[17] antiferromagnet/superconductor,[18] and ferromagnet/metal.[19] These compounds are a combination of classical magnet and molecular conductors. Miyasaka *et al.* suggests new strategy for magnetic/conducting system based on SMMs with molecular conductors (nano-dot network). In this review, three types of compounds they prepared are exhibited. Although these compounds are indeed not our final target that show hybridizing effects on both characteristic properties or synergic properties, these examples would give us significant information to consider the next strategy to design favorable candidates.

References

1. (a) Christou, G., Gatteschi, D., Hendrickson, D. N., Sessoli, R. (2000). Theme article — single-molecule magnets, *MRS Bull.*, **25**, 66–71. (b) Gatteshi, D., Sessoli, R. (2003). Quantum tunneling of magnetization and related phenomena in molecular materials, *Angew. Chem. Int. Ed.*, **42**, 268–297.

2. (a) Boyd, P. D. W., Li, Q., Vincent, J. B., Folting, K., Chang, H.-R., Streib, W. E., Huffman, J. C., Christou, G., Hendrickson, D. N. (1988). Potential building blocks for molecular ferromagnets: $[Mn_{12}O_{12}(O_2CPh)_{16}(H_2O)_4]$ with a $S = 14$ ground state, *J. Am. Chem. Soc.*, **110**, 8537–8540. (b) Caneschi, A., Gatteschi, D., Sessoli, R. (1991). Alternating current susceptibility, high field magnetization, and millimeter band EPR evidence for a ground $S = 10$ state in $[Mn_{12}O_{12}(Ch_3COO)_{16}(H_2O)_4]\cdot 2CH_3COOH\cdot 4H_2O$, *J. Am. Chem. Soc.*, **113**, 5873–5874. (c) Sessoli, R., Tsai, H.-L., Schake, A. R., Wang, S., Vincent, J. B., Folting, K., Gatteschi, D., Christou, G., Hendrickson, D. N. (1993). High-spin molecules: $[Mn_{12}O_{12}(O_2CR)_{16}(H_2O)_4]$, *J. Am. Chem. Soc.*, **115**, 1804–1816.

3. (a) Aubin, S. M. J., Wemple, M. W., Adams, D. M., Tsai, H., Christou, G., Hendrickson, D. N. (1996). Distorted $Mn^{IV}Mn^{III}_3$ cubane complexes as single-molecule magnets, *J. Am. Chem. Soc.*, **118**, 7746–7754. (b) Yoo, J., Brechin, E. K., Yamaguchi, A., Nakano, M., Huffman, J. C., Maniero, A. L., Brunel, L.-C., Awaga, K., Ishimoto, H., Christou, G., Hendrickson, D. N. (2000). Single-molecule magnets: A new class of tetranuclear manganese magnets, *Inorg. Chem.*, **39**, 3615–3623. (c) Yoo, J., Yamaguchi, A., Nakano, M., Krzystek, J., Streib, W. E., Brunel, L.-C., Ishimoto, H., Christou, G., Hendrickson, D. N. (2001). Mixed-valence tetranuclear manganese single-molecule magnets, *Inorg. Chem.*, **40**, 4604–4616. (d) Hendrickson, D. N., Christou, G., Ishimoto, H., Yoo, J., Brechin, E. K., Yamaguchi, A., Rumberger, E. M., Aubin, S. M. J., Sun, Z., Aromi, G. (2001). Magnetization tunneling in single-molecule magnets, *Polyhedron*, **20**, 1479–1488. (e) Boskovic, C., Brechin, E. K., Streib, W. E., Folting, K., Hendrickson, D. N., Christou, G. (2002). Single-molecule magnets: A new family of Mn_{12} clusters of formula $[Mn_{12}O_8X_4(O_2CPh)_8L_6]$, *J. Am. Chem. Soc.*, **124**, 3725–3736. (f) Brechin, E. K., Boskovic, C., Wernsdorfer, W., Yoo, J., Yamaguchi, A., Sanudo, E. C., Concolino, T., Rheingold, A. L., Ishimoto, H., Hendrickson, D. N., Christou, G. (2002). Quantum tunneling of magnetization in a new $[Mn_{18}]^{2+}$ single-molecule magnet with $S = 13$, *J. Am. Chem. Soc.*, **124**, 9710–9711. (g) Brechin, E. K., Soler, M., Davidson, J., Hendrickson, D. N., Parsons, S., Christou, G. (2002). A new class of single-molecule magnet: $[Mn_9O_7(OAc)_{11}(thme)(py)_3(H_2O)_2]$ with an $S = 17/2$ ground state, *Chem. Commun.*, 2252–2253.

4. (a) Delfs, C., Gatteschi, D., Pardi, L., Sessoli, R., Wieghardt, K., Hanke, D. (1993). Magnetic properties of an octanuclear iron(III) cation, *Inorg. Chem.*, **32**, 3099–3103. (b) Barra, A. L., Caneschi, A., Cornia, A., Fabrizi de Biani, F., Gatteschi, D., Sangregorio, C., Sessoli, R., Sorace, L. (1999). Single-molecule magnet behavior of a tetranuclear Iron(III)

complex. The origin of slow magnetic relaxation in Iron(III) clusters, *J. Am. Chem. Soc.*, **121**, 5302–5310. (c) Gatteschi, D., Sessoli, R., Cornia, A. (2000). Single-molecule magnets based on iron(III) oxo clusters, *Chem. Commun.*, 725–732. (d) Oshio, H., Hoshino, N., Ito, T. (2000). Superparamagnetic behavior in an alkoxo-bridged Iron(II) Cube, *J. Am. Chem. Soc.*, **122**, 12602–12603. (e) Benelli, C., Cano, J., Journaux, Y., Sessoli, R., Solan, G. A., Winpenny, R. E. P. (2001). A decanuclear Iron(III) single molecule magnet: Use of Monte Carlo methodology to model the magnetic properties, *Inorg. Chem.*, **40**, 188–189. (f) Goodwin, J. C., Sessoli, R., Gatteschi, D., Wernsdorfer, W., Powell, A. K., Heath, S. L. (2000). Towards nanostructured arrays of single molecule magnets: New Fe_{19} oxyhydroxide clusters displaying high ground state spins and hysteresis, *J. Chem. Soc. Dalton Trans.*, 1835–1840.

5. (a) Cadiou, C., Murrie, M., Paulsen, C., Villar, V., Wernsdorfer, W., Winpenny, R. E. P. (2001). Studies of a nickel-based single molecule magnet: Resonant quantum tunnelling in an S = 12 molecule, *Chem. Commun.*, 2666–2667. (b) Andres, H., Basler, R., Blake, A. J., Cadiou, C., Chaboussant, G., Grant, C. M., Güdel, H.-U., Murrie, M., Parsons, S., Paulscn, C., Semadini, F., Villar, V., Wernsdorfer, W., Winpenny, R. E. P. (2002). Studies of a Nickel-based single-molecule magnet, *Chem. Eur. J.*, **8**, 4867–4876. (c) Yang, E.-C., Wernsdorfer, W., Hill, S., Edwards, R. S., Nakano, M., Maccagnano, S., Zakharov, L. N., Rheingold, A. L., Christou, G., Hendrickson, D. N. (2003). Exchange bias in Ni_4 single-molecule magnets, *Polyhedron*, **22**, 1727–1733.
6. Castro, S. L., Sun, Z., Grant, C. M., Bollinger, J. C., Hendrickson, D. N., Christou, G. (1998). Single-molecule magnets: Tetranuclear Vanadium(III) complexes with a butterfly structure and an S = 3 ground state, *J. Am. Chem. Soc.*, **120**, 2365–2375.
7. Yang, E., Hendrickson, D. N., Wernsdorfer, W., Nakano, M., Zakharov, L. N., Sommer, R. D., Rheingold, A. L., Ledezma-Gairaud, M., Christou, G. (2002). Cobalt single-molecule magnet, *J. Appl. Phys.*, **91**, 7382–7384.
8. (a) Schake, A. R., Tsai, H., Webb, R. J., Folting, K., Christou, G., Hendrickson, D. N. (1994). High-Spin molecules: Iron(III) Incorporation into $[Mn_{12}O_{12}(O_2CMe)_{16}(H_2O)_4]$ to yield $[Mn_8Fe_4O_{12}(O_2CMe)_{16}(H_2O)_4]$ and its influence on the S = 10 ground state of the former, *Inorg. Chem.*, **33**, 6020–6028. (b) Sokol, J. J., Hee, A. G., Long, J. R. (2002). A cyano-bridged single-molecule magnet: Slow magnetic relaxation in a trigonal prismatic $MnMo_6(CN)_{18}$ cluster, *J. Am. Chem. Soc.*, **124**, 7656–7657.
9. Lecren, L., Li, Y.-G., Wernsdorfer, W., Roubeau, O., Miyasaka, H., Clérac, R. (2005). $[Mn_4(hmp)_6(CH_3CN)_2(H_2O)_4]^{4+}$: A new single-molecule magnet with the highest blocking temperature in the Mn_4/hmp family of compounds, *Inorg. Chem. Commun.*, **8**, 626–630.

10. Miyasaka, H, Clérac, R., Wernsdorfer, W., Lecren, L., Bonhomme, C., Sugiura, K., Yamashita, M. (2004). A Dimeric Manganese(III) tetradentate Schiff base complex as a single-molecule magnet, *Angew. Chem. Int. Ed.*, **43**, 2801–2805.
11. (a) Wernsdorfer, W., Sessoli, R. (1999). Quantum phase interference and parity effects in magnetic molecular clusters, *Science*, **284**, 133–135. (b) Leuenberger, M. N., Loss, D. (2001). Quantum computing in molecular magnets, *Nature*, **410**, 789–793.
12. Miller, J. S., Epstein, A. J. (1976). *Prog. Inorg. Chem.*, **20**, 1.
13. Kato, R. (2004). Conducting metal dithiolene complexes: Structural and electronic properties, *Chem. Rev.*, **104**, 5319–5346.
14. Underhill, A.E., Ahmad, M. M. (1981). A new one-dimensional "metal" with conduction through bis(dicyanoethylenedithiolato) platinum anions, *J. Chem. Soc., Chem. Commun.*, 67–68.
15. Canadell, E., Ravy, S., Pouget, J. P., Brossard, L. (1990). Concerning the band structure of $D(M(dmit)_2)_2$ (D = TTF, Cs, NMe_4); M = Ni, Pd) molecular conductors and superconductors: Role of the $M(dmit)_2$ Homo and Lumo, *Sold State Commun.*, **75**, 633–638.
16. Coronado, E., Day, P. (2004). Magnetic molecular conductors, *Chem. Rev.*, **104**, 5419–5448. Enoki, T., Miyazaki, A. (2004). Magnetic TTF-based charge-transfer complexes, *Chem. Rev.*, **104**, 5449–5478. Coronado, E., Galán-Mascarós, J., R. (2005). Hybrid molecular conductors, *J. Mater. Chem.*, **15**, 66–74. Ouahab, L. (1997). Organic/inorganic supramolecular assemblies and synergy between physical properties, *Chem. Mater.*, **9**, 1909–1926.
17. Kurmoo, M., Graham, A. W., Day, P., Coles, S. J., Hursthouse, M. B., Caulfield, J. L., Singleton, J., Pratt, F. L., Hayes, W., Ducasse, L., Guionneau, P. (1995). Superconducting and semiconducting magnetic charge transfer salts: $(BEDT\text{-}TTF)_4AFe(C_2O_4)_3$·cntdot·$C_6H_5CN$ (A = H_2O, K, NH_4), *J. Am. Chem. Soc.*, **117**, 12209–12217.
18. Ojima, E., Fujiwara, H., Kato, K., Kobayashi, H., Tanaka, H., Kobayashi, A., Tokumoto, M., Cassoux, P. (1999). Antiferromagnetic organic metal exhibiting superconducting transition, κ-$(BETS)_2FeBr_4$ [BETS = Bis(ethylenedithio)tetraselenafulvalene], *J. Am. Chem. Soc.*, **121**, 5581–5582.
19. Coronado, E., Galán-Mascarós, J. R., Gómez-Garcia, C. J., Laukhin, V. (2000). Coexistence of ferromagnetism and metallic conductivity in a molecule-based layered compound, *Nature*, **408**, 447–449. Alberola, A., Coronado, E., Galán-Mascarós, J. R., Giménez-Saiz, C., Gómez-García, C. J. (2003). A molecular metal ferromagnet from the organic donor Bis(ethylenedithio)tetraselenafulvalene and bimetallic oxalate complexes, *J. Am. Chem. Soc.*, **125**, 10774–10775.

20. (a) Wernsdorfer, W., Aliaga-Alcalde, N., Hendrickson, D. N., Christou, G. (2002). Exchange-biased quantum tunnelling in a supramolecular dimer of single-molecule magnets, *Nature*, **416**, 406–409. (b) Tiron, R., Wernsdorfer, W., Foguet-Albiol, D., Aliaga-Alcalde, N., Christou, G. (2003). Spin quantum tunneling via entangled states in a dimer of exchange-coupled single-molecule magnets, *Phys. Rev. Lett.*, **91**, 227203(1)–(4). (c) Hill, S., Edwards, R. S., Aliaga-Alcalde, N., Christou, G. (2003). Quantum coherence in an exchange-coupled dimer of single-molecule magnets, *Science*, **302**, 1015–1018. (d) Park, K., Pederson, M. R., Richardson, S. L., Aliaga-Alcalde, N., Christou, G. (2003). Density-functional theory calculation of the intermolecular exchange interaction in the magnetic Mn_4 dimer, *Phys. Rev. B*, **68**, 020405(1)–(4). (e) Wernsdorfer, W., Bhaduri, S., Tiron, R., Hendrickson, D. N., Christou, G. (2002). Spin–spin cross relaxation in single-molecule magnets, *Phys. Rev. Lett.*, **89**, 197201(1)–(4). (f) Edwards, R. S., Hill, S., Bhaduri, S., Aliaga-Alcalde, N., Bolin, E., Maccagnano, S., Christou, G., Hendrickson, D. N. (2003). *Polyhedron*, **22**, 1911.
21. Tiron, R., Wernsdorfer, W., Aliaga-Alcalde, N., Christou, G. (2003). Quantum tunneling in a three-dimensional network of exchange-coupled single-molecule magnets, *Phys. Rev. B*, **68**, 140407 (1)–(4).
22. Boskovic, C., Bircher, R., Tregenna-Piggott, P. L. W., Güdel, H. U., Paulsen, C., Wernsdorfer, W., Barra, A.-L., Khatsko, E., Neels, A., Stoeckli-Evans, H. (2003). Ferromagnetic and antiferromagnetic intermolecular interactions in a new family of Mn_4 complexes with an energy barrier to magnetization reversal, *J. Am. Chem. Soc.*, **125**, 14046–14058.
23. Miyasaka, H., Nakata, K., Sugiura, K., Yamashita, M., Clérac, R. (2004). A three-dimensional ferrimagnet composed of mixed-valence Mn_4 clusters linked by an $\{Mn[N(CN)_2]_6\}^{4-}$ unit, *Angew. Chem., Int. Ed.*, **43**, 707–711.
24. Miyasaka, H., Nakata, K., Lecren, L., Coulon, C., Nakazawa, Y., Fujisaki, T., Sugiura, K., Yamashita, M., Clérac, R. (2006). Two-dimensional networks based on Mn_4 complex linked by dicyanamide anion: from single-molecule magnet to classical magnet behavior, *J. Am. Chem. Soc.*, **128**, 3770–3783.
25. Lecren, L., Roubeau, O., Coulon, C., Li, Y.-G., Le Goff, X. F., Wernsdorfer, W., Miyasaka, H. amd Clérac, R. (2005). Slow relaxation in a one-dimensional rational assembly of antiferromagnetically coupled $[Mn_4]$ single-molecule magnets, *J. Am. Chem. Soc.*, **127**, 17353–17363.
26. Lecren, L., Wernsdorfer, W., Li, Y.-G., Vindigni, A., Miyasaka, H., Clérac, R. (2007). One-dimensional supramolecular organization of single-molecule magnets, *J. Am. Chem. Soc.*, **129**, 5045–5051.

27. Miyasaka, H., Yamashita, M. (2007). A look at molecular magnets from the aspect of inter-molecular interactions, *Dalton Trans.*, 399–406.

28. (a) Clérac, R., Miyasaka, H., Yamashita, M., Coulon, C. (2002). Evidence for single-chain magnet behavior in a Mn^{III}–Ni^{II} chain designed with high spin magnetic units: A route to high temperature metastable magnets, *J. Am. Chem. Soc.*, **124**, 12837–12844. (b) Miyasaka, H., Clérac, R., Mizushima, K., Sugiura, K., Yamashita, M., Wernsdorfer, W., Coulon, C. (2003). $[Mn_2(saltmen)_2Ni(pao)_2(L)_2](A)_2$ with L = Pyridine, 4-Picoline, 4-*tert*-Butylpyridine, *N*-Methylimidazole and A = ClO_4^-, BF_4^-, PF_6^-, ReO_4^-: A family of single-chain magnets, *Inorg. Chem.*, **42**, 8203–8213. (c) Ferbinteanu, M., Miyasaka, H., Wernsdorfer, W., Nakata, K., Sugiura, K., Yamashita, M., Coulon, C., Clérac, R. (2005). Single-chain magnet $(NEt_4)[Mn_2(5\text{-}MeOsalen)_2Fe(CN)_6]$ made of Mn^{III}–Fe^{III}–Mn^{III} trinuclear single-molecule magnet with an S_T = $^9/_2$ spin ground state, *J. Am. Chem. Soc.*, **127**, 3090–3099. (d) Lecren, L., Roubeau, O., Coulon, C., Li, Y.-G., Le Goff, X. F., Wernsdorfer, W., Miyasaka, H., Clérac, R. (2005). Slow relaxation in a one-dimensional rational assembly of antiferromagnetically coupled $[Mn_4]$ single-molecule magnets, *J. Am. Chem. Soc.*, **127**, 17353–17363. (e) Miyasaka, H., Madanbashi, T., Sugimoto, K., Nakazawa, Y., Wernsdorfer, W., Sugiura, K., Yamashita, M., Coulon, C., Clérac, R. (2006). Single-chain magnet behavior in an alternated one-dimensional assembly of a MnIII Schiff-base complex and a TCNQ radical, *Chem. Eur. J.*, **12**, 7028–7040. (f) Saitoh, A., Miyasaka, H., Yamashita, M., Clérac, R. (2007). Direct evidence of exchange interaction dependence of magnetization relaxation in a family of ferromagnetic-type single-chain magnets, *J. Mater. Chem.*, **17**, 2002–2012. (g) Lecren, L., Roubeau, O., Li, Y.-G., Le Goff, X. F., Miyasaka, H., Richard, F., Wernsdorfer, W., Coulon, C., Clérac, R. (2008). One-dimensional coordination polymers of antiferromagnetically-coupled $[Mn_4]$ single-molecule magnets, *Dalton Trans.*, 755–766.

29. (a) Miyasaka, H. Nakata, K., Sugiura, K., Yamashita, M., Clérac, R. (2004). A three-dimensional ferrimagnet composed of mixed-valence Mn_4 clusters linked by an $\{Mn[N(CN)_2]_6\}^{4-}$ Unit, *Angew. Chem., Int. Ed. Engl.*, **43**, 707–711. (b) Miyasaka, H., Nakata, K., Lecren, L., Coulon, C., Nakazawa, Y., Fujisaki, T., Sugiura, K., Yamashita, M., Clérac, R. (2006). Two-dimensional networks based on Mn_4 complex linked by Dicyanamide anion: From single-molecule magnet to classical magnet behavior, *J. Am. Chem. Soc.*, **128**, 3770–3783.

30. Hiraga, H., Miyasaka, H., Nakata, K., Kajiwara, T., Takaishi, S., Oshima, Y., Nojiri, H., Yamashita, M. (2007). Hybrid molecular material exhibiting single-molecule magnet behavior and molecular conductivity, *Inorg. Chem.*, **46**, 9661–9671.

31. Davison, A., Edelstein, N., Holm, R. H., Maki A. H. (1963). The preparation and characterization of four-coordinate complexes related by electron-transfer reactions, *Inorg. Chem.*, **2**, 1227–1232.

32. Hiraga, H., Miyasaka, H., Takaishi, S., Kajiwara, T., Yamashita, M. (2008). Hybridized complexes of $[Mn^{III}_2]$ single-molecule magnets and Ni dithiolate complexes, *Inorg. Chim. Acta*, **361**, 3863–3872.

33. Hiraga, H., Miyasaka, H., Clérac, R., Fourmigué, Yamashita M. (2009). $[M^{III}(dmit)_2]^-$-coordinated Mn^{III} salen-type dimers (M^{III} = Ni^{III}, Au^{III}; $dmit^{2-}$ = 1,3-Dithiol-2-thione-4,5-dithiolate): Design of single-component conducting single-molecule magnet-based materials, *Inorg. Chem.*, **48**, 2887–2898.

34. Clemenson, P. I. (1990). The chemistry and solid state properties of nickel, palladium and platinum bis(maleonitriledithiolate) compounds, *Coord. Chem. Rev.*, **106**, 171–203.

35. Clemenson, P. I., Underhill, A. E., Hursthouse, M. B., Short, R. L. (1989). Comparison of the structures and properties of $[N(C_2H_5)_4][Pt\{\{S_2C_2(CN)_2\}_2]$ and $[H_3O]_x[NH_4]_{1-x}[Pt\{S_2C_2(CN)_2\}_2]\cdot 2 - xH_2O$, *J. Chem. Soc. Dalton Trans.*, 61–65.

36. (a) Ahmad, M. M., Underhill, A. E. (1982). A study of the one-dimensional conductor $Li_{0.75}[Pt\{S_2C_2(CN)_2\}_2]\cdot 2H_2O$, *J. Chem. Soc., Dalton Trans.*, 1065–1068. (b) Kobayashi, A., Mori, T., Sasaki, Y., Kobayashi, H., Ahmad, M. M., Underhill, A. E. (1984). Peierls structure of $(H_3O)_{0.33}Li_{0.8}[Pt(mnt)_2]\cdot 1.67H_2O$, *Bull. Chem. Soc. Jpn.*, **57**, 3262–3268. (c) Kobayashi, A., Sasaki, Y., Kobayashi, H., Underhill, A. E., Ahmad, M. M. (1984). Crystal structure and electrical conductivity of a new conducting bisdithiolate complex $Li_{0.5}[Pt(mnt)_2]\cdot 2H_2O$, *Chem. Lett.*, 305–308. (d) Ahmad, M. M., Turner, D. J., Underhill, A. E. (1984). Physical properties and the Peierls instability of $Li_{0.82}[Pt(S_2C_2(CN)_2)_2]\cdot 2H_2O$, *Phys. Rev. B*, **29**, 4796–4799.

37. Henriques, R. T., Alcácer, L., Almeida, M., Tomic. S. (1985). Transport and magnetic properties on the family of perylene–dithiolate conductors, *Mol. Cryst. Liq. Cryst.*, **120**, 237–241.

38. Takahashi, K., Cui, H.-B., Okano, Y., Kobayashi, H., Einaga, Y., Sato, O. (2006). Electrical conductivity modulation coupled to a high-spin–low-spin conversion in the molecular system $[Fe^{III}(qsal)_2][Ni(dmit)_2]_3\cdot CH_3CN\cdot H_2O$, *Inorg. Chem.*, **45**, 5739–5741.

39. (a) Ako, A. M., Mereacre, V., Hewitt, I. J., Clérac, R.; Lecren, L., Anson, C. E., Powell, A. K. (2006). Enhancing single molecule magnet parameters. Synthesis, crystal structures and magnetic properties of mixed-valent Mn_4 SMMs, *J. Mater. Chem.*, **16**, 2579–2586. (b) Kachi-Terajima, C.,

Miyasaka, H., Saitoh, A., Shirakawa, N., Yamashita, M., Clérac, R. (2007). Single-molecule magnet behavior in heterometallic M^{II}–Mn^{III}_2–M^{II} Tetramers (M^{II} = Cu, Ni) containing Mn^{III} Salen-type dinuclear core, *Inorg. Chem.*, **46**, 5861–5872.

40. (a) Coronado, E., Galán-Mascarós, J. R., Gómez-Garcia, C. J., Laukhin, V. (2000). Coexistence of ferromagnetism and metallic conductivity in a molecule-based layered compound, *Nature*, **408**, 447–449. (b) Uji, S., Shinagawa, H., Terashima, T., Yakabe, T., Terai, Y., Tokumoto, M. Kobayashi, A., Tanaka, H., Kobayashi, H. (2001). Magnetic-field-induced superconductivity in a two-dimensional organic conductor *Nature*, **410**, 908–910. (c) Bai, Y.-L., Tao, J., Wernsdorfer, W., Sato, O., Huang, R.-B., Zhang, L.-S. (2006). Coexistence of magnetization relaxation and dielectric relaxation in a single-chain magnet, *J. Am. Chem. Soc.*, **128**, 16428–16429. (d) Takahashi, K., Cui, H.-B., Okano, Y., Kobayashi, H., Mori, H., Tajima, H., Einaga, Y., Sato, O. (2008). Evidence of the chemical uniaxial strain effect on electrical conductivity in the spin-crossover conducting molecular system: $[Fe^{III}(qnal)_2][Pd(dmit)_2]_5$·Acetone, *J. Am. Chem. Soc.*, **130**, 6688–6689.

41. (a) Kobayashi, A., Fujiwara, E., Kobayashi, H. (2004). Single-component molecular metals with extended-TTF dithiolate ligands, *Chem. Rev.*, **104**, 5243–5264. (b) Tanaka, H., Okano, Y., Kobayashi, H., Suzuki, W., Kobayashi, A. (2001). A three-dimensional synthetic metallic crystal composed of single-component molecules, *Science*, **291**, 285–287.

42. Miyasaka, H., Clérac, R., Ishii, T., Chang, H.-C., Kitagawa, S., Yamashita, M. (2002). Out-of-plane dimers of Mn(III) quadridentate Schiff-base complexes with $saltmen^{2-}$ and $naphtmen^{2-}$ ligands: structure analysis and ferromagnetic exchange, *J. Chem. Soc., Dalton Trans.*, 1528–1534.

43. Kachi-Terajima, C., Miyasaka, H., Sugiura, K., Clérac, R., Nojiri, H. (2006). From an S_T = 3 single-molecule magnet to diamagnetic ground state depending on the molecular packing of Mn^{III}salen-type dimers decorated by *N,N′*-Dicyano-1,4-naphthoquinonediiminate radicals, *Inorg. Chem.*, **45**, 4381–4390.

44. Lü, Z., Yuan, M., Pan, F., Gao, S., Zhang, D., Zhu, D. (2006). Syntheses, crystal structures, and magnetic characterization of five new dimeric Manganese(III) tetradentate Schiff Base complexes exhibiting single-molecule-magnet behavior, *Inorg. Chem.*, **45**, 3538–3548.

45. Miyasaka, H., Saitoh, A., Abe, S. (2007). Magnetic assemblies based on Mn(III) salen analogues, *Coord. Chem. Rev.*, **251**, 2622–2664.

46. Cortés, R., Drillon, M., Solans, X., Lezama, L., Rojo, T. (1997). Alternating ferromagnetic–antiferromagnetic interactions in a

Manganese(II)–Azido one-dimensional compound: [Mn(bipy)$(N_3)_2$], *Inorg. Chem.*, **36**, 677–683.

47. Fisher, M. E. (1964). Magnetism in one-dimensional systems — The Heisenberg Model for infinite spin, *Am. J. Phys.*, **32**, 343–346.

48. Kahn, O. (1993). *Molecular Magnetism*, VCH Publishers, Inc., Weinheim, Germany.

49. Thomos, L., Lionti, F., Ballou, R., Gatteschi, D., Sessoli, R., Barbara, B. (1996). Macroscopic quantum tunnelling of magnetization in a single crystal of nanomagnets, *Nature*, **383**, 145–147.

50. Ako, A. M., Mereacre, V., Hewitt, I. J., Clérac, R., Lecren, L., Anson, C. E., Powell, A. K. (2006). Enhancing single molecule magnet parameters. Synthesis, crystal structures and magnetic properties of mixed-valent Mn_4 SMMs, *J. Mater. Chem.*, **16**, 2579–2589.

51. (a) Faulmann, C., Dorbes, S., Bonneval, B.G., Molnar, G., Bousseksou, A., Gomez-Garcia, C.J., Coronado, E., Valade, L. (2005). Towards molecular conductors with a spin-crossover phenomenon: Crystal structures, magnetic properties and Mössbauer spectra of [Fe(salten)Mepepy] [M$(dmit)_2$] complexes, *Eur. J. Inorg. Chem.*, **16**, 3261–3270. (b) Dorbes, S., L. Valade, S., Real, J. A., Faulmann, C. (2005). [Fe(sal_2-trien)][Ni$(dmit)_2$]: Towards switchable spin crossover molecular conductors, *Chem. Commun.*, **1**, 69–71. (c) Faulmann, C., Jacob, K., Dorbes, S. Lampert, S., Malfant, I., Doublet, L., Valade, M.-L., Real, J. A., (2007). Electrical conductivity and spin crossover: A new achievement with a metal bis dithiolene complex, *Inorg. Chem.*, **46**, 8548–8559. (d) Takahashi, K., Cui, H.-B., Kobayashi, H., Einaga, Y., Sato, O. (2005). The light-induced excited spin state trapping effect on Ni$(dmit)_2$ salt with an Fe(III) spin-crossover cation: [Fe$(qsal)_2$][Ni$(dmit)_2$]·$2CH_3CN$, *Chem. Lett.* **34**, 1240–1241.

Chapter 3

Multifunctional Single-Molecule Magnets and Single-Chain Magnets

Ling-Chen Kang and Jing-Lin Zuo

State Key Laboratory of Coordination Chemistry, School of Chemistry and Chemical Engineering, Nanjing University, Hankou Road 22, Nanjing 210093, Jiangsu Province, People's Republic of China

zuojl@nju.edu.cn

3.1 Introduction

During the past 30 years, molecule-based magnetic materials have attracted much attention from both fundamental scientific views and technological applications. Currently, one of the hottest topics in magnetic materials is molecular nanomagnet, which comprises both *single-molecule magnets* (SMMs)[1] and *single-chain magnets* (SCMs).[2]

SMMs are molecular species that combine intramolecular properties of a high-spin ground state and large easy-axis-type magnetic anisotropy and exhibit slow relaxation of the magnetization at low temperatures.[3] They can retain magnetization in the absence of a magnetic field below a blocking temperature for days, behaving like a tiny magnet. Analogous to the situation in SMMs, slow relaxation of the magnetization is also observed in one-dimensional

Multifunctional Molecular Materials
Edited by Lahcène Ouahab

ISBN 978-981-4364-29-4 (Hardcover), 978-981-4364-30-0 (eBook)
www.panstanford.com

(1-D) magnetic systems without any interchain interaction, and this type of materials are called SCMs.[2] Compared with classic bulk 3-D magnets, these molecular materials possess advantages of molecular dimension, enormous monodispersivity, high processing ability, and low cost, and they are potentially useful in high-density magnetic storage devices,[4] memories,[5] sensors,[6] quantum computing applications,[7] and more recently spintronics.[8]

Since the discovery of SMMs and SCMs, a great number of SMMs and SCMs, with various structures and magnetic properties, have been reported.[9] In spite of the rapid growth in basic scientific research and promises of application, several challenges remain in the development of SMM and SCM chemistry. First, the blocking temperature (T_B) is still too low for application. It is imperative to achieve a higher blocking temperature (at least above the boiling point of liquid nitrogen) in new SMMs or SCMs. Second, SMMs and SCMs are crystalline materials and fragile. Developing new fabrication technologies that are compatible with the current semiconductor electronic industries will challenge not only chemists but also material scientists and engineers. Although breakthroughs have been rare in recent years, some important developments representing the new trends in the research of SMMs and SCMs have been made, and one of them is known as *multifunctional single-molecule magnets and single-chain magnets*.

With the development of science and technology, the need for new materials that have more diversified and more sophisticated properties is continuously increasing. One of the goals is to prepare materials that not only possess one expected property or function but also combine two or more of them in a multifunctional system. The synthesis of materials combining magnetism with another physical property is an emerging trend in molecular magnetism. Multifunctional SMMs and SCMS are new multifunctional molecular materials, where, in addition to slow relaxation of the magnetization, physical properties such as conductivity, fluorescence, photochromism, optical activity, and ferroelectricity are realized within the same molecular system.

3.2 Conductive Single-Molecule Magnets

Due to the important applications in molecular spintronics, the synthesis of new materials with both metallic conductivity and ferromagnetism has attracted increasing attention in the last few years.[10] Several strategies to realize conductive single-molecular magnets have been developed, where sulfur-rich ligands or related metal complexes are always utilized. These sulfur-rich units may lead to molecular conductivity through interchalcogen–atom interactions, which are often observed in molecule-based conductors, even superconductors.

With the use of the tetranuclear [Mn_4] SMM, [$Mn^{II}_2Mn^{III}_2(hmp)_6(MeCN)_2(H_2O)_4](ClO_4)_4\cdot 2MeCN$ (hmp^- = 2-hydroxymethyl-pyridinate), as building block and the conductive [$Pt^{III}(mnt)_2]^-$ (mnt^{2-} = maleonitriledithiolate) anion as bridging ligand, two 1-D chain complexes [$\{Mn^{II}_2Mn^{III}_2(hmp)_6(MeCN)_2\}\{Pt(mnt)_2\}_2][Pt(mnt)_2]_2 \cdot 2MeCN$ (**1**) and [$\{Mn^{II}_2Mn^{III}_2(hmp)_6(MeCN)_2\}\{Pt(mnt)_2\}_4][Pt(mnt)_2]_2$ (**2**) were obtained by Miyasaka and Yamashita *et al.*[11a] As shown in Fig. 3.1, in complexes **1** and **2**, [$Mn^{II}_2Mn^{III}_2(hmp)_6(MeCN)_2]^{4+}$ double-cuboidal cluster units are linked by [$Pt(mnt)_2]^-$ anions, forming 1-D infinite chain structures. The magnetic properties of **1** and **2** (Fig. 3.2) are almost consistent with the isolated [Mn_4] SMMs. The magnetic interactions between the [Mn_4] units are

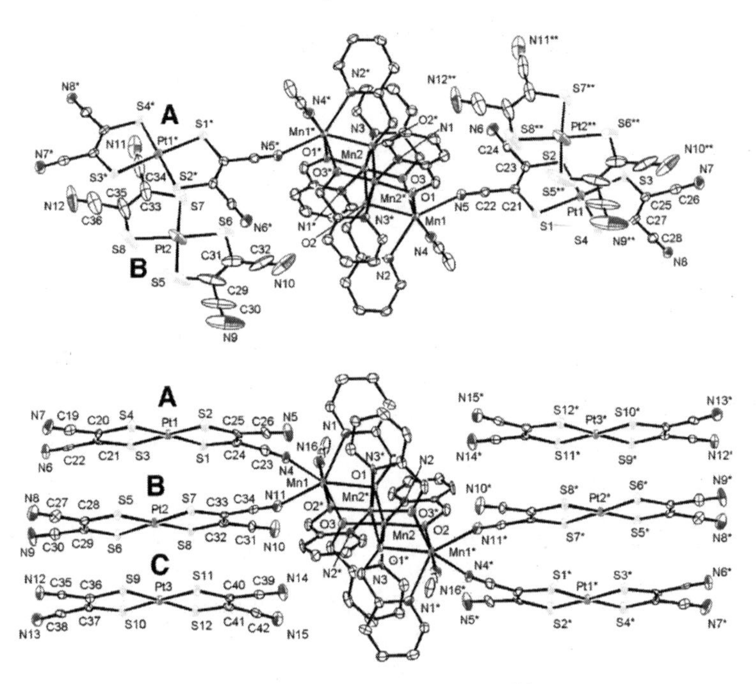

Figure 3.1 ORTEP view of complexes **1** (a) and **2** (b).

negligible and no detectable coupling interaction is observed between the spins of the $[Mn_4]$ units and the localized spins of Pt^{III}. Furthermore, due to the close packing of the $[Pt^{III}(mnt)_2]^-$ units (Fig. 3.3, left), complex **2** exhibits semiconductor behavior (Fig. 3.3, right). On the other hand, complex **1** is an insulator because of the insufficient interchalcogen–atom interactions. Complex **2** represents the first example of a hybridized material exhibiting both SMM behavior and electronic conductivity.

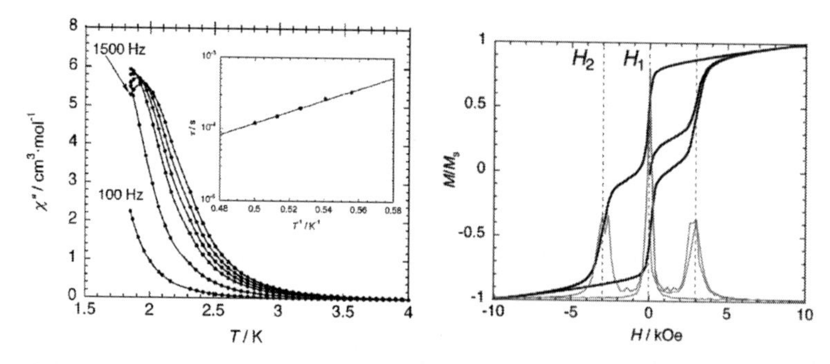

Figure 3.2 **Left:** Temperature dependence of the out-of-phase ac susceptibilities of **2** measured at several frequencies. **Inset:** Arrhenius fitting. **Right:** Field dependence of magnetization measured on field-oriented single crystals of **2**.

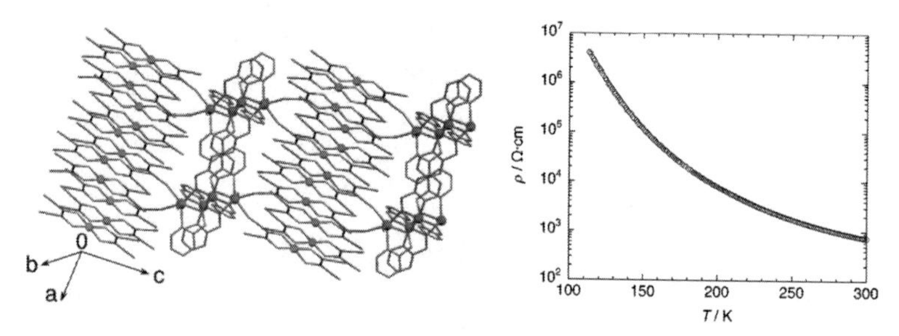

Figure 3.3 **Top:** Packing view of **2**. **Bottom:** Temperature dependence of resistivity for a single crystal of **2** along the $[Pt(mnt)_2]^{n-}$ column.

The same group also used some cationic $[Mn_2]$ SMMs, $[Mn_2(5\text{-MeOsaltmen})_2(H_2O)_2](PF_6)_2$ and $[Mn_2(5\text{-Mesaltmen})_2(H_2O)_2](PF_6)_2$ (5-MeOsaltmen^{2-} = *N,N'*-(1,1,2,2-tetramethylethylene)-bis(5-methylsalicylideneiminate), 5-MeOsaltmen^{2-} = *N,N'*-(1,1,2,2-tetramethylethylene)-bis(5-methoxysalicylideneiminate)) to react with $[M(dmit)_2]^-$ anion (dmit^{2-}=1,3-dithiol-2-thione-4,5-dithiolate,

M = Ni and Au). A series of new complexes, $[Mn_2(5\text{-MeOsaltmen})_2(Acetone)_2][Ni(dmit)_2]_7\cdot 4Acetone$ (**3**), $[Mn_2(5\text{-MeOsaltmen})_2(MeCN)_2][Ni(dmit)_2]_7\cdot 4MeCN$ (**4**), $[Mn(5\text{-MeOsaltmen})\{Ni(dmit)_2\}]_2$ (**5**), $[Mn(5\text{-MeOsaltmen})\{Au(dmit)_2\}]_2$ (**6**), $[Mn(5\text{-Mesaltmen})\{Ni(dmit)_2\}]_2$ (**7**), and $[Mn(5\text{-Mesaltmen})\{Au(dmit)_2\}]_2$ (**8**), were prepared as shown in Figs. 3.4 and 3.5.[11b,c]

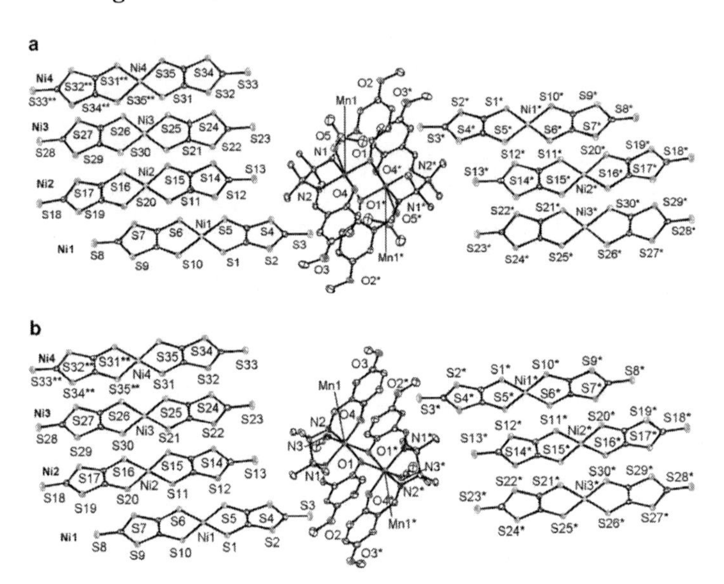

Figure 3.4 ORTEP drawings of **3** (a) and **4** (b).

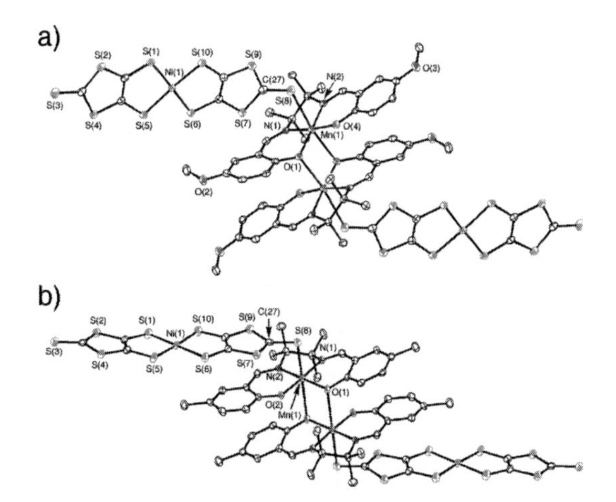

Figure 3.5 ORTEP drawings of **5** (a) and **6** (b).

Complexes **3** and **4** were isolated by electrochemical oxidation method and composed of $[Mn_2(5\text{-MeOsaltmen})_2(Acetone)_2]^{2+}$ cations (SMM units) and noninteger charged $[Ni(dmit)_2]^{n-}$ anions (conductive units). They show layered structures similar to alternative magnetic layers and conductive layers (Fig. 3.6, left), and exhibit both semiconductor behavior (Fig. 3.6, right) and slow relaxation of the magnetization.

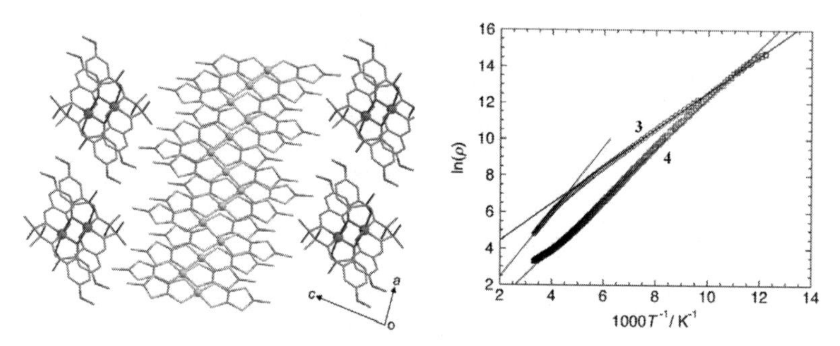

Figure 3.6 **Left:** Packing diagrams of **3**. **Right:** Ln(ρ) vs 1/T plots for a single crystal of **3** and **4**.

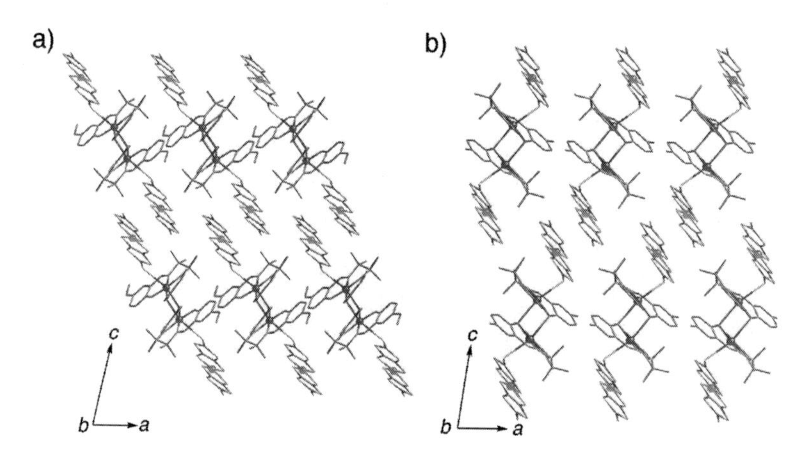

Figure 3.7 Packing diagrams of **5** (a) and **6** (b).

On the other hand, complexes **5–8** were isolated by evaporating the solutions. They are composed of neutral single-component $[Mn(5\text{-Rsaltmen})\{M(dmit)_2\}]_2$ molecules. Complexes **5**, **6**, and **8** exhibit SMM behavior, while complex **7** is antiferromagnet at low temperature. Though interchalcogen–atom interactions and close-packing structures are found within the crystal structures of complexes **5–8** (Fig. 3.7), their electrical conductivities are poorer than complexes **3** and **4.** Complexes **5** and **6** show measurable conductivity as $\sigma = 7\times10^{-4}$ S·cm^{-1} and 1×10^{-4} S·cm^{-1} (Fig. 3.8) while complexes **7** and **8** are insulators. Compared with **3** and **4**, the decreased quantities of conductive components ([Ni(dmit)]$^{n-}$ units) may be the main reason for the decreased conductivity.

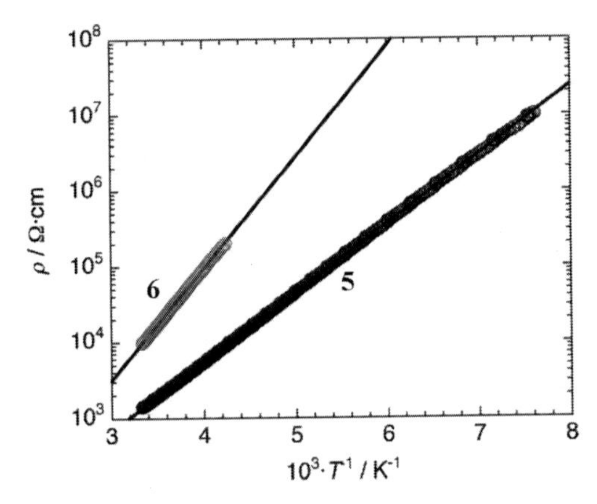

Figure 3.8 Temperature dependence of electrical resistivity on single crystals of **5** and **6.**

Tricyanometalate-based clusters are widely studied magnetic systems and these well-tailored cyano-bridged clusters sometimes exhibit SMMs behaviors at low temperatures. Zuo *et al.* introduced different π-conjugated sulfur-rich ligands containing coupled bis(2-pyridine) group and 1,3-dithiol-2-ylidene as auxiliary ligands in the assembly of tricyanometalate-based clusters.[12] Two new tetranuclear clusters, $[(Tp)Fe(CN)_3Ni(L_1)_2]_2\cdot(ClO_4)_2\cdot6H_2O$ (**9**, Tp = hydridotris(pyrazolyl)borate, L_1 = 4,5-[1′,4′]dithiino[2′,3′-*b*] quinoxaline-2-bis(2-pyridyl)-methylene-1,3-dithiole) and $[(i\text{-}BuTp)Fe(CN)_3Ni(L_3)_2]_2\cdot(ClO_4)_2\cdot6H_2O$ (**10,** *i*-BuTp = 2-methylpropyltris (pyrazolyl)borate, L_3 = dimethyl-2-[di(pyridin-2-yl)-methylene]-1,3-dithiole-4,5-dicarboxylate), show SMM behavior with the effective spin-reversal barriers of 8.7 and 13.5 K, respectively, as shown in Fig. 3.9. Unfortunately, these complexes are insulators due to the lack of effective close packing and interchalcogen–atom interactions. However, it is still one of the few attempts to fabricate conductive SMMs.

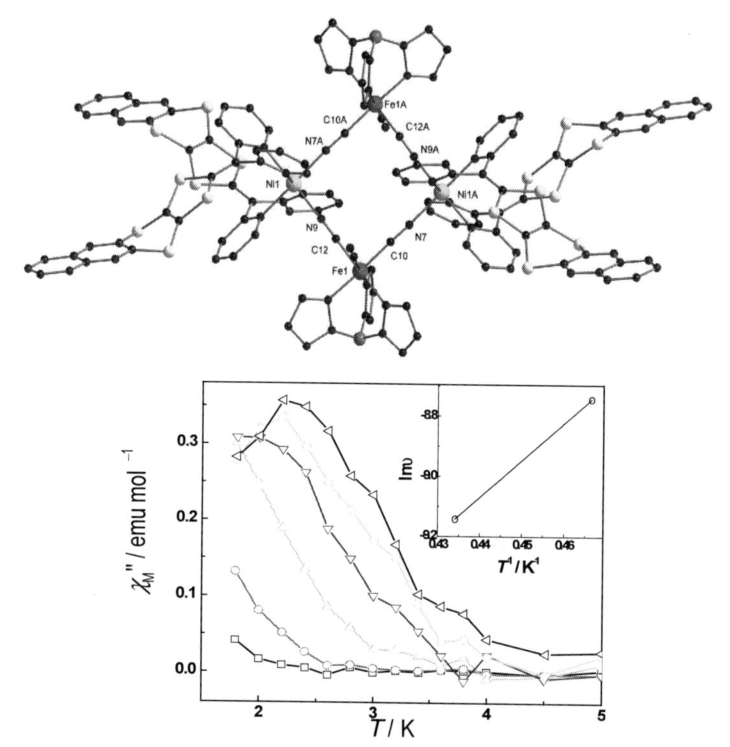

Figure 3.9 **Top:** Structure of complex **9**. **Bottom:** Frequency dependencies of the out-of-phase χ_M'' signals of **9**. Inset: Arrhenius fitting. See also Color Insert.

Tetrathiafulvalene (TTF) and its derivatives are well-known sulfur-rich organic molecules. A variety of TTF-contained ligands with different donating groups have been prepared.[13] Ouahab *et al.* fabricated pyridine-donating TTF ligands with paramagnetic clusters, and some new interesting complexes were obtained,[14] pointing to another promising approach for new conductive SMMs.

3.3 Photoluminescent Single-Molecule Magnets

Recently, efforts have been made to study SMMs at the single-molecule level by placing them on surfaces to examine their conductive properties.[15] However, it is always difficult to determine the exact position or dispersion of species deposited on surfaces. When SMMs with photoluminescent properties are employed, it may be possible to precisely ascertain the positions and concentration of molecules on surfaces. Therefore, the presence of photoluminescent SMMs could prove valuable in the detection of single molecules, which is an issue of vital importance in the field of molecular electronics.

On the basis of emissive ligands, 9-anthracenecarboxylic acid, Hendrickson *et al.* synthesized a new $[Mn_4]$ single-molecular magnet, $[Mn_4(anca)_4(Hmdea)_2(mdea)_2]\cdot 2CHCl_3$ (**11**, anca = 9-anthracenecarboxylate anion, Hmdea = *N*-methyldiethanolamine), as shown in Fig. 3.10.[16] Complex **11** exhibits obvious magnetic hysteresis loops below 1.3 K (Fig. 3.11). Besides, as shown in Fig. 3.12, the emission spectrum of **11** exhibits the same line shape and peak position (460 nm) as those of Hanca and NH_4-anca, indicating that the photoluminescent property is arising from the surrounding 9-anthracenecarboxylic acid ligand. Compared with the free ligand, significant quenching of the emission intensity for **11** is also observed, which may be due to the paramagnetic effect. Theoretically, below blocking temperature, the fluorescent lifetimes of the emissive ligands may be dramatically affected. Thus, the introduction of such emissive ligands to the synthetic routes of SMMs may also provide a new tool for studying the fundamental quantum behavior exhibited by SMMs on a 10^{-9}–10^{-12} s time scale.

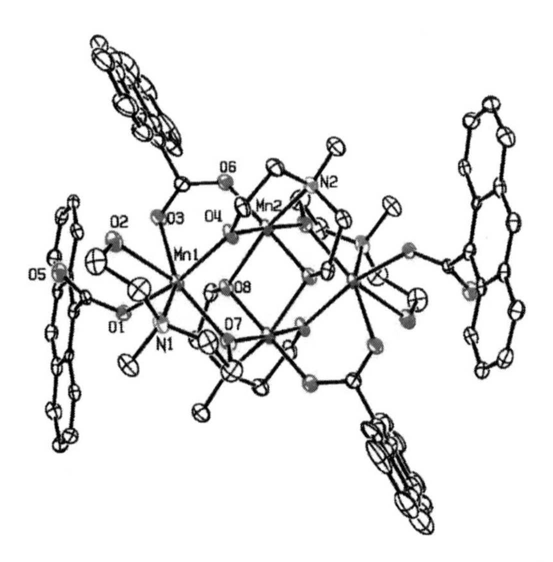

Figure 3.10 ORTEP drawing of **11** with thermal ellipsoids at 30% probability.

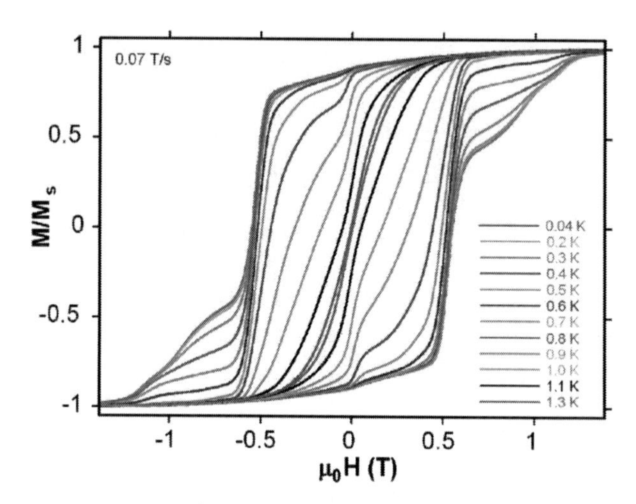

Figure 3.11 Plot of temperature-dependent magnetization versus field hysteresis loops for complex 11. See also Color Insert.

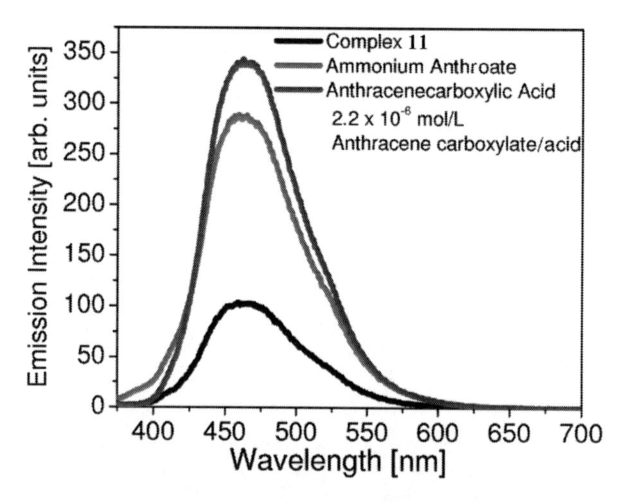

Figure 3.12 Room temperature fluorescence emission spectra of complex 11. See also Color Insert.

3.4 Photochromic Single-Molecule Magnets

Diarylethene compounds are well-established photoswitching and photochromic molecules due to photoisomerization reaction.[17]

Various photochromic compounds have been synthesized and investigated for applications in molecular memory and switching devices.[18]

The biscarboxylate substituted diarylethene ligand (dae) was used as bridging ligand to connect [Mn_4] SMMs, [$Mn_4(hmp)_6Br_2(H_2O)_2$] $Br_2 \cdot 4H_2O$ (Hhmp = 2-hydroxymethylpyridine) by Yamashita *et al.* The open-form and close-form chain complexes, [$Mn_4(hmp)_6(dae\text{-}o)_2$ $(ClO_4)_2$]$\cdot 6H_2O$ (**12o**) and [$Mn_4(hmp)_6(dae\text{-}c)_2(H_2O)_2$]$(ClO_4)_2 \cdot CH_3C$ $N \cdot 4H_2O$ (**12c**), were isolated separately (Fig. 3.13).[19] Despite being bridged by the dae ligand, magnetic measurements on **12o** and **12c** showed that they still behaved as the isolated double-cuboidal [Mn_4] SMMs (Fig. 3.14). The photochromic reactivity of **12o** and **12c** in the solid state was examined through visual color changes using KBr-diluted pellet samples of both compounds as shown in Fig. 3.15.

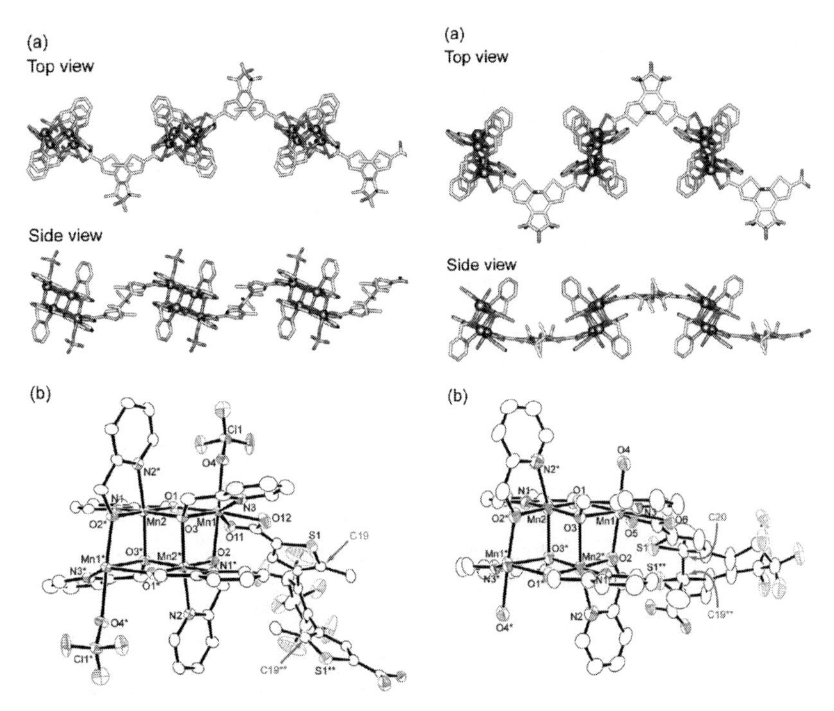

Figure 3.13 **Left:** (a) 1-D chain structure and (b) ORTEP drawing of a repeating unit of **12o**. **Right:** (a) 1-D chain structure and (b) ORTEP drawing of a repeating unit of **12c**.

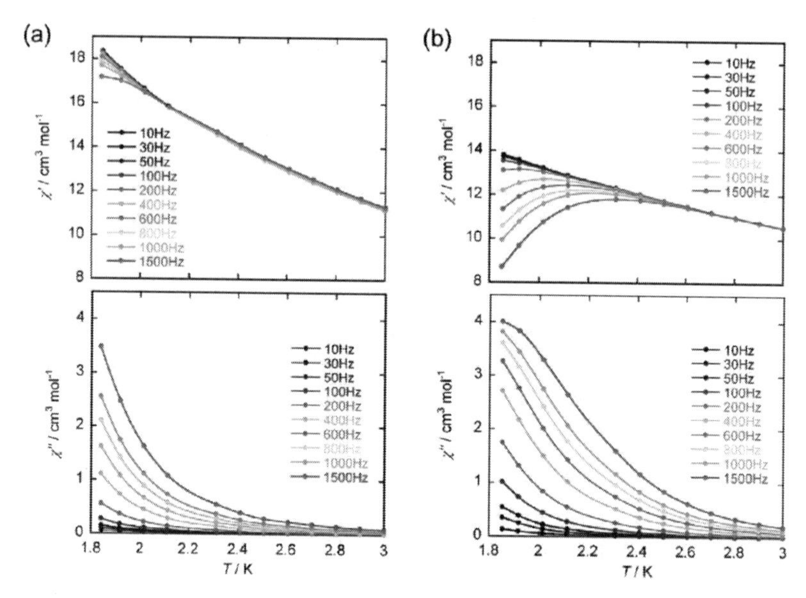

Figure 3.14 Temperature dependence of in-phase (χ') and out-of-phase (χ'') ac susceptibilities for (a) **12o** and (b) **12c** measured at several frequencies. See also Color Insert.

The effects of photoinduced isomerization reactions of dae on the magnetic properties were demonstrated with microcrystalline powder samples of **12o** and **12c** (Fig. 3.15). The open-ring form **12o** was irradiated with UV light (λ = 313 nm) for 26 h (denoted as **12o-UV**). No significant change in the magnetic properties of **12o-UV** was observed when **12o** was photocyclized, although dae-c in **12o-UV** should have a well-conjugated pathway. In contrast, the closed-ring form **12c** was irradiated with visible light (λ > 480 nm) for 3 h (denoted as **12c-Vis**). XRPD data revealed significant structural change of the crystal-lattice parameters from **12c** to **12c-Vis,** which was caused by a change in the [Mn_4] arrangement between the neighboring 1-D chains due to the geometrical change in the dae moiety during the photoreaction. Such structural change led to a dramatic change in magnetic properties as reflecting on field-dependent magnetization plot and ac magnetic susceptibility (Fig. 3.16). Finally, **12c-Vis** was irradiated with UV again to give **12c-Vis-UV** and the magnetic data of **12c-Vis-UV** nearly returned to those of **12c**, verifying the reversibility of such photoinduced changes in magnetic properties.

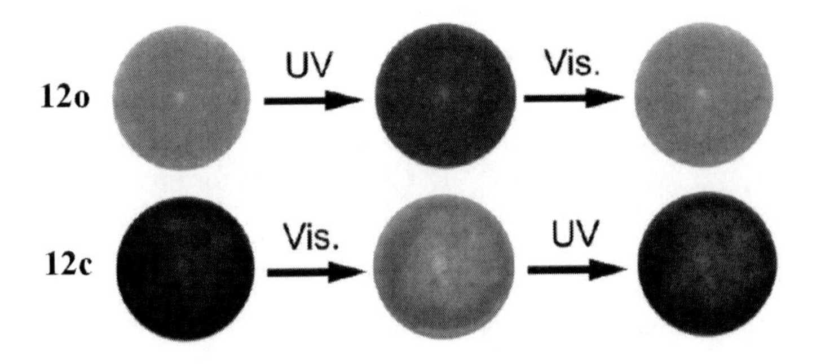

Figure 3.15 Photochromic effects between **12o** and **12c**. See also Color Insert.

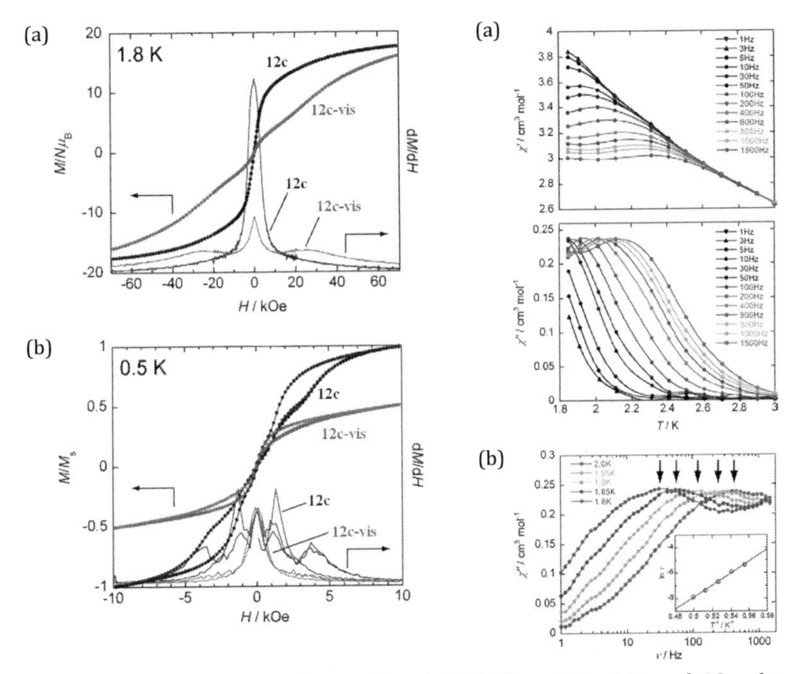

Figure 3.16 **Left:** Plots of M vs. H and $\Delta M/\Delta H$ vs. H for **12c** and **12c-vis** using (a) a nonorientation polycrystalline sample at 1.8 K and (b) field-oriented one at 0.5 K. **Right:** Plots of χ' and χ'' as a function of (a) temperature and (b) frequency for **12c-vis**. See also Color Insert.

This study provides the first example of introducing photochromic and photoswitching molecules into SMM systems, which lead to new multifunctional and stimulus-responsive SMMs.

3.5 Chiral Single-Molecule Magnets and Ferroelectric Single-Chain Magnets

Chirality has played an important role in the research of molecule-based magnets. Since the first observation of magneto-chiral dichroism in 1997,[20] the exploration of enantiopure chiral magnets, hybrid of natural optical activity and long-range ferromagnetic or ferrimagnetic ordering, has attracted much attention.[21] Analogously, SMMs or SCMs with optical activity are acquired by assembly of chirality into magnetic clusters or nanowires.

With ligand substitution reaction, Veciana and collaborators introduced different chiral carboxylate groups into the classic [Mn_{12}] SMMs and several chiral SMMs were obtained. Among them, enantiomers of (*R*)- and (*S*)-[$Mn_{12}O_{12}(O_2CCHMeCl)_{16}(H_2O)_4$] (**13R**, **13S**) possess good mirror-image Cotton effects in CD spectra and slow relaxation of magnetization in ac magnetic susceptibility (Fig. 3.17).[22]

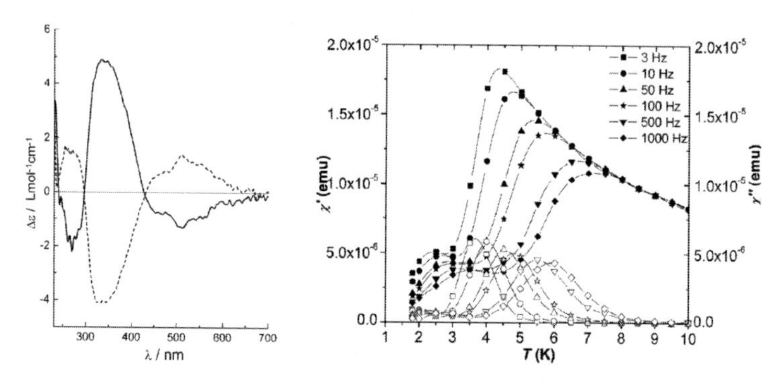

Figure 3.17 **Left:** Room temperature natural CD spectrum of the enantiomers of **13S** (solid line) and **13R** (dashed line). **Right:** Ac magnetic susceptibility of complex **13S** as a function of temperature at different frequencies.

MCD measurements were performed on samples of **13R** and **13S** in a 1:1 toluene–dichloromethane glass. The resulting spectra (Fig. 3.18, left) are very similar to those of the achiral cluster [$Mn_{12}O_{12}(O_2CMe)_{16}(H_2O)_4$] recorded in a DMF–MeCN glass, which are composed of a series of Faraday C terms. At lower temperatures, hysteresis was observed, manifested as a large remnant MCD that persists after the removal of the applied magnetic field. When the field direction is reversed, the signs of the C terms and the subsequent zero-field remnant bands are also reversed. More generally, cycling of the applied magnetic field generated a symmetric hysteresis loop (Fig. 3.18, right) with a coercive field of approximately 1.25 T. Although no detectable magneto-chiral effect was observed under these conditions, these chiral SMMs displayed optical bistability when monitored using MCD.

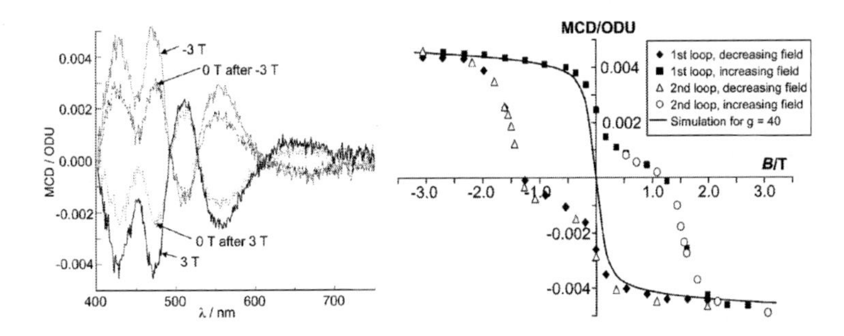

Figure 3.18 **Left:** 1.65 K MCD spectra of **13S** in 1:1 dichloromethane–toluene glass at 3, −3, and 0 T after removing the applied fields. **Right:** MCD hysteresis loop at 1.65 K for the 472 nm band of **13S** in a 1:1 dichloromethane–toluene glass.

Besides optical activity, the presence of chiral ligands may reduce the symmetry of the crystal lattice. And particularly, the molecular assemblies with chiral ligands may crystallize in polar point group, which sometimes exhibits ferroelectric ordering. Multiferroic materials, exhibiting simultaneous ferroelectric and magnetic properties, have attracted increasing attention over the past few years because of their great potential for applications as memory devices that can be electrically written and magnetically read.[23] While much of the research has been directed toward pure inorganic compounds, some success has also been achieved in molecular materials.[24]

By introducing chiral enantiomers (1*S*, 2*S*)-(+)-1,2-diaminocyclohexane((1*S*,2*S*)-chxn)and(1*R*,2*R*)-(−)-1,2-diaminocyclohexane ((1*R*, 2*R*)-chxn) as auxiliary ligands to react with $Ni(ClO_4)_2·6H_2O$ and $(Bu_4N)[(meTp)Fe(CN)_3]$, (meTp = methyltris(pyrazolyl)borate), heterotrinuclear enantiomers $\{(MeTp)_2Fe_2(CN)_6Ni[(1R, 2R)\text{-}chxn]_2\}$ (**14R**) and $\{(MeTp)_2Fe_2(CN)_6Ni[(1S, 2S)\text{-}chxn]_2\}$ (**14S**) were synthesized (Fig. 3.19).[25]

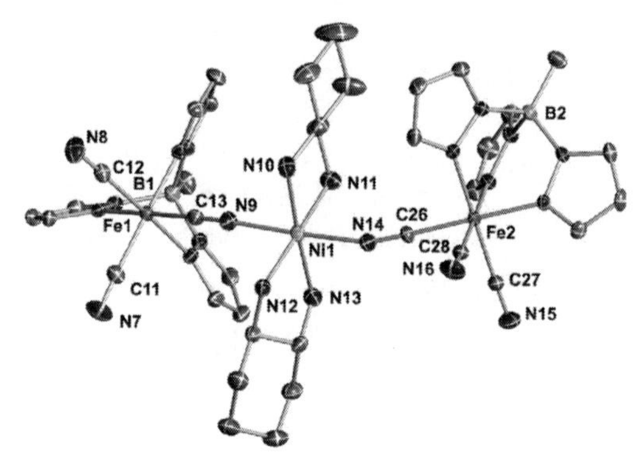

Figure 3.19 Thermal ellipsoids of complex **14R**.

Complexes **14R** and **14S** crystallize in polar point group C_2 (space group $P2_1$) and display both optical activity and ferroelectricity at room temperature (Fig. 3.20). Furthermore, these clusters exhibit intramolecular ferromagnetic interactions and significant zero-field splitting (Fig. 3.21) and provide good examples for the coexistence of ferromagnetism and ferroelectricity.

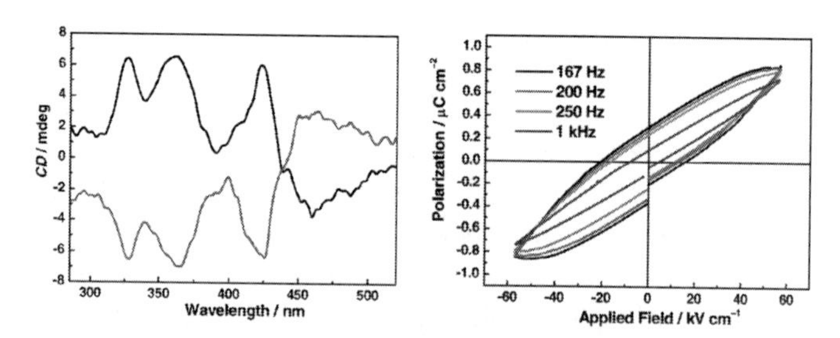

Figure 3.20 **Left:** CD spectra of **14R** (gray) and **14S** (black) in KBr pellets. **Right:** Polarization versus applied electric field curve at room temperature. See also Color Insert.

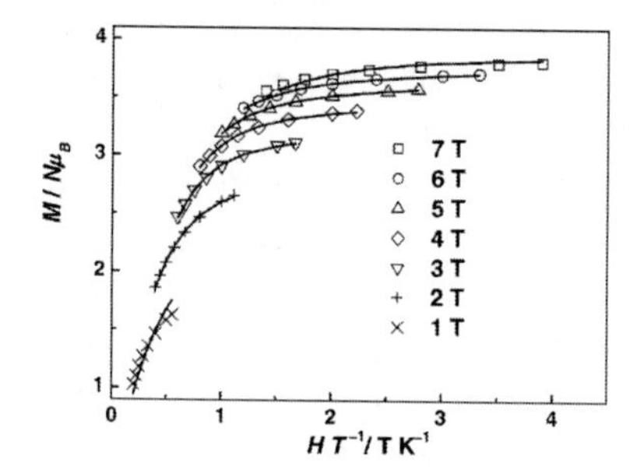

Figure 3.21 Plot of reduced magnetization for **14R**.

When $(Bu_4N)[(Tp)Fe(CN)_3]$ (Tp = hydridotris(pyrazolyl)borate) was used instead of $(Bu_4N)[(meTp)Fe(CN)_3]$ in the above-mentioned assembly, 1-D chain complexes $\{[(Tp)_2Fe_2(CN)_6Ni_3((1S,2S)\text{-chxn})_6](ClO_4)_4\cdot 2H_2O\}_n$ (**15S**) and $\{[(Tp)_2Fe_2(CN)_6Ni_3((1R,2R)\text{-chxn})_6](ClO_4)_4\cdot 2H_2O\}_n$ (**15R**) were isolated (Fig. 3.22).[26]

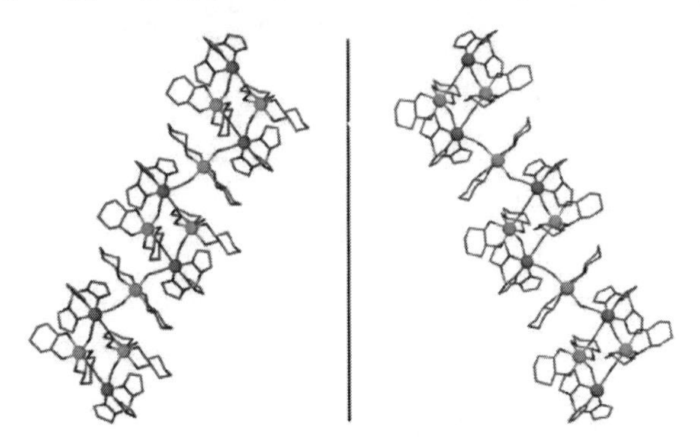

Figure 3.22 1-D 3,2-chain structures of **15S** (left) and **15R** (right).

Complexes **15S** and **15R** crystallize in polar point group C_1 (space group $P1$) with highly anisotropic 1-D structure. As a result, they show ferroelectric ordering at room temperature as well as optical activity (Fig. 3.23). Importantly, these highly magnetic anisotropic chains exhibit typical slow relaxation of the magnetization below 5 K (Fig. 3.24), typical of SCMs, with an effective spin-reversal barrier of U_{eff} = 19.5 K. These chain complexes provide the first example of metal–organic compounds bearing both slow magnetization relaxation and ferroelectricity. These results demonstrate that multiferroic materials may be achieved by a new synthetic strategy based on the assembly of chirality into magnetic clusters or nanowires.

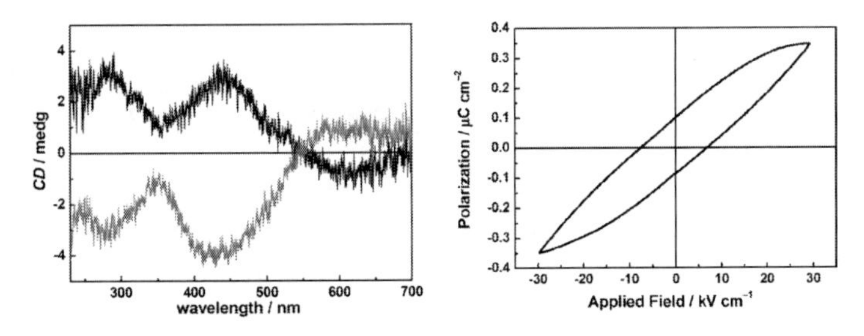

Figure 3.23 **Left:** Circular dichroism spectra of **15S** (black) and **15R** (gray) in KBr pellets. **Right:** Polarization versus applied electric field curve at room temperature.

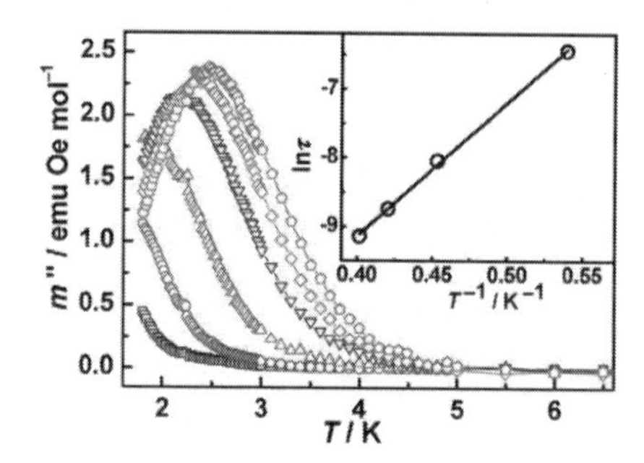

Figure 3.24 Out-of-phase component of the ac magnetization data of **15S**. Inset: Arrhenius law fit.

3.6 Functionalized Single-Molecule Magnets

All of the said potential applications are based on the use of intrinsic properties of the individual molecules of SMMs. Traditional measurements, such as magnetic susceptibility, are focused on bulk solid samples, which reflect the macroscopical properties of molecular assemblies rather than single molecules. Before SMMs stimulate the realization of a prospective molecular memory or computational device, some basic studies that help us to better understand and control their intricate characteristics are highly required. One of the challenges that has attracted much attention over the last few years is the development of new techniques of nanostructuration and deposition of SMMs on surfaces. Focused on the classic $[Mn]_{12}$ SMM family, several different methods have been developed, including thin-film growth and micro- and nanopatterning.[27] Furthermore, detection of materials properties at single-molecule level has enabled a crucial evolution in the research of SMM-based spintronics.

Theoretical analyses of the interplay of current, especially spin-polarized current, and magnetic anisotropic molecules have been investigated.[28] Experimentally, Heersche *et al.* measured the electron transport through single $[Mn_{12}]$ molecules that are weakly coupled to gold electrodes (Fig. 3.25).[29] Current suppression and negative differential conductance on the energy scale of the anisotropy barrier were observed, relating the conducting characteristics to the spin properties of single molecules. These results provide the conductivity of SMM molecules and may lead to the electronic control of SMMs.

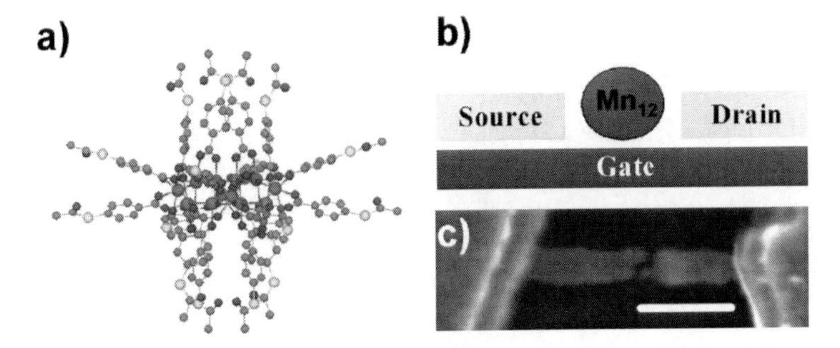

Figure 3.25 (a) Side view of a Mn_{12} molecule with tailormade ligands. (b) Schematic drawing of the Mn_{12} molecule trapped between electrodes. (c) Scanning electron microscopy image of the electrodes.

Although the research of the self-assembly monolayers (SAMs) of SMMs has been active for decades, direct magnetic measurement using the bulk measurement techniques, given the small quantities of materials in a SAM, can be difficult. Since SMMs have been shown to be very dependent on structural and crystalline restrictions, it is also a concern whether the material will retain its intrinsic magnetic property when immobilized on surface or in a device.

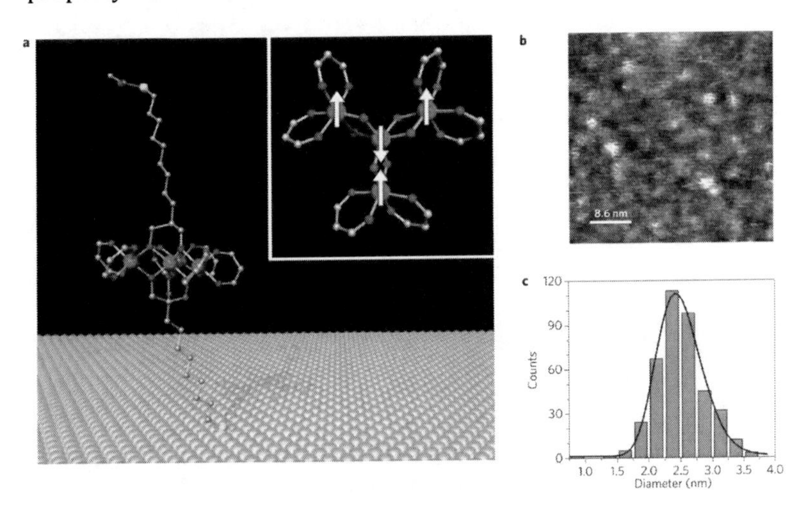

Figure 3.26 Monolayer of Fe_4 on gold. (a) Schematic diagram of the anchoring on a gold surface of Fe_4. (b) Room temperature constant-current STM image of the Fe_4 monolayer. (c) Statistical distribution of molecular diameters. See also Color Insert.

Using the newly developed synchrotron-based X-ray absorption spectroscopy and X-ray magnetic circular dichroism techniques, Sessoli *et al.* recently measured the magnetic property of the SAM of [Fe_4] SMMs deposited on the Au(111) surface (Fig. 3.26), and the magnetic hysteresis loop was directly observed for the first time (Fig. 3.27). These results demonstrate that isolated SMMs do retain their intrinsic magnetic properties when they are deposited on a conductive metal surface. The investigation of the interactions between electron transport and magnetism of SMMs in their blocked magnetization state at the molecular scale is possible.[30]

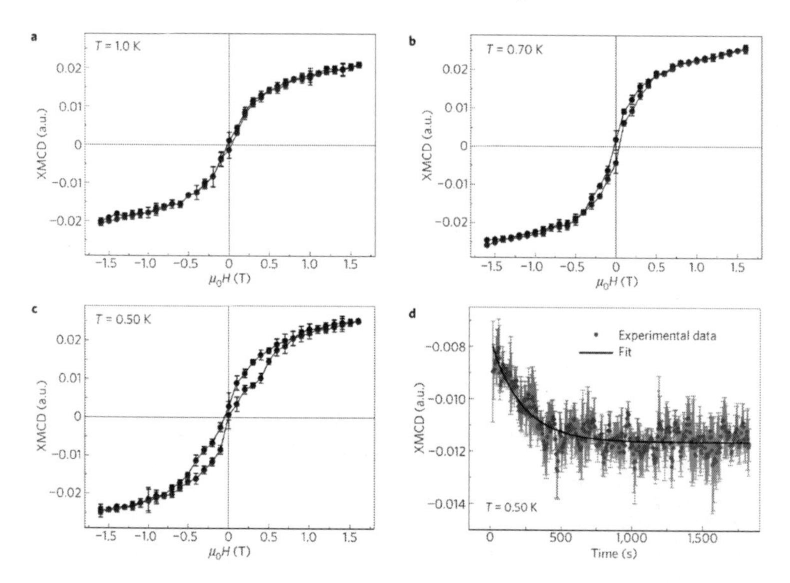

Figure 3.27 Hysteresis at different temperatures and magnetization dynamics of Fe_4 monolayer.

3.7 Other Emerging Areas of Single-Molecule Magnets and Single-Chain Magnets

Most of the early examples of SMMs and SCMs focus on complexes containing the first-row transition metals, such as the first SMM (Mn^{III}/Mn^{IV})[1] and the first SCM (Co^{II}).[2a] As mentioned above, SMMs combines large spin ground state and large magnetic anisotropy. In order to further increase the blocking temperature, both factors are seriously considered. Due to the limitation of the magnetic moment of individual 3-D metal ions, chemists have made great effort to increase the nuclearity of SMMs so as to further increase the total spins of the ground state. A typical example is the discovery of the giant $[Mn_{84}]$ SMMs, $[Mn_{84}O_{72}(O_2CMe)_{78}(OMe)_{24}(MeOH)_{12}(H_2O)_{42}(OH)_6]\cdot xH_2O\cdot yCHCl_3$.[9a] On the other hand, the second- or third-row transition metal ions possess greater spin-orbital coupling constants than the first-row transition metals, which may lead to larger magnetic anisotropy. In recent years, it has also become an important strategy to introduce the second or third row, even the actinide transition metal ions, into nanomagnet systems. Moreover, the enhancement of intrachain ferromagnetic coupling interactions is also a key factor in theory in spite of the scarcity of successful SCM examples.[31]

Introduction of heavy metal ions with great magnetic anisotropy is another new trend in the development of SMMs and SCMs. One of the typical examples is the discovery of single-ion magnets, which usually refer to some mononuclear rare earth complexes.[32] Recently, a mononuclear actinide metal complex was also found to exhibit slow relaxation of magnetization.[33] Another utility of heavy metal ions is to construct heavy metal polycyanide building blocks.[34] On the basis of these new building blocks, clusters or 1-D chain complexes were prepared that display slow relaxation of magnetization, and some of them indeed exhibit greater spin-reversal barriers.[35]

3.8 Conclusion

Since the discovery of SMMs in 1993, a great deal of SMMs and SCMs have been intensively studied. Several trends of recent years in the design and syntheses of SMMs and SCMs are noteworthy. Combinations of slow relaxation of magnetization behavior and other physical properties have led to a new class of molecular materials named multifunctional SMMs and SCMs. The crossover of SMM chemistry with modern nanoscience makes it possible to control magnetization reversal by conduction electrons, which paves the way for information storage applications. Additionally, heavy metal ions, possessing greater magnetic anisotropy than first-row metals, become popular in the synthesis of SMMs and SCMs, thereby providing a chance for increasing the relaxing barrier and blocking temperature.

References

1. Sessoli, R., Tsai, H. L., Schake, A. R., Wang, S., Vincent, J. B., Folting, K., Gatteschi, D. Christou, G., Hendrickson, D. N. (1993). High-spin molecules $[Mn_{12}O_{12}(O_2CR)_{16}(H_2O)_4]$, *J. Am. Chem. Soc.*, **115**, 1804–1816. (b) Sessoli, R., Gatteschi, D., Caneschi, A., Novak, M. A. (1993). Magnetic bistability in a metal–ion cluster, *Nature*, **365**, 141–143.
2. Caneschi, A., Gatteschi, D., Lalioti, N., Sangregorio, C., Sessoli, R., Venturi, A., Vindigni, A., Rettori, A., Pini, M. G., Novak, M. A. (2001). Cobalt(II)–nitronyl nitroxide chains as molecular magnetic nanowires, *Angew. Chem. Int. Ed.*, **40**, 1760–1763. (b) Bogani, L., Vindigni, A., Sessoli, R., Gatteschi, D. (2008). Single chain magnets: Where to from here?, *J. Mater. Chem.*, **18**, 4750–4758.
3. (a) Gatteschi, D., Sessoli, R. (2003). Quantum tunneling of magnetization and related phenomena in molecular materials, *Angew. Chem. Int. Ed.*, **42**, 268–297. (b) Christou, G., Gatteschi, D., Hendrickson, D. N., Sessoli, R. (2000). Single-molecule magnets, *Mater. Res. Bull.*, **25**, 66–71.
4. Krusin-Elbaum, L., Shibauchi, T., Argyle, B., Gignac, L., Weller, D. (2001). Stable ultrahigh-density magnetooptical recordings using introduced linear defects, *Nature*, **410**, 444–446.
5. Cavallini, M., Gomez-Segura, J., Ruiz-Molina, D., Massi, M., Albonetti, C., Rovira, C., Veciana, J., Biscarini, F. (2005). Magnetic information storage on polymers by using patterned single-molecule magnets, *Angew. Chem., Int. Ed.*, **44**, 888–892.
6. Chang, C.-C., Sun, K. W., Lee, S.-F., Kan, L.-S. (2007). Self-assembled molecular magnets on patterned silicon substrates: Bridging biomolecules with nanoelectronics, *Biomaterials*, **28**, 1941–1947.
7. Chudnovsky, E. M., Tejada, J. (1998). *Macroscopic Quantum Tunneling of the Magnetic Moment*, Cambridge University Press, Cambridge.
8. (a) Rocha, A. R., Garcia-Suarez, V. M., Bailey, S. W., Lambert, C. J., Ferrer, J., Sanvito, S. (2005). Towards molecular spintronics, *Nat. Mater.*, **4**, 335–339. (b) Bogani, L., Wernsdorfer, W. (2008). Molecular spintronics using single-molecule magnets, *Nat. Mater.*, **7**, 179–186.
9. (a) Tasiopoulos, A. J., Vinslava, A., Wernsdorfer, W., Abboud, K. A., Christou, G. (2004) Giant single-molecule magnets: A $\{Mn_{84}\}$ torus and its supramolecular nanotubes, *Angew. Chem. Int. Ed.*, **43**, 2117–2121. (b) Mishra, A., Wernsdorfer, W., Abboud, K. A., Christou, G. (2004). Initial observation of magnetization hysteresis and quantum tunneling in mixed manganese-Lanthanide single-molecule magnets, *J. Am. Chem. Soc.*, **126**, 15648–15649. (c) Song, Y., Zhang, P., Ren,

X. M., Shen, X. F., Li, Y. Z., You, X. Z. (2005). Octacyanometallate-based single-molecule magnets: $Co^{II}_9M^V_6$ (M = W, Mo), *J. Am. Chem. Soc.*, **127**, 3708–3709. (d) Pardo, E., Ruiz-García, R., Lloret, F., Faus, J., Julve, M., Journaux, Y., Delgado, F., Ruiz-Pérez, C. (2004). Cobalt(II)-copper(II) bimetallic chains as a new class of single-chain magnets, *Adv. Mater.*, **16**, 1597–1600. (e) Kajiwara, T., Nakano, M., Kaneko, Y., Takaishi, S., Ito, T., Yamashita, M., Igashira-Kamiyama, A., Nojiri, H., Ono, Y., Kojima, N. (2005). A single-chain magnet formed by a twisted arrangement of ions with easy-plane magnetic anisotropy, *J. Am. Chem. Soc.*, **127**, 10150–10151. (f) Wang, S., Zuo, J. L., Gao, S., Song, Y., Zhou, H. C., Zhang, Y. Z., You, X. Z. (2004). The observation of superparamagnetic behavior in molecular nanowires, *J. Am. Chem. Soc.*, **126**, 8900–8901.

10. (a) Coronado, E., Galán-Mascarós, J. R., Gómez-García, C. J., Laukhin, V., (2000). Coexistence of ferromagnetism and metallic conductivity in a molecule-based layered compound, *Nature*, **408**, 447–449. (b) Kosaka, Y., Yamamoto, H. M., Nakao, A., Tamura, M., Kato, R. (2007). Coexistence of conducting and magnetic electrons based on molecular π-electrons in the supramolecular conductor (Me-3,5-DIP)[Ni(dmit)$_2$]$_2$, *J. Am. Chem. Soc.*, **129**, 3054–3055. (c) Day, P., Kurmoo, M. (1997). Molecular magnetic semiconductors, metals and superconductors: BEDT-TTF salts with magnetic anions, *J. Mater. Chem.*, **7**, 1291–1295. (d) Motokawa, N., Miyasaka, H., Yamashita, M., Dunbar, K. R. (2008). An electron-transfer ferromagnet with T_c = 107 K based on a three-dimensional [Ru$_2$]$_2$/TCNQ System, *Angew. Chem. Int. Ed.*, **47**, 7760–7763.

11. (a) Hiraga, H., Miyasaka, H., Nakata, K., Kajiwara, T., Takaishi, S., Oshima, Y., Nojiri, H., Yamashita, M. (2007). Hybrid molecular material exhibiting single-molecule magnet behavior and molecular conductivity, *Inorg. Chem.*, **46**, 9661–9671. (b) Hiraga, H., Miyasaka, H., Takaishi, S., Kajiwara, T., Yamashita, M. (2008). Hybridized complexes of [Mn^{III}_2] single-molecule magnets and Ni dithiolate complexes, *Inorg. Chim. Acta*, **361**, 3863–3872. (c) Hiraga, H., Miyasaka, H., Clérac, R., Fourmigué, M., Yamashita, M. (2009). [M^{III}(dmit)$_2$]$^-$-coordinated Mn^{III} salen-type dimers (M^{III} = Ni^{III}, Au^{III}; dmit^{2-} = 1,3-dithiol-2-thione-4,5-dithiolate): design of single-component conducting single-molecule magnet-based materials, *Inorg. Chem.*, **48**, 2887–2898.

12. Peng, Y. H., Meng, Y. F., Hu, L., Li, Q. X., Li, Y. Z., Zuo, J. L., You, X. Z. (2010). Syntheses, structures, and magnetic properties ofheterobimetallic clusters with tricyanometalate and π-conjugated ligands containing 1,3-dithiol-2-ylidene, *Inorg. Chem.*, **49**, 1905–1912.

13. (a) Bouguessa, S., Herve, K., Golhen, S., Ouahab, L., Fabre, J. M. (2003). Synthesis, redox behaviour and X-ray structure of bis-TTF containing

a pyridine unit as a potential building block in the construction of conducting magnetic materials, *New J. Chem.*, **27**, 560–564. (b) Goze, C., Liu, S. X., Leiggener, C., Sanguinet, L., Levillain, E., Hauser, A., Decurtins, S. (2008). Synthesis of new ethynylbipyridine–linked mono- and bis-tetrathiafulvalenes: Electrochemical, spectroscopic, and Ru(II)-binding studies, *Tetrahedron*, **64**, 1345–1350. (c) Madalan, A. M., Rethore, C., Avarvari, N. (2007) Copper (II) and cobalt (II) complexes of chiral tetrathiafulvalene-oxazoline (TTF-OX) and tetrathiafulvalene-thiomethyl-oxazoline (TTF-SMe-OX) derivatives, *Inorg. Chim. Acta*, **360**, 233–240. (d) Herve, K., Liu, S. X., Cador, O., Golhen, S., Le Gal, Y., Bousseksou, A., Stoeckli-Evans, H., Decurtins, S., Ouahab, L. (2006). Synthesis of a BEDT-TTF bipyridine organic donor and the first Fe^{II} coordination complex with a redox-active ligand, *Eur. J. Inorg. Chem.*, **2006**, 3498–3502. (e) Chahma, M., Wang, X. S., van der Est, A., Pilkington, M. (2006). Synthesis and characterization of a new family of spin bearing TTF ligands, *J. Org. Chem.*, **71**, 2750–2755. (f) Zhu, Q. Y., Liu, Y., Lu, W., Zhang, Y., Bian, G. Q., Niu, G. Y., Dai, J. (2007). Effects of protonation and metal coordination on intramolecular charge transfer of tetrathiafulvalene compound, *Inorg. Chem.*, **46**, 10065–10070. (g) Lorcy, D., Bellec, N., Fourmigué, M., Avarvari, N. (2009). Tetrathiafulvalene-based group XV ligands: Synthesis, coordination chemistry and radical cation salts, *Coord. Chem. Rev.*, **253**, 1398–1438. (h) Liu, W., Wang, R., Zhou, X. H., Zuo, J. L., You, X. Z. (2008). Syntheses, structures, and properties of tricarbonyl rhenium(I) heteronuclear complexes with a new bridging ligand containing coupled bis(2-pyridyl) and 1,2-dithiolene units, *Organometallics*, **27**, 126–134. (i) Pointillart, F., Le Gal, Y., Golhen, S., Cador, O., Ouahab, L. (2008). First paramagnetic 4d transition-metal complex with a redox-Active tetrathiafulvalene derivative, [Ru(salen)(PPh_3)(TTF-CH=CH-Py)] BF_4 [$salen^{2-}$ = *N,N′*-Ethan-1,2-diylbis(salicylidenamine), PPh_3= Tri-phenylphosphine,TTF-CH=CH-Py)$_4$-(2-Tetrathiafulvalenylethenyl)p yridine], *Inorg. Chem.*, **47**, 9730–9732. (j) Liu, W., Xiong, J., Wang, Y., Zhou, X. H., Wang, R., Zuo, J. L., You, X. Z. (2009). Syntheses, structures, and properties of tricarbonyl (chloro) Rhenium(I) complexes with redox-active tetrathiafulvalene-pyrazole Ligands, *Organometallics*, **28**, 755–762.

14. (a) Setifi, F., Ouahab, L., Golhen, S., Yoshida, Y, Saito, G. (2003) First radical cation salt of paramagnetic transition metal complex containing TTF as ligand, [Cu^{II}(hfac)$_2$(TTF-py)$_2$](PF_6)·2CH_2Cl_2 (hfac = hexafluoroacetylacetonate and TTF-py)$_4$-(2-tetrathiafulvalenyl-ethenyl)pyridine), *Inorg. Chem.*, **42**, 1791–1793. (b) Benbellat, N., Gavrilenko, K. S., Gal, Y. L., Cador, O., Golhen, S., Gouasmia, A., Fabre, J. M., Ouahab, L. (2006). Co(II)-Co(II) paddlewheel complex with a

redox-active ligand derived from TTF, *Inorg. Chem.*, **45**, 10440–10442. (e) Gavrilenko, K. S., Gal, Y. L., Cador, O., Golhen, S., and Ouahab, L. (2007). First trinuclear paramagnetic Transition metal complexes with redox active ligands derived from TTF: $Co_2M(PhCOO)_6$(TTF-CH=CHpy)$_2$·$2CH_3CN$, M = Co^{II}, Mn^{II}, *Chem. Commun.*, 280–282.

15. (a) Jo, M. H., Grose, J. E., Baheti, K., Deshmukh, M. M., Sokol, J. J., Rumberger, E. M., Hendrickson, D. N., Long, J. R., Park, H., Ralph, D. C. (2006). Signatures of molecular magnetism in single-molecule transport spectroscopy, *Nano Lett.*, **6**, 2014–2020. (b) Ni, C., Shah, S., Hendrickson, D., Bandaru, P. R. (2006). Enhanced differential conductance through light induced current switching in Mn_{12} acetate molecular junctions, *Appl. Phys. Lett.*, **89**, 212104. (c) Bogani, L., Wernsdorfer, W. (2008). Molecular spintronics using single-molecule magnets, *Nat. Mater.*, **7**, 179–186. (d) Coronado, E., Martí-Gastaldo, C., Tatay, S. (2007). Magnetic molecular nanostructures: Design of magnetic molecular materials as monolayers, multilayers and thin films, *Appl. Surf. Sci.*, **254**, 225–235. (e) Abdi, A. N., Bucher, J. P., Rabu, P., Toulemonde, O., Drillon, M., Gerbier, P. (2004). Magnetic properties of bulk Mn_{12}Pivalates$_{16}$ single molecule magnets and their self assembly on functionnalized gold surface, *J. Appl. Phys.*, **95**, 7345–7347.
16. Beedle, C. C., Stephenson, C. J., Heroux, K. J., Wernsdorfer, W., Hendrickson, D. N. (2008). Photoluminescent Mn_4 Single-Molecule Magnet, *Inorg. Chem.*, **47**, 10798–10800.
17. Brown, G. H. (1971). *Photochromism*, Wiley-Interscience, New York.
18. Irie, M. (2000). Diarylethenes for memories and switches, *Chem. Rev.*, **100**, 1685–1716.
19. Morimoto, M., Miyasaka, H., Yamashita, M., Irie, M. (2009). Coordination assemblies of [Mn_4] single-molecule magnets linked by photochromic ligands: Photochemical control of the magnetic properties, *J. Am. Chem. Soc.*, **131**, 9823–9835.
20. Rikken, G. L. J. A., Raupach, E. (1997). Observation ofmagneto-chiral dichroism, *Nature*, **390**, 493–494.
21. (a) Barron, L. D. (2008). Chirality and magnetism shake hands, *Nat. Mater.*, **7**, 691–692. (b) Train, C., Gheorghe, R., Krstic, V., Chamoreau, L. M., Ovanesyan, N. S., Rikken, G. L. J. A., Vergaguer, M. (2008). Strong magneto-chiral dichroism in enantiopure chiral ferromagnets, *Nat. Mater.*, **7**, 729–734.
22. Gerbier, P., Domingo, N., Gómez-Segura, J., Ruiz-Molina, D., Amabilino, D. B., Tejada, J., Williamson, B. E., Veciana, J. (2004). Chiral, single-molecule nanomagnets: synthesis, magnetic characterization

and natural and magnetic circular dichroism, *J. Chem. Mater.*, **14**, 2455–2460.

23. (a) Ramesh, R., Spaldin, N. A. (2007). Multiferroics: Progress and prospects in thin films, *Nat. Mater.*, **6**, 21–29. (b) Scott, J. F. (2007). Data storage-multiferroic memories, *Nat. Mater.*, **6**, 256–257. (c) Eerenstein, W., Mathur, N. D., Scott, J. F. (2006). Multiferroic and magnetoelectric materials, *Nature*, **442**, 759–765. (d) Spaldin, N. A., Fiebig, M. (2005). The renaissance of magnetoelectric multiferroics, *Science*, **309**, 391–392.

24. (a) Cui, H. B., Wang, Z., Takahashi, K., Okano, Y., Kobayashi, H., Kobayashi, A. (2006). Ferroelectric porous molecular crystal, $[Mn_3(HCOO)_6](C_2H_5OH)$, exhibiting ferrimagnetic transition, *J. Am. Chem. Soc.*, **128**, 15074–15705. (b) Ohkoshi, S., Tokoro, H., Matsuda, T., Takahashi, H., Irie, H., Hashimoto, K. (2007). Coexistence of ferroelectricity and ferromagnetism in a rubidium manganese hexacyanoferrate, *Angew. Chem., Int. Ed.*, **46**, 3238–3241.

25. Wang, C. F., Gu, Z. G., Lu, X. M., Zuo, J. L., You, X. Z. (2008). Ferroelectric heterobimetallic clusters with ferromagnetic interactions, *Inorg. Chem.*, **47**, 7957–7959.

26. Wang, C. F., Li, D. P., Chen, X., Li, X. M., Li, Y. Z., Zuo, J. L., You, X. Z. (2009). Assembling chirality into magnetic nanowires: cyano-bridged iron(III)-nickel(II) chains exhibiting slow magnetization relaxation and ferroelectricity, *Chem. Commun.*, 6940–6942.

27. (a) Gómez-Segura, J., Veciana, J., Ruiz-Molina, D. (2007). Advances on the nanostructuration of magnetic molecules on surfaces: the case of single-molecule magnets (SMM), *Chem. Commun.*, 36, 3699–3707. (b) Cavallini, M., Facchini, M., Albonetti, C., Biscarini, F. (2008). Single molecule magnets: from thin films to nano-patterns, *Phys. Chem. Chem. Phys.*, **10**, 784–793.

28. (a) Timm, C., Elste, F. (2006). Spin amplification, reading, and writing in transport through anisotropic magnetic molecules, *Phys. Rev. B*, **73**, 235304. (b) Misiorny, M., Barnaś., J. (2007). Magnetic switching of a single molecular magnet due to spin-polarized current, *Phys. Rev. B*, **75**, 134425.

29. Heersche, H. B., de Groot, Z., Folk, J. A., van der Zant, H. S. J. (2006). Electron transport through single Mn_{12} molecular magnets, *Phys. Rev. Lett.*, **96**, 206801.

30. Mannini, M., Pineider, F., Sainctavit, P., Danieli, C., Otero, E., Sciancalepore, C., Talarico, A. M., Arrio, M. A., Cornia, A., Gatteschi, D., Sessoli, R. (2009). Magnetic memory of a single-molecule quantum magnet wired to a gold surface, *Nat. Mater.*, **8**, 194–197.

31. Miyasaka, H., Julve, M., Yamashita, M., Clérac, R. (2009). Slow dynamics of the magnetization in one-dimensional coordination polymers: Single-chain magnets, *Inorg. Chem.*, **48**, 3420–3437.

32. Ishikawa, N., Sugita, M., Ishikawa, T., Koshihara, S., Kaizu, Y. (2003). Lanthanide double-decker complexes functioning as magnets at the single-molecular level, *J. Am. Chem. Soc.*, **125**, 8694–8695.

33. Rinehart, J. D., Long, J. R. (2009). Slow magnetic relaxation in a trigonal prismatic uranium(III) Complex, *J. Am. Chem. Soc.*, **131**, 12558–12559.

34. (a) Schelter, E. J., Bera, J. K., Bacsa, J., Galn-Mascars, J. R., Dunbar, K. R. (2003). New paramagnetic Re(II) compounds with nitrile and cyanide ligands prepared by homolytic Scission of dirhenium complexes, *Inorg. Chem.*, **2003**(42), 4256–4258. (b) Bennett, M. V., Long, J. R. (2003). New cyanometalate building units: synthesis and characterization of $[Re(CN)_7]^{3-}$ and $[Re(CN)_8]^{3-}$, *J. Am. Chem. Soc.*, **125**, 2394–2395. (c) Harris, T. D., Bennett, M. V., Clérac, R., Long, J. R. (2010). $[ReCl_4(CN)_2]^{2-}$: A high magnetic anisotropy building unit giving rise to the single-chain magnets $(DMF)_4MReCl_4(CN)_2$ (M = Mn, Fe, Co, Ni), *J. Am. Chem. Soc.*, **132**, 3980–3988. (d) Yeung, W. F., Man, W. L., Wong, W. T., Lau, T. C., Gao, S. (2001). Ferromagnetic ordering in a diamond-like cyano-bridged $Mn^{II}Ru^{III}$ bimetallic coordination polymer, *Angew. Chem. Int. Ed.*, **40**, 3031–3033. (e) Yeung, W. F., Lau, P. H., Lau, T. C., Wei, H. Y., Sun, H. L., Gao, S., Chen, Z. D., Wong, W. T. (2005). Heterometallic $M^{II}Ru^{III}_2$ compounds constructed from trans-$[Ru(Salen)(CN)_2]^-$ and trans-$[Ru(acac)_2(CN)_2]^-$. Synthesis, structures, magnetic properties, and density functional theoretical study, *Inorg. Chem.*, **44**, 6579–6590. (f) Guo, J. F., Yeung, W. F., Lau, P. H., Wang, X. T., Gao, S., Wong, W. T., Chui, S. S. Y., Che, C. M., Wong, W. Y., Lau, T. C. (2010). Trans-$[Os^{III}(salen)(CN)_2]^-$: A new paramagnetic building block for the construction of molecule-based magnetic materials, *Inorg. Chem.*, **49**, 1607–1614.

35. (a) Schelter, E. J., Prosvirin, A. V., Dunbar, K. R. (2004). Molecular cube of Re^{II} and Mn^{II} that exhibits single-molecule magnetism, *J. Am. Chem. Soc.*, **126**, 15004–15005. (b) Schelter, E. J., Karadas, F., Avendano, C., Prosvirin, A. V., Wernsdorfer, W., and Dunbar, K. R.(2007). A family of mixed-metal cyanide cubes with alternating octahedral and tetrahedral corners exhibiting a variety of magnetic behaviors including single molecule magnetism, *J. Am. Chem. Soc.*, **129**, 8139–8149. (c) Freedman, D. E., Jenkins, D. M., Iavarone, A. T., Long, J. R. (2008). A redox-switchable single-molecule magnet incorporating $[Re(CN)_7]^{3-}$, *J. Am. Chem. Soc.*, **130**, 2884–2885.

Chapter 4

Magnetism and Chirality

Katsuya Inoue[a] **and Jun-ichiro Kishine**[b]

[a]*Department of Chemistry, Faculty of Science, Hiroshima University, 1-3-1 Kagamiyama, Higashi-hiroshima 739-8526, Japan*

[b]*The Open University of Japan, Chiba 261-8586, Japan*

kxi@hiroshima-u.ac.jp

This chapter provides an outline of chiral magnetism along with the hierarchy of space groups of crystals, magnetic representation theory, and some examples of reported materials. In the field of solid state chemistry and/or physics, the physical properties of symmetric materials or crystals are well understood nowadays, but those of asymmetric materials or crystals, such as polar or chiral materials, are not clear. In this chapter, we focus on the magnetic chirality of materials.

4.1 Introduction

The concept of chirality plays an essential role in biochemistry, organic chemistry, and biosciences. Chirality means an asymmetric shape of matter. The term was coined by Load Kelvin (1824–1907)

Multifunctional Molecular Materials
Edited by Lahcène Ouahab

ISBN 978-981-4364-29-4 (Hardcover), 978-981-4364-30-0 (eBook)
www.panstanford.com

in 1904 in his Baltimore lectures on "Molecular Dynamics and the Wave Theory of Light." In this lecture, he stated: "I call any geometrical figure, or any group of points, chiral, and say it has chirality, if its image in a plane mirror, ideally realized, cannot be brought to coincide with itself." A good example is that our hands have chirality, which is often called handedness: the left hand is a nonsuperimposable mirror image of the right hand; it is impossible for all the major features of both hands in any orientations to coincide by simple rotation (Fig. 4.1). The sense of handedness can be easily understood by shaking hands with each other using the left and right hands. A chiral object and its mirror image are said to be enantiomorphs. The word chirality is derived from the Greek word for palm (hand), *χειρ* (cheir) and the word enantiomorph stems from the Greek word for opposite, *εναντιοζ* (enantios) and form *μορφη* (morphe).

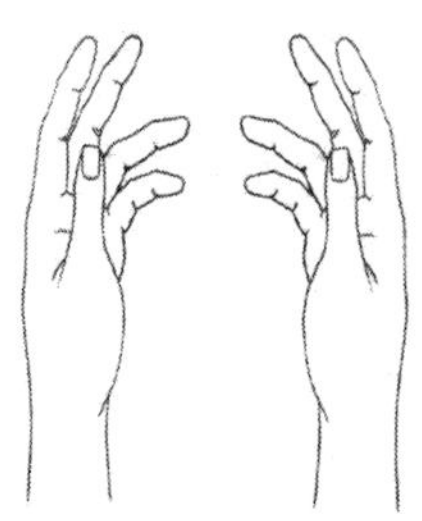

Figure 4.1 Example of chiral shape (Dr. Hiroyuki Higashikawa, 2005).

Table 4.1 Types of chirality

	Electronic chirality (P)	Magnetic chiral (P&T)
Microscopic Chirality	Molecular Chirality	Spin Chirality
Crystal Chirality	Maroscopic Chirality	Chiral Helimagnetic Order

Because the chirality is tied to the geometrical shape of matter, it is also applicable to objects in microscopic scales. From this point of view, we can categorize the concept of chirality as shown in Table 4.1. For instance, chiral molecules are categorized to micro-scale and chiral crystals such as quartz crystals are categorized to macroscale chirality. In these chiral structures, because each atom has different electronic negativity, finite electronic fields remain asymmetric in space. These features can be observed as natural optical activities. Magnetic spins also generate a magnetic field, which can possibly be defined as magnetic chirality. In this case, we need at least three noncollinear spins.

4.2 Space Groups of Crystal

The space groups of crystals are categorized by the symmetries of the crystal and can be divided into 230 types. The space groups in 3-D space are made from combinations of the 32 crystallographic point groups with the 14 Bravais lattice, which belong to one of the seven lattice systems. This results in a space group having some combination of the translational symmetry of a unit cell; the point group symmetry operations of reflection, rotation, and improper rotation; and the screw axis and glide plane symmetry operations. The combination of all these symmetry operations results in 230 unique space groups and can describe all possible crystal symmetries. The crystallographic point group is the symmetry that is common to all of its macroscopic physical properties. It follows that the symmetry group of any property of a crystal must include the symmetry operations of the crystallographic point group. This is the so-called "Neumann's principle" or "principle of symmetry," which can be used to derive information about the symmetry of a crystal from its physical properties. Twenty-one non-centrosymmetric crystallographic point groups, $1(C_1)$, $2(C_2)$, $m(C_S)$, $222(D_2)$, $2mm(C_{2v})$, $4(C_4)$, $-4(S_4)$, $422(D_4)$, $4mm(C_{4v})$, $-42m(D_{2d})$, $3(C_3)$, $32(D_3)$, $3m(C_{3v})$, $6(C_6)$, $-6(C_{3h})$, $622(D_6)$, $6mm(C_{6v})$, $-62m(D_{3h})$, $23(T)$, $432(O)$, and $-43m(T_d)$, are divided from al_l of the 32 crystallographic point groups. The crystals belong to non-centrosymmetric crystallographic point groups and are primarily properties represented by polar tensors of odd rank (e.g., pyroelectricity, piezoelectricity) or axial tensors of second rank (e.g., optical activity). Eleven chiral crystallographic point groups, 1, 2, 222, 4, 422, 3, 32, 6, 622, 23, and 432, are divided from 21 of the non-centrosymmetirc crystallographic point groups. The term of "chirality" refer to the same symmetry restriction, the absence of improper rotations (rotoinversions, rotoreflections) in a crystal or molecule. As a consequence, such chiral crystals or molecules can occur in two different shapes, the so-called left-handed and right-handed shapes. These two shapes of a crystal or molecule are mirror-related and never superimposable. Thus, the only symmetry operations of proper rotations are allowed for chiral crystals or molecules.

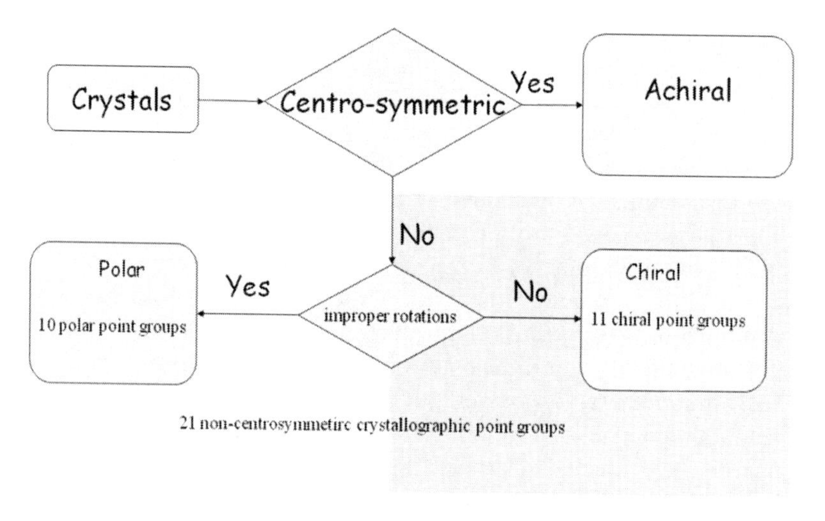

Figure 4.2 Point groups.

4.3 Symmetry and Complex Magnetic Structures

In this section, we briefly review basic properties of chiral helimagnetic structure from the viewpoints of crystallographic symmetry. As described so far, recent progress of material synthesis promotes systematic researches on a series of magnets belonging to chiral space group without any rotoinversion symmetry elements.[1] In the chiral magnets, the crystallographic chirality possibly gives rise to the asymmetric Dzyaloshinskii interaction that stabilizes the chiral helimagnetic structure, where either left-handed or right-handed magnetic chiral helix is formed.[2]

The chiral helimagnetic structure is a noncolinear incommensurate magnetic structure with a single propagation vector $\mathbf{Q}_0 = (0, 0, Q_0)$. The space group G consists of the elements $\{g_i\}$. Among them, some elements leave the propagation vector $\mathbf{Q}_0$ invariant, i.e., these elements form the little group G_{Q_0}. The magnetic representation[3] Γ_{mag} is written as $\Gamma_{\text{mag}} = \Gamma_{\text{perm}} \otimes \Gamma_{\text{axial}}$, where Γ_{perm} and Γ_{axial} represent the Wyckoff permutation representation and the axial vector representation, respectively. Then, Γ_{mag} is decomposed into the nonzero irreducible representations of G_{Q_0}. The incommensurate magnetic structure is determined by a "magnetic basis frame" of an axial vector space and the propagation vector k. In specific magnetic ion, the decomposition becomes $\Gamma_{\text{mag}} = \sum_i n_i \Gamma_i$, where Γ_i is the irreducible representations of G_{Q_0}. Then, we have two cases leading to the chiral helimagnetic magnetic structure. Case I: The magnetic moments are described by two independent 1-D real representations that form 2-D basis frames, or Case II: The magnetic moments are described by a single 2-D real or 1-D complex representations that form 2-D basis frames. Now, we summarize these characteristics.

Case I: This case is realized even in crystal classes possessing symmetry lower than tetragonal, if the unit cell has more than two magnetic ions. So, this case is not subject to severe symmetry restrictions.

Case II: This case is realized only for the following 52 space groups:

- *Tetragonal:* $P4$, $P4_1$, $P4_2$, $P4_3$, $I4$, $I4_1$, $P422$, $P42_12$, $P4_122$, $P4_12_12$, $P4_222$, $P4_22_12$, $P4_322$, $P4_32_12$, $I422$, $I4_122$
- *Trigonal:* $P3$, $P3_1$, $P3_2$, $R3$, $P3_12$, $P3_21$, $P3_112$, $P3_121$, $P3_212$, $P3_221$, $R32$
- *Hexagonal:* $P6$, $P6_1$, $P6_5$, $P6_2$, $P6_4$, $P6_3$, $P622$, $P6_122$, $P6_522$, $P6_222$, $P6_422$, $P6_322$
- *Cubic:* $P23$, $F23$, $I23$, $P2_13$, $I2_13$, $P432$, $P4_232$, $F432$, $F4_132$, $I432$, $P4_332$, $P4_132$, $I4_132$

From this classification scheme, it is interesting to compare inorganic and molecular-based crystals. Generally speaking, in inorganic crystals, magnetic ions tend to occupy rather high symmetry site. So, the number of equivalent position is small and the unit cell tends to be more closely packed. Consequently, many-body correlations work well and noncolinear magnetic structures are realized easily. On the other hand, in molecular-based crystals, magnetic ions tend to occupy rather low symmetry sites. So, the number of equivalent position is large and the unit cell tends to be more loosely packed. Consequently, many-body correlations do not work well and noncolinear magnetic structures are not so easily realized. Of course, this scheme is an extreme. From the viewpoints of material synthesis, it is interesting to locate the boundary between the two. By doing so, we may control the commensurability of the magnetic structure based on the symmetry arguments.

4.4 Magnetic Structure for Asymmetric Materials

In terms of the magnetic structure of chiral or non-centrosymmetric magnetic materials, the following aspects must be considered: (i) all spins are sited in asymmetric magnetic positions, and (ii) the materials exhibit an electronic dipole moment. On the basis of (i), the spins experience an asymmetric magnetic dipole field. The asymmetric magnetic dipole moments operate the spins aligned asymmetrically through spin–orbital interaction. In order to facilitate matters, we assume that the spins are sited on a helix and the ferro- or ferrimagnetic interaction occurs between the nearest neighbor spins. In this case, the spins are aligned in helical or conical spin alignment by an asymmetric magnetic dipole field and ferromagnetic interaction.[1] Moreover, chiral crystals display an internal asymmetric electronic dipole field, thus, all magnetic spins are tilted by Dzyaloshinsky–Moriya (DM) interaction.

4.5 Nonlinear Magneto-Optical Effects of Chiral Magnets

Chiral magnets can exhibit nonlinear magneto-optical effects due to their non-centrosymmetric structure and spontaneous magnetization. Over the last 20 years, the nonlinear magneto-optical phenomena have been intensively studied[5–13] and magnetization-induced second harmonic generation (MSHG) has received special attention due to its large magnetic effects, e.g., the second-order nonlinear Kerr rotation on the Fe/Cr magnetic film surface is close to 90°.[7] In the electric dipole approximation, second harmonic generation (SHG) is allowed in media with broken inversion symmetry.[14] MSHG effects are mainly reported from the surface of magnetic materials.[6,7] MSHG observations from the bulk crystals are limited. For antiferromagnetic materials, only chromium oxide Cr_2O_3[8] and yttrium-manganese oxide $YMnO_3$[9] are reported to exhibit MSHG. The reports of bulk MSHG for ferromagnetic materials are limited to bisubstituted yttrium iron garnet (Bi:YIG) magnetic film[10–12] and a ternary-metal Prussian blue analog $\{(Fe^{II}_x Cr^{II}_{1-x})_{1.5}[Cr^{III}(CN)_6]\cdot 7.5H_2O\}$-based magnetic film.[13] The MSHG effect is also useful for the topography of magnetic domains.[15] Furthermore, applying a magnetic field can control the intensity of MSHG signal.[11–13] Hence, an attractive method for studying nonlinear optics is to prepare ferromagnetic materials, which display second harmonic activity. Chiral magnets are advantageous when compared to conventional metal or metal oxide magnetic materials because the space-inversion and time-reversal symmetry are simultaneously broken.

4.6 Magnetic Structure of Molecule Based Chiral Magnet

To obtain non-centrosymmetric (polar) or chiral molecule-based magnets, the geometric symmetry such as chirality must be controlled in the molecular structure as well as in the entire crystal structure. In this section, recent results regarding the construction, structure, magnetic, and optical properties of molecule-based magnets with chiral structures will be described.

The major strategy relating to crystal design for magnetic materials exhibiting long-range magnetic ordering and spontaneous magnetization involves generation of an extended array of paramagnetic metal ions (M) with bridging ligands (L). The high-spin nitroxide or nitronylnitroxide radicals, cyanide ions, or oxalate dianions are used often for bridging ligands. The cyanide-bridged Prussian-blue and oxalate-bridged systems are generally obtained as bimetallic assemblies with 2- or 3-D networks by the reaction of hexacyanometalate $[M^{III}(CN)_6]^{3-}$ with a metallic ion M^{II} or tris(oxalato) metalate $[M^{III}(ox)_3]^{3-}$ with a metallic ion M^{III}, respectively. Extensive research has led to the production of a material displaying magnetic ordering at T_C as high as 372 K[4] for a

Prussian-blue analogues. The incorporation of such a ligand leads to the blockade of some coordinated linkages to M^{II} of cyanide groups in $[M^{III}(CN)_6]^{3-}$. It follows that various novel structures have been obtained in cyanide-bridged system, depending on the organic molecule. In oxalate-bridged systems, several different shapes of counter ions are often used to control crystal structures. These methods afford the possibility of crystal design in this system. In this section, we will describe the crystal design toward a chiral magnet utilizing these systems; additionally, several examples are presented.

Some chiral diamines,[18,21,23,24,27] nitroxide radicals,[16,26] nitonylnitroxide radicals,[17] and amino acid derivatives,[22,28] serve as ligands for the chiral source in the entire crystal structures of these systems. For oxalate-bridged systems, some chiral counter cations are used.[19,20] For instance, in cyanide-bridged systems, a target magnetic compound can be generated by the reaction between a hexacyanometalate $[M^{III}(CN)_6]^{3-}$ and a mononuclear complex $[M^{II}(L)_n]$. However, the combination of M^{II} and M^{III} in the stage of crystallization, which generates ferromagnetic interaction through M^{III}–CN–M^{II}, must be known in order to obtain a ferromagnet. These methods hold many possibilities with respect to obtaining various chiral magnets via alteration of the component substances. Reported chiral molecule-based magnets are shown in Table 4.2.

DM interaction is an antisymmetric interaction and it tends to align neighboring spins perpendicular to one another. The antiparallel DM vectors arrangement along the crystal axis stabilize canted spin structure (weak ferromagnetism) (Fig. 4.3, left), while the monodirectional parallel DM vectors stabilize chiral helical spin structures (incommensurate helical spin structures) (Fig. 4.3, center) or chiral conical spin structures (Fig. 4.3, right). Therefore, the magnetic chirality is made out of the crystallographic chirality. The magnetic materials have chiral conical magnetic structures, with both space-inversion and time-reversal symmetry simultaneously broken, and have new properties.[29–30] For molecule-based magnets, non-centrosymmetric examples have already been reported: weak ferromagnetism of the manganese alkylphosphonate hydrates ($Pnm2_1$),[29] etc., and chiral helical/conical spin structures for cyanide-bridged systems.[24,32] These canted and/or chiral helical/conical spin structures are stabilized by DM interactions. Normally, the DM interactions are small compared to exchange and magnetic dipole–dipole interactions. Also, the canting angles are small and the pitches of helix become very long. Due to this, neutron diffraction and muon experiments are used to determine the magnetic structure of these materials. Chiral spin solitons are expected for chiral helical spin structures and have been reported for some cases.[1,32]

Table 4.2 Reported chiral molecule-based magnets

Formula	Type of sequence	Space group	T_C or T_N (magnetism)	Ref.
Mn(hfac)$_2$(BNO R*) R* = CMe$_2$O CH$_2$C*H(Me)CH$_2$CH$_3$ hfac=fexafluoroacety lacetonato	---Mn-BNOR*-Mn- BNOR* ----	*P*1	T_N = 5.4 K	16
Mn(hafc)$_2$(NITR*) R = OCH$_2$C*H(Me)CH$_2$CH$_3$	---Mn-NITR* -Mn- NITR* ----	$P2_12_12_1$	T_C = 4.6 K	17
K$_{0.4}$[Cr(CN)$_6$][Mn(*S*)-pn](*S*)-pnH$_{0.6}$((S)-pn = (S)-1,2-diaminopropane)	3-D Helical	$P6_1$, $P6_5$	T_C = 53 K	18
[Z^{II}(bpy)$_3$][ClO$_4$][M^{II}Cr^{III} (ox)$_3$] (Z^{II} = Ru, M^{II} = Mn, Fe, Co, Ni, Cu; Z^{II} = Fe, M^{II} = Mn, Fe, Co; Z^{II} = Co, M^{II} = Mn, and Z^{II} = Ni; M^{II} = Mn, Fe; ox = Oxalate Dianion)	3-D 3-connected 10-gon oxalate-based anionic network (10,3). Single Crystal; Z^{II} = Ru, M^{II} = Mn	$P4_132$	T_C (Z^{II} M^{II}) K T_C (Ru Mn) < 2.0 T_C (Ru Fe) = 2.5 T_C (Ru Co) = 2.8 T_C (Ru Ni) = 6.4 T_C (Ru Cu) = 1.9 T_C (Fe Mn) = 3.9 T_C (Fe Fe) = 4.7 T_C (Fe Co) = 6.6 T_C (Ni Mn) = 2.3 T_C (Ni Fe) = 4.0 T_C (Co Mn) = 2.2	19
[Ru(bpy)$_3$][ClO$_4$][MnCr (ox)$_3$]	3-D 3-connected 10-gon oxalate-based anionic network (10,3).	$P2_13$	T_C = 4.2(Δ), 5.8 (Λ)	20
[{Cr(CN)$_6$}{Mn(*S*, *R*)-pnH(H$_2$O)}](H$_2$O); ((*S*, *R*)-pn =(*S*, *R*)-1,2-diaminopropane).	2-D	$P2_12_12_1$	T_C = 38 K	21
[{Cr(CN)$_6$}(Mn (*S*), (*R*)-NH$_2$ala)$_3$]·3H$_2$O	3-D triple helical strand structure	$P6_3$	T_C = 35 K	22

(Continued)

Table 4.2 (*Continued*)

Formula	Type of sequence	Space group	T_C or T_N (magnetism)	Ref.
$[Cr(CN)_6][Mn(S, R)$-$pnH(DMF)]\cdot 2H_2O$	2-D	$P2_12_12_1$	$T_N = 28$ K (metamagnet)	23
$[W(CN)_8]_4[Cu\{(S$ or $R)$-$pn\}H_2O]_4[Cu\{(S$ or $R)$-$pn\}]_2 2.5H_2O$	2-D	$P2_1$	$T_N = 8.5$ K (metamagnet)	24
$[NH_4][M(HCOO)_3]$, where (M) divalent Mn, Co, or Ni	3-D $4^9, 6^6$ topology	$P6_322$	$T_N = 8.4$ (Mn), 9.8 (Co), and 29.5 (Ni) K (metamagnet)	25
Co(hfac)$_2$(BNO R*) R* = CMe$_2$O CH$_2$C*H(Me)CH$_2$CH$_3$	---Co-BNOR*-Co-BNOR* ----	$P1$	$T_N = 20$ K	26
$[\{Cr(CN)_6\}\{Mn(S, R)$-$pnH(H_2O)\}](H_2O)$; ((S, R)-pn =(S, R)-1,2-diaminopropane).	2-D	$P2_12_12_1$	$T_C = 38$ K	27
$[\{Cr(CN)_6\}\{Mn(S, R)$-$pnH(H_2O)\}](H_2O)$; ((S, R)-pn = (S, R)-1,2-diaminopropane).	3-D	$P2_12_12_1$	$T_C = 73$ K	27
$[Cu(H_2O)_2]_{0.5}[Cu(R$ or $S)$-$pn]\ [Cr(CN)_6]\cdot H_2O$	3-D	$C2$	$T_C = 48.0$ K	28
$K[Mn(L)$-$ser]_2[Cr(CN)_6]\cdot 2H_2O$	3-D	$P2_12_12$	$T_C = 48.0$ K	28
$[Mn(NH_3$-$ala)_2]_3[Cr(CN)_6]_2$	3-D	$R3$	$T_C = 21$ K	28
$[Cu_3(S$ or R-$phea)_2(N_3)_6]_n$; phea = 1-phenylethylamine	2-D	$P2_12_12_1$	$T_C = 5.5$ K	34
$[Cu_3(S$ or R-$chea)_2(N_3)_6]_n$; chea = 1-cyclohexylethlamine	2-D	$P2_1$	$T_C = 5.0$ K	35
$[Cu_3(S$-$phpa)_2(N_3)_6]_n$; phpa = 1-phenylpropylamine	2-D	$P2_12_12_1$	$T_C = 5.0$ K	35
$[(S)$-$[PhCH(CH_3)N(CH_3)_3]]\ [Mn(CH_3CN)_{2/3}Cr(ox)_3]\ (CH_3CN)$(solvate)	2-D	$P3$	$T_C = 5.6$ K	36

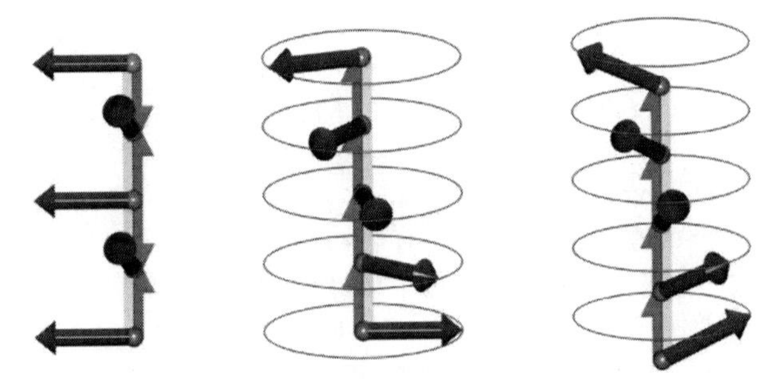

Figure 4.3 Spin structures stabilized by DM interactions. (left) Canted spin structure (weak ferromagnetism); (center) chiral helical spin structures (incommensurate helical spin structures); (right) chiral conical spin structures.

4.7 Magnetic Chirality and Crystal Chirality

Notable features of molecule-based magnets are that they can be designed and are transparent. A novel category of materials suitable for chiral or non-centrosymmetric magnets has been successfully fabricated for application in the field of molecule-based magnetic materials. These materials display new optical and/or spin dynamics phenomena, such as magnetization-induced nonlinear optical effect, magneto-chiral optical effect, and nonlinear magnetic susceptibility due to their non-centrosymmetric or chiral magnetic structure. These materials are of keen scientific interest and show the possibility for use in new types of devices. In this chapter, we described how to stabilize noncollinear and/or chiral magnetic structures for several chiral and non-centrosymmetric molecule-based magnets. These magnetic structures are explained by magnetic representation analysis theories and show the possibility of being used in new fields of magnetism.

4.8 Epilogue

The basic strategy of natural science is "reductionism," which means that all materials can be divided into point-like building blocks, e.g., atomic particles. However, there are some physical phenomena, such as optical activity, which cannot be completely understood until we go beyond reductionism and take into account the geometrical shape of the atomic cluster. Chirality becomes the starting concept to capture the sense of geometrical shape. As mentioned in this chapter, behind the concept of chirality lies the contrasts between chemistry and physics, molecule and crystals, and geometry and magnetism. Naively speaking, even the simple reason why spherical atoms are arranged in a chiral manner is a nontrivial issue.

In this chapter, we emphasized notable features of molecule-based magnets, i.e., the ability to be designed and transparency. Then, we discussed the interplay of crystallographic and magnetic chirality, which may play a key role to provide new optical phenomena, such as magnetization-induced nonlinear optical effects and magneto-chiral optical effects. Such materials are not only of keen scientific interest, but they may also open a possible window for new device synthesis and fabrication. We believe that the chiral magnets certainly carve out new facets of a long-standing history of magnetism and magnetic materials.

References

1. Kishine, J., Inoue, K, Yoshida, Y. (2005). Synthesis, structure and magnetic properties of chiral molecule-based magnets, *Prog. Theoret. Phys., Suppl.*, **159**, pp. 82–95.
2. Dzyaloshinskii, I. E. (1964). Theory of helicoidal structures in antiferromagnets. 1. Nonmetals, *Soviet Phys. JETP-USSR*, **19**, 960–971.
3. Kovalev, O. V. (1993). *Representations of the Crystallographic Space Groups*, 2nd ed., Gordon and Breach Science Publishers, Switzerland.
4. Ferlay, S., Mallah, T., Ouahes, R., Veillet, P., Verdaguer, M. (1995). A room temperature organometallic magnet based on prussian blue, *Nature*, **378**, 701–703.
5. Pan, R. P., Wei, H. D., Shen, Y. R. (1989). Optical 2nd-harmonic generation from magnetized surfaces, *Phys. Rev. B*, **39**, 1229–1234.
6. Pustogowa, U., Hubner, W., Bennemann, K. H. (1994). Enhancement of the magneto-optical Kerr angle in nonlinear-optical response, *Phys. Rev. B*, **49**, 10031–10034.
7. Rasing, Th., Koerkamp, M. G., Koopmans, B. (1996). Giant nonlinear magneto-optical Kerr effects from Fe interfaces, *J. Appl. Phys.*, **79**, 6181–6185.
8. Fiebig, M., Frohlich, D., Krichevtsov, B. B., Pisarev, R. V. (1994). 2nd-harmonic generation and magnetic-dipole-electric-dipole interference in antiferromagnetic Cr_2O_3, *Phys. Rev. Lett.*, **73**, 2127–2730.

9. Frohlich, D., Leute, St., Pavlov, V. V., Pisarev, R. V. (1998). Nonlinear optical spectroscopy of the two-order-parameter compound $YMnO_3$, *Phys. Rev. Lett.*, **81**, 3239–3242.

10. Aktsipetrov, O. A., Braginskii, O. V., Esikov, D. A. (1990). Nonlinear optics of gyrotropic media: SHG in rare-earth iron garnets, *Sov. J. Quant. Electron*, **20**, 259–263.

11. Pavlov, V. V., Pisarev, R. V., Kirilyuk, A., Rasing, Th. (1997). Observation of a transversal nonlinear magneto-optical effect in thin magnetic garnet films, *Phys. Rev. Lett.*, **78**, 2004–2007.

12. Gridnev, V. N., Pavlov, V. V., Pisarev, R. V., Kirikyuk, A., Rasing, Th. (2001). Second harmonic generation in anisotropic magnetic-films, *Phys. Rev. B*, **63**, 184407.

13. Ikeda, K., Ohkoshi, S., Hashimoto, K. (2001). Second harmonic generation from ternary metal Prussian blue analog films in paramagnetic and ferromagnetic regions, *Chem. Phys. Lett.*, **349**, 371–375.

14. Shen, Y. R. (1984). *The Principles of Nonlinear Optics*, Wiley, New York.

15. Fiebig, M., Frohlich, D., Leute, St., Pisarev, R. V. (1998). Topography of antiferromagnetic domains using second harmonic generation with an external reference, *Appl. Phys. B*, **66**, 265–270.

16. Kumagai, H., Inoue, K. (1999). A chiral molecular based metamagnet prepared from manganese ions and a chiral triplet organic radical as a bridging ligand, *Angew. Chem. Int. Ed.*, **38**, 1601–1603.

17. Kumagai, H., Markosyan, A. S., Inoue, K. (2000). Synthesis, structure and magnetic properties of a chiral one-dimensional molecular-based magnet, *Mol. Cryst. Liq. Cryst.*, **343**, 415–420.

18. Inoue, K., Imai, H., Ghalsasi, P. S., Kikuchi, K., Ohba, M., Okawa, H., Yakhmi. J. V., (2001). A three-dimensional ferrimagnet with a high magnetic transition temperature (T_C) of 53 K based on a chiral molecule, *Angew. Chem. Int. Ed.*, **48**, 4242–4245.

19. Coronado, E., Galan-Mascaros, J. R., Gómez-García, C. J., Martinez-Agugo, J. M. (2001). Molecule-based magnets formed by bimetallic three-dimensional oxalate networks and chiral tris(bipyridyl) complex cations. The series $[Z^{II}(bpy)_3][ClO_4][M^{II}Cr^{III}(ox)_3]$ (Z^{II}) (Ru, Fe, Co, and Ni; M^{II}) Mn, Fe, Co, Ni, Cu, and Zn; ox) Oxalate Dianion), *Inorg. Chem.*, **40**, 113–120.

20. Andrés, R., Bissard, M., Gruselle, M., Train, C., Vaissermann, J., Malézieux, B., Jamet, J.-P., Verdaguer, M., (2001). Rational design of three-dimensional (3-D) optically active molecule-based magnets: Synthesis, structure, optical and, magnetic properties of $\{[Ru(bpy)_3]^{2+}$,

ClO_4^-, $[(MnCr^{III})\text{-}Cr^{II}(ox)_3]^-\}_n$ and $\{[Ru(bpy)_2ppy]^+$, $[(MCr^{III})\text{-}Cr^{II}(ox)_3]^-\}_n$, with M-II = Mn-II, Mn-II. X-ray structure of $\{[\Delta\ Ru(bpy)_3]^{2+}$, ClO_4^-, $[\Delta$ Mn-II Δ Cr-III$(ox)_3]^-\}_n$ and $\{[\Lambda$ Ru(bpy)$_2$ppy]+, $[\Delta$ Mn-II Λ Cr-III$(ox)_3]^-\}_n$, *Inorg. Chem.*, **40**, 4633–4640.

21. Inoue, K., Kikuchi, K., Ohba, M., Okawa, H. (2003). Structure and magnetic properties of a chiral two-dimensional ferrimagnet with T_C of 38 K, *Angew. Chem. Int. Ed.*, **42**, 4810–4813.
22. Imai, H., Inoue, K., Kikuchi, K., Yoshida, Y., Ito, M., Sunahara, T, Onaka, S. (2004). Three-dimensional chiral molecule-based ferrimagnet with triple-helical-strand structure, *Angew. Chem. Int. Ed.*, **43**, 5618–5621.
23. Imai, H., Inoue, K., Kikuchi, K., (2005). Crystal structures and magnetic properties of chiral and achiral cyanide-bridged bimetallic layered compounds, *Polyhedron*, **24**, 2808–2812.
24. Higashikawa, H., Okuda, K., Kishine, J., Masuhara, N, Inoue, K. (2007). Chiral effects on magnetic properties for chiral and racemic W-V-Cu-II Prussian blue analogues, *Chem. Lett.*, **36**, 1022–1023.
25. Wang, Z., Zhang, B., Inoue, K., Fujiwara, H., Otsuka, T., Kobayashi, H, Kurmoo, M. (2007). Occurrence of a rare $4_9\ 6_6$ structural topology, chirality, and weak ferromagnetism in the $[NH_4][M^{II}(HCOO)_3]$ (M = Mn, Co, Ni) frameworks, *Inorg. Chem.*, **46,** 437–445.
26. Numata, Y., Inoue, K., Baranov, N., Kurmoo, M, Kikuchi, K. (2007). Field-induced ferrimagnetic state in a molecule-based magnet consisting of a Co-II ion and a chiral triplet bis(nitroxide) radical, *J. Am. Chem. Soc.*, **129**, 9902–9909.
27. Yoshida, Y, Inoue, K., unpublished data.
28. Imai, H, Inoue, K. unpublished data.
29. Rikken, G. L. J. A, Raupach, E. (1997). Observation of magneto-chiral dichroism, *Nature*, **390**, 493–494.
30. Muhlbauer, S., Binz, B., Jonietz, F., Pfleiderer, C., Rosch, A., Neubauer, A., Georgii, R., Boni, P., (2009). Skyrmion lattice in a chiral magnet, *Science*, **323**, 915–919.
31. Yu, X. Z., Onose, Y., Kanazawa, N., Park, J. H., Han, J. H., Matsui, Y., Nagaosa, N., Tokura, Y. (2010). Real-space observation of a two-dimensional skyrmion crystal, *Nature*, **465**, 901–904.
32. Carling, S. G., Day, P., Visser, D, Kremer, R. K. (1993). Weak Ferromagnetic Behavior of the Manganese Alkylphosphonate Hydrates $MnC_nH_{2n+1}PO_3 \cdot H_2O$, $n = 1$–4, *J. Solid State Chem.*, **106**, 111–119.
33. Mito, M., Iriguchi, K., Deguchi, H., Kishine, J., Kikuchi, K., Ohsumi, H., Yoshida, Y., Inoue, K. (2009). Giant nonlinear magnetic response in a molecule-based magnet, *Phys. Rev. B*, **79**, 012406.

34. Gu, Z. G., Song, Y., Zuo, J. L, You, X. Z. (2007). Chiral molecular ferromanets based on copper(II) polymers with end-on bridges, *Inorg. Chem.*, **46**, 9522–9524.

35. Gu, Z. G., Xu, Y. F., Yin, X. J., Zhou, X. H., Zuo, J. L, You, X. Z. (2008). Cluster-based copper(II) coordination polymers with azido bridges and chiral magnets, *Dalton Trasactions.*, 5593–5602.

36. Clemente-Leon, M., Coronado, E., Dias, JC., Soriano-Portillo, A., Willett, R. D. (2008). Synthesis, structure, and magnetic properties of [(*S*)-[PhCH(CH_3)N(CH_3)$_3$]][Mn(CH_3CN)$_{2/3}$Cr(ox)$_3$](CH_3CN)(solvate), a 2-D chiral magnet containing a quaternary ammonium chiral cation, *Inorg. Chem.*, **47**, 6458–6463.

Chapter 5

Toward Bifunctional Materials with Conducting, Photochromic, and Spin Crossover Properties

Lydie Valade, Isabelle Malfant, and Christophe Faulmann

CNRS, LCC (Laboratoire de Chimie de Coordination), 205 route de Narbonne, BP 44099, F-31077 Toulouse Cedex 4, France

Université de Toulouse, UPS, INPT, F-31077 Toulouse Cedex 4, France

lydie.valade@lcc-toulouse.fr

5.1 Introduction

There has been an increasing interest during the past 30 years in the preparation of materials exhibiting multiple properties. Many reasons justify conducting research programs along this line. First, materials are present in numerous personal or industrial equipments. As demand increases, production means have to take into account environmental concerns (clean production and recyclability) together with reduction of costs. Therefore, reduction in weight and size, search for higher performances at smaller size (nanomaterials), and multifunctionality are active lines that research programs follow to produce materials fulfilling these requirements.

These programs largely target materials for electronic equipments. Reduction in weight is studied through the use of organic

Multifunctional Molecular Materials
Edited by Lahcène Ouahab

ISBN 978-981-4364-29-4 (Hardcover), 978-981-4364-30-0 (eBook)
www.panstanford.com

materials as polymers, largely applied as matrix materials in composites incorporating glass or carbon fibers for mechanical applications or carbon nanotubes or fibers for conductive applications. A composite is a typical bifunctional material, both the matrix and the embedded material affording two different properties. Another method to afford bifunctionality is the use of single-component materials. Inorganic bifunctional materials are known and studied (ferroelectric ceramics, for example), but considerable efforts have been made for about 20 years to produce molecular multifunctional materials.

Molecular materials are built from molecular building blocks. In the solid state, these organic or metal-organic building blocks are organized in such a way that interactions occur between them and give rise to physical properties such as conductivity or magnetism. On the other hand, an optical property such as photochromism is rather a molecular property. The property results either from intermolecular interactions, from the molecule itself, from molecular materials, or from reduction in size such as nano-sized components for devices.

Molecular materials can be built from the association of two different building blocks, each having a different property. In this type of bifunctional materials, in addition to the combination of properties, interplay between them may occur and lead to intelligent materials. As mentioned above, the properties of molecular materials are directly connected to their structural organization. In an intelligent photomagnetic molecular material, for example, irradiation would change the structure of the photochromic entities, modify their interactions with the magnetic entities, and induce a modification of spin interactions. Such a material may act as an optical switch. Although a majority of molecular materials are built from the association of two components, complexes such as $[Cu(hfac)_2(TTF\text{-}py)_2](X)_2{\cdot}2CH_2Cl_2$ ($X = PF_6$, BF_4),[1,2] and $[(ppy)Au(C_8H_4S_8)]_2[PF_6]$[3] include both a conductive and magnetic moiety chemically bonded within a single molecule. The most spectacular multiproperty switchable organic material was reported by R. C. Haddon's group in 2002.[4] Haddon's material is based on an organic molecular conductor, a spiro-biphenalenyl neutral radical (Fig. 5.1) that simultaneously exhibits bistability in three physical properties: electrical, optical, and magnetic.

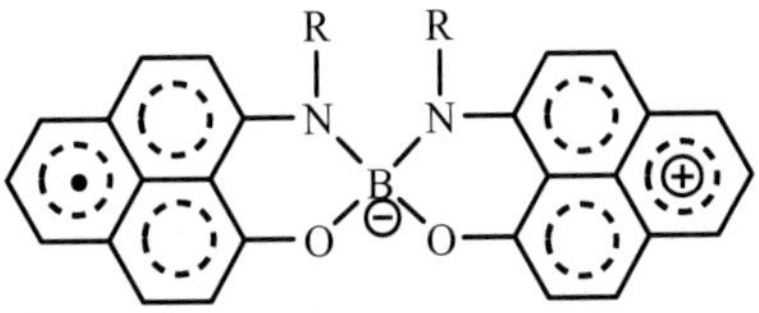

$R_2C_{26}H_{14}O_2N_2B$
R = Bu, Et

Figure 5.1 The spiro-biphenalenyl derivative.

This type of multifunctional material has the potential to be used for optoelectronics and spintronics devices, where multiple physical properties are used for writing, reading, and transferring information.

With the aim of reducing the size of components in devices, the preparation of molecular materials as thin films or nanoparticles is actively studied. Precursors to molecular materials are either soluble or volatile molecules that can be engaged in various processing techniques[5,6]: gas phase methods as chemical vapor deposition,[7,8] evaporation under ultra-high vacuum,[9] or solution-based methods such as dipping,[10] spin coating, ink-jet printing,[11] electrodeposition,[12,13] and polymeric casting.[14] Micro- and nanowires of molecular conductors[15,16] and nanoparticles of magnetic,[17–19] conductive,[20] and photochromic[21,22] molecular materials have also been isolated. These nanoparticles can find applications in many domains as active thin layers or as charges into polymer matrix for composite fabrication. All these methods of thin film or nanoparticle growth can be easily applied to multifunctional materials.

A development in the research on molecular conductors concerns the possibility to create solids in which conductivity coexists with other physical phenomena. This chapter focuses on two classes of bifunctional molecular materials built from transition metal complexes: conductive–photochromic materials and conductive–spin crossover materials.

Photochromic materials are based on molecules that can be switched between two discrete states upon irradiation of light. Spin crossover (SCO) materials exhibit a spin transition between two electronic states. In both types of materials, excited metastable states are produced.[23] They differ from the ground state in structural and electronic properties. If the metastable states have long-enough lifetimes, if they can be easily detected and differentiated by optical or magnetic methods, and if they can reversibly switch back to ground states without fatigue, the material may find applications in switching or display devices.

Photochromic materials have appeared to be materials that could overcome the upper capacity limit reached by data recording using compact and magneto-optical disks. What renders photochromic systems very powerful for optical data storage is the possibility of recording data within the volume of the material and not only on the surface, as for other technologies.[24]

In SCO materials, spin transition may be induced by temperature variation (thermoswitching), irradiation (photoswitching), or application of pressure (piezoswitching). Examples have demonstrated the potential for practical applications in switching[25] and display devices.[26,27]

Photochromic molecules can take advantage of being associated with conductive molecules in a bifunctional material. SCO molecules may also be combined with molecules exhibiting other physical or chemical properties[28]: magnetic, liquid crystalline, nonlinear optics, electrical conductivity.

This class of multifunctional materials should provide a unique opportunity to study the competition and mutual influence, and even better, a possible interplay between two properties in the same crystal lattice. In the resulting bifunctional materials, synergic effects may enhance spin transition signals, switching, and sensing. Examples of materials including photochromic or SCO and conductive building blocks will be described.

5.2 Conductive and Photochromic Salts

An approach involving conducting radical ion salts with photochromic ions has been explored for several years. The interest in the synthesis and characterization of molecular conductors containing photochromic ions is to analyze the influence of the electronic excitation of the ions over the conducting electrons. For example, one can imagine the design of a photochromic molecular conductor in which the conducting properties can be tuned by light. Both cationic and anionic photochromic ions have been isolated. Therefore they can be combined with conductive bricks of donor or acceptor type (Fig. 5.2).

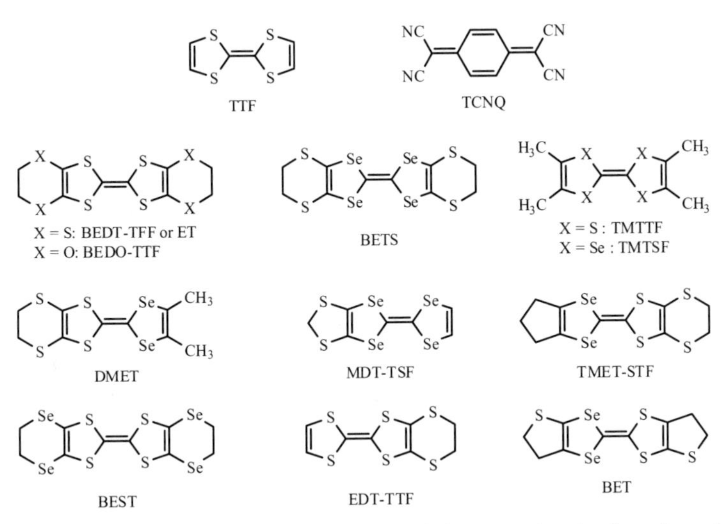

Figure 5.2 TTF donor, TCNQ acceptor, and donor molecules based on the TTF core.

In this context, the use of the photochromic nitroprusside $[Fe(CN)_5NO]^{2-}$ anion and related transition-metal mononitrosyl complexes, such as counter-anions in radical cation salts based on the TTF molecules (TTF = $C_6H_4S_4$, tetrathiafulvalene), is particularly interesting.

Donor molecules that have been particularly studied are the BEDT-TTF (BEDT-TTF = bis(ethylenedithio) tetrathiafulvalene) and analogues; BEDO-TTF, which is the oxygen-substituted analogue of BEDT-TTF (BEDO-TTF = bis(ethylene dioxy)tetrathiafulvalene); and the BETS molecule, which is a modification of BEDT-TTF obtained by substituting selenium for sulfur in the central tetrathiafulvalene fragment (BETS = bis(ethylenedithio)tetraselenafulvalene) (Fig. 5.2). These three donors are also well known to have given rise to numerous superconductors[29] exhibiting several different structural arrangements: α, β, κ...[30]

The photochromic nitroprusside $[Fe(CN)_5NO]^{2-}$ anion and related transition-metal mononitrosyl complexes are of high interest because they possess extremely long-living metastable excited states that can be generated by laser irradiation.[31–36] $[Fe(CN)_5NO]^{2-}$ presents two light-induced long-lived metastable states due to a bond isomerization of the NO group that comes from the nitrosyl N-bond (GS ground state) to (η^2) N–O (MS2 metastable state) and to isonitrosyl O–N bond (MS1 metastable state) coordination (Fig. 5.3).

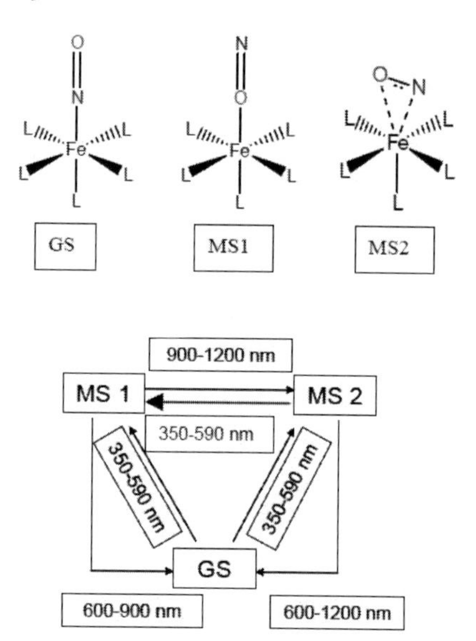

Figure 5.3 Irradiation conditions for the two metastables states MS1 and MS2 of the nitrosyl complexes.

Several systems combining the photochromic moieties and conductive bricks have been synthesized and characterized. Table 5.1 lists the compounds which will be presented in this section.

Table 5.1 Association of conductive and photochromic compounds

Compounds	References
$(BEDT\text{-}TTF)_4M[Fe(CN)_5NO]_2$ with M = Na^+, K^+, NH^{4+}, Tl^+, Rb^+, Cs^+	37, 38
$(BEDT\text{-}TTF)_n[OsCl_5NO]$	39
$(BEDO\text{-}TTF)_4[Fe(CN)_5NO]$	40, 41
$(TTF)_7[Fe(CN)_5NO]_2$	42
$(BEST)_2[Fe(CN)_5NO]$	43
$(BET)_2[Fe(CN)_5NO]\cdot CH_2Cl_2$	44
θ-$(BETS)_4[Fe(CN)_5NO]$ and $(BETS)_2[RuX_5NO]$ (X = Cl, Br)	45, 46
$(EDT\text{-}TTF)_3[Fe(CN)_5NO]$	47, 48
$(EDX\text{-}DMTTF)_2[Fe(CN)_5NO]$ with X = O, Se	49
$(EDXIY\text{-}TTF)_4[Fe(CN)_5NO]$	50
{*trans*-$[Ru(ox)(en)_2NO]\}_2(TCNQ)_3$	51

5.2.1 Association Donor/Photochromic Anions

All the photochromic/conductive salts based on the TTF-like molecules are obtained by electrocrystallization[52] with the donor molecule in the anodic compartment and the photochromic counter-anions as supporting electrolytes.

(BEDT-TTF)$_4$M[Fe(CN)$_5$NO]$_2$ with M = Na^+, K^+, NH^{4+}, Tl^+, Rb^+, Cs^+

These salts are isostructural and behave as stable metals down to helium temperature. Their structures are characterized by radical cation layers of the β''-type alternating with layers of complex anions (Fig. 5.4). No superconducting transition was found in these crystals within the studied temperature range (>1.3 K). The theoretical calculations of the band structure showed that the Fermi surface of these crystals consist of two parts: a strongly warped open part related to hole carriers and an electronic part, which can be either open or closed, depending on minor changes in the donor–donor interactions.[37,38]

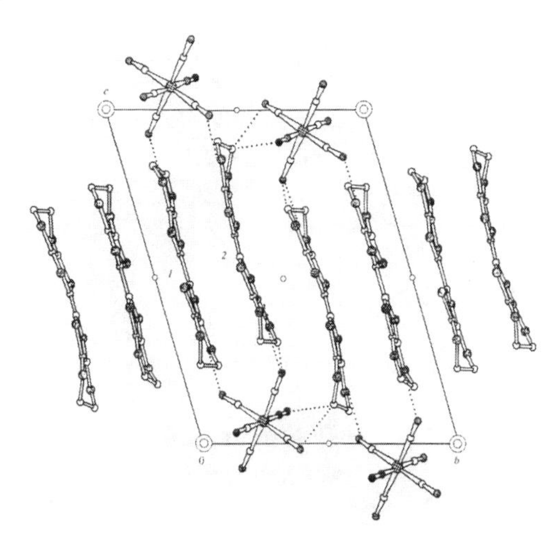

Figure 5.4 Projection of the (BEDT-TTF)$_4$M[Fe(CN)$_5$NO]$_2$ structure along the *a* direction.

(BEDT-TTF)$_n$[OsCl$_5$NO]

Four different phases have been obtained depending on the electrocrystallization conditions (Table 5.2): α'-(BEDT-TTF)$_4$

$[OsCl_5NO]$.NB, δ-(BEDT-TTF)$_4[OsCl_5NO]_{1.33}\cdot NB_{0.67}$, β-(BEDT-TTF)$_2$ $[OsCl_5NO]$, κ-(BEDT-TTF)$_4[OsCl_5NO]\cdot$BN (NB nitrobenzene, BN benzonitrile).[39] Moreover, the four phases isolated using the mononitrosyl osmium complex show different types of conducting layers.

The crystal structure of α'-(BEDT-TTF)$_4[OsCl_5NO]\cdot$NB is characterized by the presence of BEDT-TTF radical cation layers alternating with anionic layers in which the photochromic salts is disordered. The room temperature conductivity of a single crystal of α'-(BEDT-TTF)$_4[OsCl_5NO]\cdot$NB in the *ab* plane is ~0.5–1.0 S $cm^{-1}$and it behaves as a semiconductor. In parallel, the analysis of the calculated band structure shows that this compound is a band gap semiconductor.

Table 5.2 Electrocrystallization conditions

Salt	Reagents and solvents	I(μA)	T(°C)	Time/days
α'-(BEDT-TTF)$_4$ $[OsCl_5NO]\cdot NB^a$	BEDT-TTF (1.3 × 10^{-3} M)	0.25	25	21
δ-(BEDT-TTF)$_4$ $[OsCl_5NO]_{1.33}\cdot NB_{0.67}{}^a$	$(Ph_4P)_2[OsCl_5NO]$ (1.4 × 10^{-3} M)			
β-(BEDT-TTF)$_2[OsCl_5NO]^a$	NB (20 mL)			
κ-(BEDT-TTF)$_4[OsCl_5NO]\cdot$ BN	BEDT-TTF (1.3 × 10^{-3} M) $K_2[OsCl_5NO]$ (1.4 × 10^{-3} M) 18-crown-6 (9.8 × 10^{-3} M) BN (20 mL) EtOH (2 mL)	0.25	10–14	7

[a]α', β, δ-phase were formed together in the electrocrystallization process, the δ product was obtained as the major product.

In β-(BEDT-TTF)$_2[OsCl_5NO]$, the room temperature conductivity is 0.1 S cm^{-1}. The room temperature conductivity of δ-(BEDT-TTF)$_4[OsCl_5NO]_{1.33}\cdot NB_{0.67}$ is 1 S cm^{-1}. It increases upon cooling down to 160 K where it undergoes a sharp metal-to-insulator transition. Semiconducting behavior is observed for κ-(BEDT-TTF)$_4[OsCl_5NO]\cdot$BN with an activation energy of 0.1 eV (room temperature conductivity ≈1–2 S cm^{-1}). These results evidence that subtle changes in the structure influence electrical behavior.

In Ref. 39, the authors mentioned the comparison of these systems with the one obtained with the ruthenium photochromic salts. No major difference was observed.

$(BEDO\text{-}TTF)_4[Fe(CN)_5NO]$

This compound is a two-dimensional metal stable down to liquid helium temperature and is not superconducting above 1.3 K.[40] It does not exhibit Shubnikov de Haas (SdH) oscillations in a magnetic field. The lack of superconductivity and SdH oscillations are probably related to disorders of the anions.

The linear correlation between the formal charge of BEDO-TTF and the experimental Raman Ag ν_2 and Ag ν_3 shift frequencies of the donor indicates that the charge transfer is around 0.5 (Fig. 5.5), which is a confirmation of the 4:1 stoichiometry of the $(BEDO\text{-}TTF)_4[Fe(CN)_5NO]$ compound.[41]

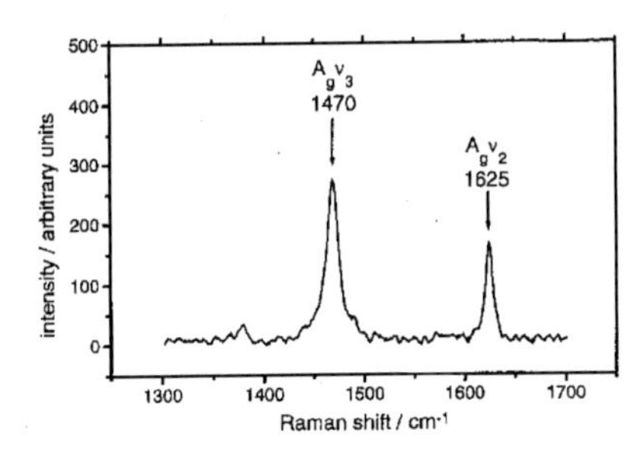

Figure 5.5 Raman spectrum of $(BEDO\text{-}TTF)_4M[Fe(CN)_5NO]$.

$(TTF)_7[Fe(CN)_5NO]_2$

The donor slabs are made of orthogonal arrangement of hexameric and monomeric TTF units to form a novel type of κ-pattern[42] (Fig. 5.6).

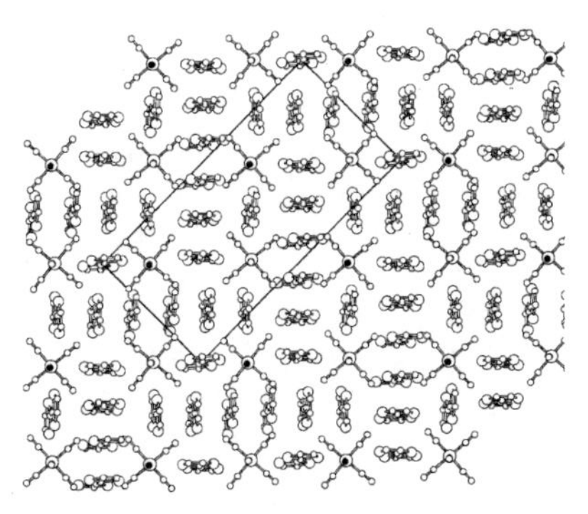

Figure 5.6 View of the structure in the *bc* plane showing the novel type of κ arrangement of TTF in $(TTF)_7[Fe(CN)_5NO]_2$.

The conductivity measurements indicate that $(TTF)_7[Fe(CN)_5NO]_2$ behaves as a semiconductor with an activation energy of 0.42 eV (Fig. 5.7). The detailed investigation of the electronic properties indicates that $(TTF)_7[Fe(CN)_5NO]_2$ can be described as a series of pairs of moderately interacting dimeric $(TTF_2)^{2+}$ units surrounded by two different types of neutral TTF molecules in such a way that there is no charge delocalization throughout the layer.

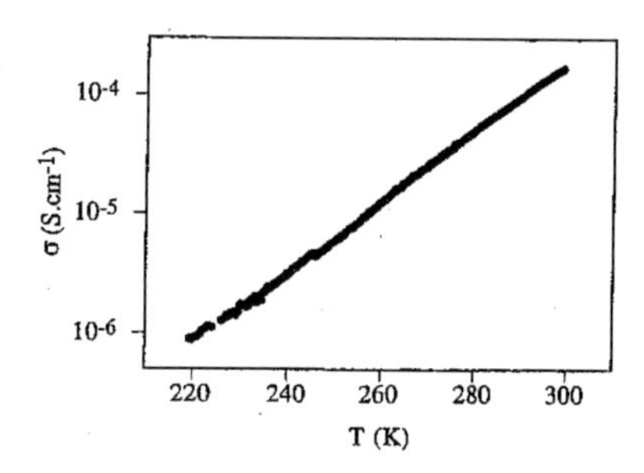

Figure 5.7 Single-crystal electrical conductivity of $(TTF)_7[Fe(CN)_5NO]_2$.

$(BEST)_2[Fe(CN)_5NO]$

The structure of this compound consists of alternating layers of the organic BEST molecules and the inorganic anions.[43] The organic layers are formed by only one independent BEST molecule forming zigzag chains packed in the β-mode (Fig. 5.8). The inorganic layer is formed by discrete nitroprusside anions that lie in the holes left by the BEST molecules. The estimation of the ionic charge on the donor molecules suggests that BEST is completely charged (+1) in agreement with the 2:1 stoichiometry.

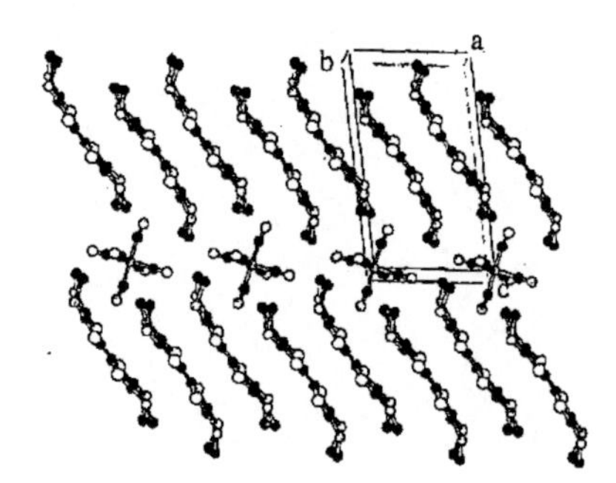

Figure 5.8 View of the alternating layers in $(BEST)_2[Fe(CN)_5NO]$.

$(BET)_2[Fe(CN)_5NO]\cdot CH_2Cl_2$

This compound presents an original structure (Fig. 5.9) where the anions occupy the tunnels formed by dimers of organic molecules.[44] The stoichiometry indicates that all the BET-TTF molecules bear a charge of +1, which is in agreement with the semiconductive behavior with a low-room temperature conductivity ($<10^{-6}$ S cm^{-1}) and a strong electron localization.

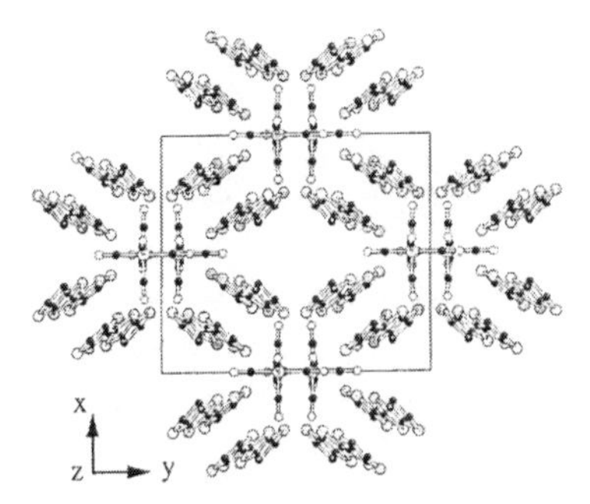

Figure 5.9 Channels created by the BET-TTF dimers in $(BET)_2[Fe(CN)_5NO]\cdot CH_2Cl_2$; view of the *ab*-plane.

θ-$(BETS)_4[Fe(CN)_5NO]$ and $(BETS)_2[RuX_5NO]$ (X = Cl, Br)

Three different phases are obtained when the BETS molecule is associated with the $[Fe(CN)_5NO]^{2-}$ anion and its substituted halide ruthenium mononitrosyl $[RuX_5NO]^{2-}$ (X = Cl, Br) complexes.[45] All these phases have been characterized by X-ray structure determination and energy-dispersive X-ray spectrometry (EDS).

The different molecular arrangements lead to different physical behaviors. The single-crystal room temperature conductivity of θ-$(BETS)_4[Fe(CN)_5NO]$ is 0.01 S cm^{-1} and the temperature-dependent resistance shows a weakly metallic behavior down to ~40 K (σ_{40} K = 0.07 S cm^{-1}). Below this temperature, the compound undergoes a metal-to-semiconductor phase transition. By contrast, $(BETS)_2[RuBr_5NO]$ is a semiconductor with a very small activation energy (E_a = 0.03 eV), and $(BETS)_2[RuCl_5NO]$ behaves as an insulator.

The different physical behaviors of $(BETS)_2[RuBr_5NO]$ and $(BETS)_2[RuCl_5NO]$, corresponding to the same 2:1 stoichiometry, have been correlated to the conformation of the $[RuX_5NO]$ (X = Cl, Br) species (Fig. 5.10). Depending on the Ru–N–O bending angle, different oxidation states are expected for the ruthenium atom and therefore different electron transfer from the BETS layers to the $[RuX_5NO]$ species.

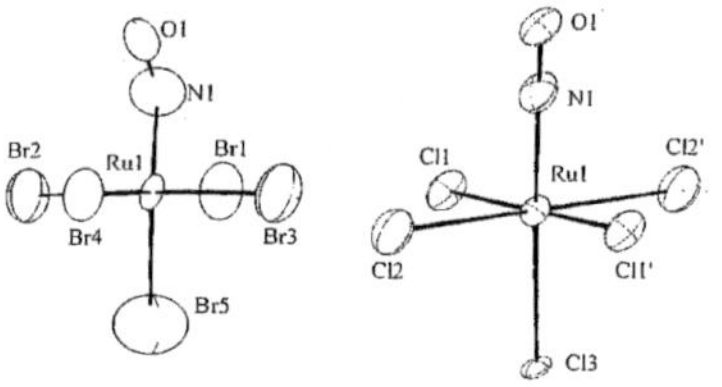

Figure 5.10 Anionic conformation in $(BETS)_2[RuBr_5NO]$ (left) and $(BETS)_2[RuCl_5NO]$ (right).

When NO is linearly bonded to the transition metal, the stabilization of a $[Ru^{II}Cl_5NO^+]^{2-}$ anions dictates an integral oxidation state for the BETS molecules, in agreement with the insulating behavior of $(BETS)_2[RuCl_5NO]$. When NO bends, a formal intermediate charge between NO^+ and NO^- may be considered, yielding a Ru^{n+} ($n > 2$) oxidation state. The lack of electron density around the transition metal is thus proposed to be retained by the BETS layers, in agreement with the small activation energy of $(BETS)_2[RuBr_5NO]$.

This hypothesis has been confirmed through the use of a detailed study of the structural data and the analysis of the Raman spectra of $(BETS)_2[RuCl_5NO]$ and $(BETS)_2[RuBr_5NO]$.[46] The fact that the degree of charge transfer is different from 1 in $(BETS)_2[RuBr_5NO]$ necessarily implies that the π-conducting electrons of the BETS molecule are sensitive to the presence of the photochromic $[RuBr_5NO]^{2-}$ anion. This feature is the signature of a real interplay between the donor molecule and the anion. Therefore, the nature of the anion leaves the possibility of an influence of irradiation on the electrical behavior of $(BETS)_2[RuBr_5NO]$.

$(EDT\text{-}TTF)_3[Fe(CN)_5NO]$

The structure of this compound can be described as a series of dimers of $EDT\text{-}TTF^+$ cations with $EDT\text{-}TTF^0$ neutral molecules in-between.[47,48] The calculated HOMO...HOMO interaction energies suggest that the strong intradimer interaction is responsible for the charge separation and the semiconducting properties of this system.

$(EDX\text{-}DMTTF)_2[Fe(CN)_5NO]$ with X = O, Se

Although $(EDSe\text{-}DMTTF)_2[Fe(CN)_5NO]$ and $(EDO\text{-}DMTTF)_2[Fe(CN)_5NO]$ possess the same stoichiometry (EDO-DMTTF = ethylenedioxodimethyltetra-thiafulvalene; (EDSe-DMTTF) ethylenediselenodimethyltetrathiafulvalene), they crystallize in two different space group[49]: the triclinic $P\bar{1}$ space group for $(EDO\text{-}DMTTF)_2[Fe(CN)_5NO]$ and the monoclinic $P2_1/a$ space group for $(EDSe\text{-}DMTTF)_2[Fe(CN)_5NO]$.

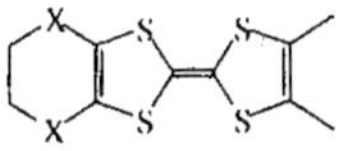

EDX-DMTTF with X= O, Se

Nevertheless, the structures consist in both cases in dimers stacked in columns. This is in agreement with the insulating behavior of these two systems. In both cases, the Ru–N–O bending angle in the anionic nitroprussiate salt is almost linear, which corresponds to NO^+. An analysis of the electron density distribution over each donor shows that EDX-DMTTF are much more polarized toward their external rings than a donor such as the BETS molecule. Moreover, the strong π-acceptor character of CN^- results in a lack of electron density around the transition metal that a linear NO^+ can compensate.

$(EDXIY\text{-}TTF)_4[Fe(CN)_5NO]$ with $\{(EDSI_2\text{-}TTF)$ X = S, Y = I, $(EDSeI_2\text{-}TTF)$ X = Se, Y = I, (EDSIH-TTF) X = S, Y = H}

Radical cation salts of iodine-substituted TTF derivatives [diiodoethylenedithiotetrathiafulvalene ($EDSI_2$-TTF), diiodoethylenedithio diselenadithiafulvalene ($EDSeI_2$-TTF) or iodoethylenedithiotetrathiafulvalene (EDSIH-TTF)] with the $[Fe(CN)_5NO]^{2-}$ nitroprusside anion, namely, $(EDSI_2\text{-}TTF)_4[Fe(CN)_5NO]$, $(EDSeI_2\text{-}TTF)_4[Fe(CN)_5NO]$ and $(EDSIH\text{-}TTF)]_3[Fe(CN)_5NO]$, were synthesized by electrocrystallization from solutions of the appropriate TTF derivative in CH_2Cl_2 with bis(tetraphenylphosphonium)nitroprussi de $[(PPh_4)_2\{Fe(CN)_5NO\}]$ as electrolyte.[50]

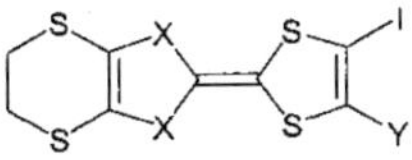

($EDSI_2$-TTF) X = S, Y = I; ($EDSeI_2$-TTF) X = Se, Y = I;
(EDSIH-TTF) X = S, Y = H

An X-ray structure analysis indicates that the crystal structures of $(EDSI_2\text{-}TTF)_4[Fe(CN)_5NO]$ and $(EDSeI_2\text{-}TTF)_4[Fe(CN)_5NO]$ are very similar. In both cases, the donor molecules are stacked along the *c* direction, with their long molecular axis parallel to each other. By contrast, in $(EDSIH\text{-}TTF)]_3[Fe(CN)_5NO]$ EDSIH-TTF molecules are stacked in triad units in which two of the molecules are arranged in a face-to-face fashion and the remaining molecule is arranged in a head-to-tail fashion with respect to its neighboring molecules. In all cases, several effective iodine–nitrogen contacts between donor molecules and $[Fe(CN)_5NO]^{2-}$ anions are observed. The room temperature electrical conductivities of $(EDSI_2\text{-}TTF)_4[Fe(CN)_5NO]$, $(EDSeI_2\text{-}TTF)_4[Fe(CN)_5NO]$ and $(EDSIH\text{-}TTF)]_3[Fe(CN)_5NO]$ are less than 10^{-6}, 2.2×10^{-2}, and 5 S cm^{-1}, respectively. The temperature dependence of the electrical conductivity of these complexes reveals semiconducting behavior with a small activation energy of 30 meV. The temperature dependence of the electrical conductivity of $(EDSeI_2\text{-}TTF)_4[Fe(CN)_5NO]$ also reveals a semiconducting behavior, but with two regimes: above 200 K the activation energy is 60 meV and becomes much smaller (0.2 meV) below this temperature.

5.2.2 Association Acceptor/Photochromic Cations

Only a few examples corresponding to the association of acceptor molecules (such as TCNQ = 7,7,8,8-tetracyanoquinodimethane) and photochromic cations have been studied.

{*trans*-[Ru(ox)(en)$_2$NO]}$_2$(TCNQ)$_3$

Single crystals of {*trans*-[Ru(ox)(en)$_2$NO]}$_2$(TCNQ)$_3$ (ox = oxalate; en = ethylenediamine) were obtained by slow interdiffusion of water solution of the complex and LiTCNQ.[51]

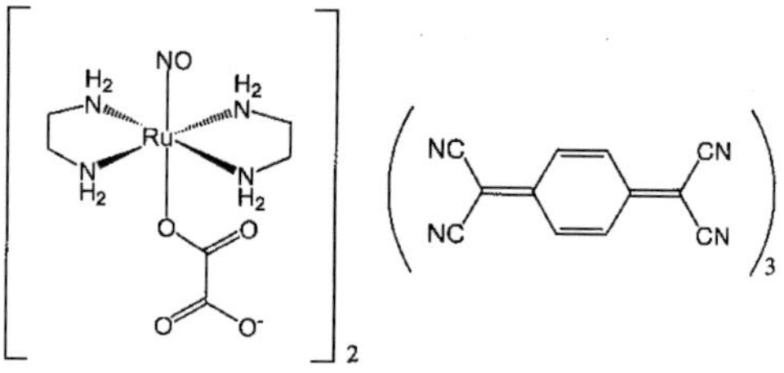

{*trans*-[Ru(ox)(en)$_2$NO]}$_2$(TCNQ)$_3$

This compound is the first example in which TCNQ molecules are connected with *trans*-$[Ru(ox)(en)_2NO]^+$ cations through hydrogen bonds.

The electric conductive property less than 10^{-6} S cm^{-1} can be understood by considering the crystal and molecular structures, despite the presence of several –N...HN– type hydrogen bonds between nitrile groups of TCNQ and ethylenediamine ligands of the photochromic cations.

In all compounds previously described, we have often noticed that the presence of a photochromic species in association either with a donor or an acceptor molecule may change conductivity properties. Unfortunately, in all systems, no clear evidence of photochromic properties have yet been clearly demonstrated. Nevertheless, a shift in the ν_{NO} band in IR spectra, which is the signature of the presence of metastable states, has been observed in the case of the association of the photochromic $[Ru(Cl)(py)_4NO]^{2+}$ complex and TCNQ molecules.[53]

5.3 Conductor and Spin Crossover

Combining electrical conductivity and magnetic properties in the same material is not a new idea, since this was already reported more than 30 years ago by Alcacer *et al.*[54] for $(perylene)_2[M(mnt)_2]$ complexes (M = Ni, Cu, Pd; mnt = maleonitrile dithiolate) (Fig. 5.11).

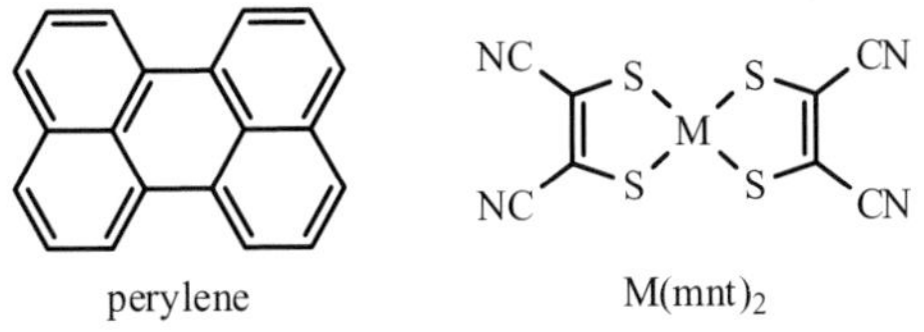

Figure 5.11 Perylene and $M(mnt)_2$ units.

In this compound, the perylene unit was used as the electrical pathway whereas the $M(mnt)_2$ moiety was used as the magnetic component.

From that date, numerous attempts have been made with the aim of obtaining materials with coexisting or coupled magnetic and electrical properties.

A general trend to reach this goal was to combine a donor molecule (such as the TTF derivatives shown in Fig. 5.2) with a paramagnetic anion (metal halogenides, metal oxalates) or to combine an acceptor entity (such as the organic compound TCNQ or a metal bisdithiolene complex, like $M(dmit)_2$; TCNQ = $C_{12}H_4N_4$, tetracyanoquinodimethane; Fig. 5.12) with a paramagnetic cation (such as metallocenium for instance).

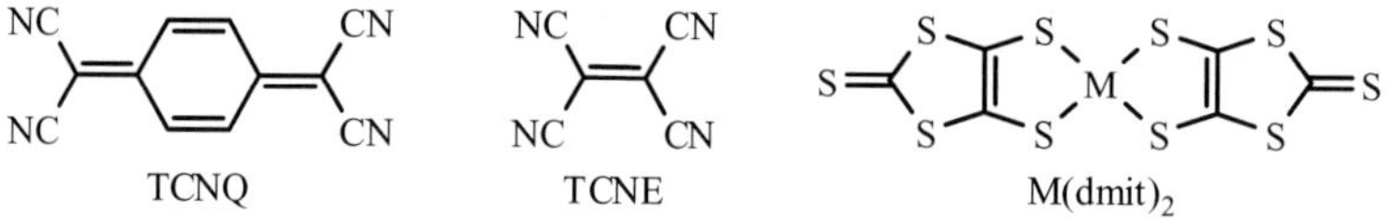

Figure 5.12 TCNQ and various acceptors.

This strategy has led to numerous compounds exhibiting interesting properties such as coexistence of paramagnetism (or antiferromagnetism) and superconductivity[55–58] or ferromagnetism and metallic conductivity.[59] A few complexes do exhibit an interplay between the itinerant electrons of the donor molecule and localized spin of the counter-ion. This is the case for some BETS-based complexes.[60–63]

In the field of molecular magnetism, one of the important phenomena is the spin crossover. This property is characterized by a bistability between a low-spin state (LS) and a high-spin state (HS). External stimuli such as temperature, pressure, or light irradiation allow switching between these two spin states.[64,65] The magnetic transition is also accompanied by structural changes (variation of the metal–ligand distances) and optical changes (different UV–vis absorption spectra). Moreover, when the structural changes are efficiently transmitted through the lattice, cooperativity can be observed, and this induces sharp transitions, sometimes accompanied with hysteresis. These characteristics make SCO complexes good candidates for memories, thermal displays, molecular switching, sensors, and medical applications.[27,28,66,67]

Combining material conductivity and SCO properties is of interest for tuning either conducting properties by acting on the SCO counter-ion or spin transition thanks to the electrical behavior.

Up to now, in the field of molecular complexes, there exist only a very few materials combining electrical properties and SCO properties. In this section, we will focus on these examples, together with their precursors, if any.

Except for six compounds, to our knowledge, all materials exhibiting (or expected to exhibit) SCO and electrical properties are based on metal bisdithiolene complexes. These compounds are listed in Table 5.3; the metal bisdithiolene complexes and the SCO compounds are represented in Figs. 5.13 and 5.14, respectively.

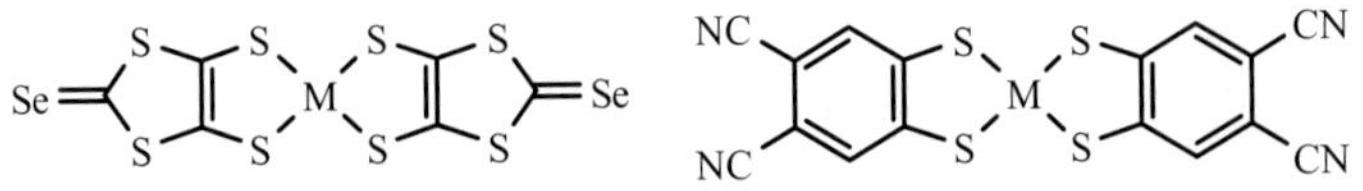

M(dmise)$_2$ M(dcbdt)$_2$

Figure 5.13 Metal bisdithiolene complexes cited in the text.

Table 5.3 List of spin crossover compounds with various acceptors

Compound	σ_{RT} (S cm^{-1})	Electrical behavior	Magnetic properties	Ref.
[Fe(abpt)$_2$(TCNQ)$_2$]	—	—	ST (400–300 K)	68
[Fe(acpa)$_2$](TCNQ)	—	—	ST (160 K)	69
[Fe(acpa)$_2$](TCNQ)$_2$	2.8×10^{-3}	—	ST (370 K)	69
[Fe(acpa)$_2$][Ni(dmit)$_2$]	—	—	ST (300–100 K)	70
[Fe(acpa)$_2$][Ni(dmit)$_2$]$_x$	0.16	SC	Possible ST	70
$R_2C_{26}H_{14}O_2N_2B$ R = Bu, Et	*ca.* 0.1	SC	ST (335 or 150 K) with hysteresis for R = Bu	4
[Fe(sal$_2$-trien)][Ni(dmit)$_2$]	—	—	Abrupt ST (245 K) with hysteresis (30 K wide)	71–73
[Fe(sal$_2$-trien)][Ni(dmit)$_2$]$_3$	0.1	SC	Paramagnet	74
[Fe(sal$_2$-trien)][M(dcbdt)$_2$] M = Ni, Au	—	—	—	75
[Fe(phen)$_3$][M(dcbdt)$_2$]$_2$ M = Ni, Au	—	—	—	75
[Fe(qsal)$_2$][Ni(dmit)$_2$]·2CH$_3$CN	—	—	ST (230–150 K), hysteresis, LIESST effect	76
[Fe(qsal)$_2$][Ni(dmit)$_2$]$_3$·CH$_3$CN·H$_2$O	2	SC	ST (300–100 K) LIESST effect	77
[Fe(qsal)$_2$][Ni(dmit)$_2$]	—	—	ST (300–10 K)	78
[Fe(qsal)$_2$][Ni(dmise)$_2$]·2CH$_3$CN	—	—	ST (260–230 K), hysteresis	79
[Fe(salten)Mepepy][M(dmit)$_2$]·CH$_3$CN M = Ni, Pd, Pt	—	—	ST (350–20 K)	80
[Fe(salten)Mepepy][Ni(dmit)$_2$]$_3$	0.1	—	AFM	80
[Fe(3-MeO-salEen)$_2$] [Ni(dmit)$_2$]·CH$_3$OH	—	—	ST (400–200 K)	81
[Fe(salEen)$_2$]$_2$[Ni(dmit)$_2$](NO$_3$)·CH$_3$CN	—	—	ST (400–300 K)	81
[Fe(salEen)$_2$]$_2$[Ni(dmit)$_2$]$_5$·6CH$_3$CN	0.12	SC	ST (350–50 K)	81
[Fe(qnal)$_2$][Pd(dmit)$_2$]$_3$·(CH$_3$)$_2$CO	1.6×10^{-2}	SC	ST (240–180 K) and LIESST effect	82
[Fe(HB(pz)$_3$)$_2$]	6×10^{-10} to 2×10^{-13}	—	ST with hysteresis	83,84
[Fe(qsal)$_2$][Co(Pc)(CN)$_2$]$_3$	10^{-1}	SC	ST with hysteresis	85

SC and ST stands for semi-conducting behavior and Spin Transition respectively.

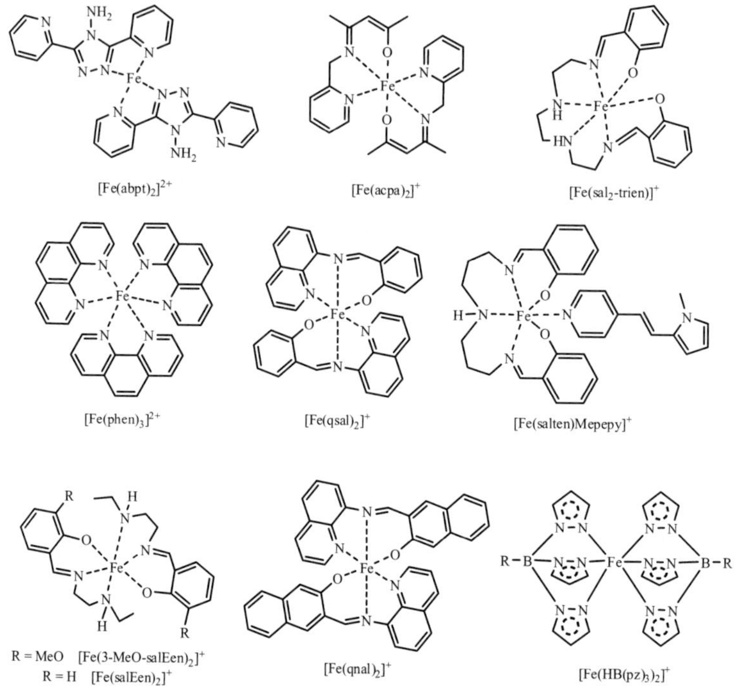

$[Fe(abpt)_2]^{2+}$ $[Fe(acpa)_2]^+$ $[Fe(sal_2\text{-trien})]^+$

$[Fe(phen)_3]^{2+}$ $[Fe(qsal)_2]^+$ $[Fe(salten)Mepepy]^+$

R = MeO $[Fe(3\text{-MeO-salEen})_2]^+$
R = H $[Fe(salEen)_2]^+$

$[Fe(qnal)_2]^+$ $[Fe(HB(pz)_3)_2]^+$

Figure 5.14 Spin crossover compounds cited in text.

The three first compounds in Table 5.3 are based on the organic acceptor TCNQ. TCNQ is a potential conductive unit: indeed, when combined with the organic donor TTF, it forms a donor–acceptor compound, with metallic properties down to 66 K,[86] and becomes an insulator because of Peierls distortion.[87]

Originally, $[Fe(abpt)_2(TCNQ)_2]$ was synthesized to promote spin–spin interaction between unpaired electron spins situated on the metal ion and on the radical (TCNQ), i.e., to combine magnetic properties arising from both units in the complex. Since the acceptor TCNQ still is an entire oxidation state (−1), no electrical property is expected and logically observed in $[Fe(abpt)_2(TCNQ)_2]$.

Compounds based on the acpa ligand are Fe(III) complexes. In this series, the lack of structural data and clear evidences of their physical properties prevent any consideration about their properties.

The case of the spiro-biphenalenyl compounds $R_2C_{26}H_{14}O_2N_2B$ (R = Bu, Et; Fig. 5.1) finds here its place although they were not originally presented as SCO compounds: their magnetic and electric behaviors were not well understood in 2001.[88]

Indeed, an increase in conductivity from the high-temperature paramagnetic form to the low-temperature diamagnetic form was not consistent with Peierls dimerization, which should lead to an insulating ground state (as in (TTF)(TCNQ)).[87] Nevertheless, this class of materials is among the only one to simultaneously exhibit bistability in optical, magnetic, and electrical properties.[4] This bistability occurs at 150 K for R = Et and around 335 K for R = Bu, with a paramagnetic, insulating, and infrared transparent state and a diamagnetic, conducting, and opaque state. This bistability finds its origin in an intramolecular electron switching. This creates an increase in the energy gap between the low-temperature and the high-temperature polymorphs, which explains the bistability of these compounds.[89]

$[Fe(sal_2\text{-trien})][Ni(dmit)_2]$ was the first complex combining an SCO cation ($[Fe(sal_2\text{-trien})]^+$) with a potentially conductive unit ($[Ni(dmit)_2]^-$) with the aim of combining electric and SCO properties in the same material.[71] It exhibits a cooperative spin transition behavior with a wide hysteresis loop (30 K) around 240 K. (Fig. 5.15).

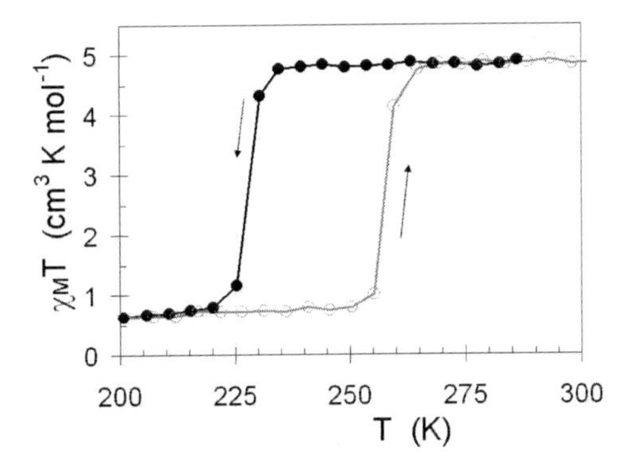

Figure 5.15 Magnetic behavior of $[Fe(sal_2\text{-trien})][Ni(dmit)_2]$ as a function of temperature.

Such behavior is quite unusual for Fe(III) complexes,[90] and even more for $[Fe(sal_2\text{-trien})]$-based compounds, which usually undergo only gradual and incomplete spin transition[91–94] or even no transition with Cl^-, I^-, and $(NO_3)^-$.[91,92] Actually, the $[Ni(dmit)_2]^-$ anions force the $[Fe(sal_2\text{-trien})]^+$ units to stack close together, favoring a strong coupling via π-stacking (Fig. 5.16).

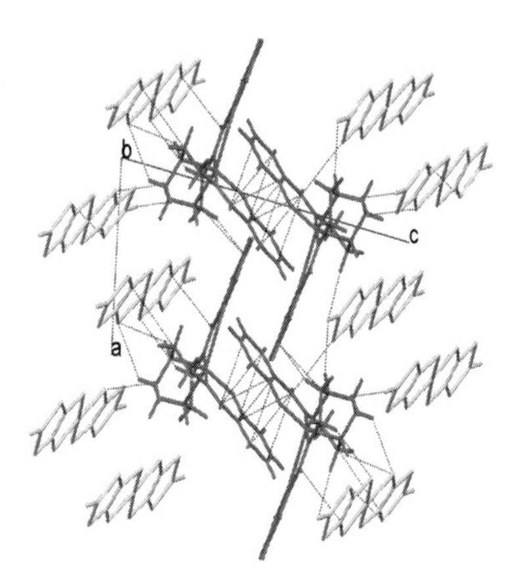

Figure 5.16 Projection of the structure onto the *ac*-plane (dotted lines represent short contacts between atoms (less than sum of the van der Waals radii).

Therefore the structural modifications due to spin transition are efficiently transmitted through the crystal lattice, which is at the origin of the large and abrupt hysteresis. The effects of temperature and pressure have been investigated on this complex[72]: under hydrostatic pressures up to 5.7 kbar, the thermal transition shifts to higher temperatures by ~16 K/kbar. Interestingly, at a low applied pressure of 500 bar the hysteresis loop becomes wider (~61 K) and the transition is blocked at ~50% upon cooling, indicating a possible (irreversible) structural phase transition under pressure. Unfortunately, the formation of several (pseudo)-polymorphs prevents further studies on $[Fe(sal_2\text{-trien})][Ni(dmit)_2]$.[73]

With the final goal of obtaining an SCO molecular conductor, $[Fe(sal_2\text{-trien})][Ni(dmit)_2]$ has been oxidized by electrocrystallization to form the fractional oxidation state complex $[Fe(sal_2\text{-trien})][Ni(dmit)_2]_3$.[74] This latter exhibits a room temperature conductivity of ~0.1 S cm^{-1}. A statistical disorder of the $[Fe(sal_2\text{-trien})]^+$ units makes the cationic network not dense enough to favor cooperative effects (Fig. 5.17). This causes the lack of any spin transition.

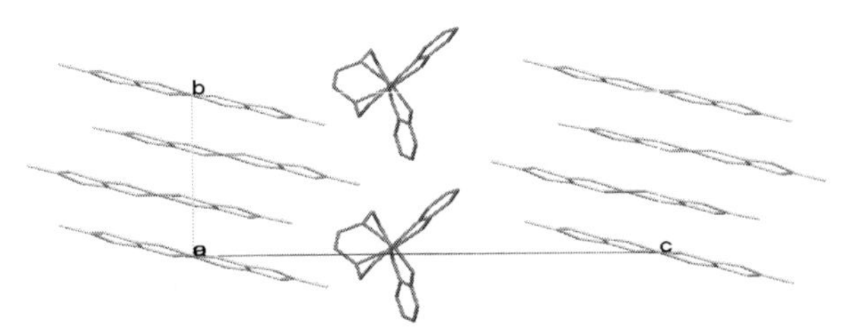

Figure 5.17 Projection of the structure of [Fe(sal$_2$-trien)][Ni(dmit)$_2$] onto the *bc*-plane.

For the same purpose, [Fe(sal$_2$-trien)]$^+$ was combined with [M(dcbdt)$_2$]$^-$ (M = Ni, Au), but none of them exhibits a spin transition.[75] The same authors also associated [M(dcbdt)$_2$]$^-$ (M = Ni, Au) with [Fe(phen)$_3$]$^+$. As for [Fe(sal$_2$-trien)][M(dcbdt)$_2$], no clear spin transition was observed, resulting probably from a very rigid crystal lattice.

[Fe(qsal$_2$)]$^+$ seems to be a good candidate to obtain cooperative effects together with electrical properties, since for instance [Fe(qsal$_2$)]SeCN displays abrupt spin transition accompanied with hysteresis thanks to π–π overlap between quinoline and phenyl rings.

Takahashi *et al.* then combined [Ni(dmit)$_2$]$^-$ with [Fe(qsal$_2$)]$^+$, to synthesize [Fe(qsal)$_2$][Ni(dmit)$_2$]·2CH$_3$CN.[76] During the first cooling–heating cycle process, this complex exhibited an apparent hysteresis centered around 212 K. This is due to a reversible molecular slipping occurring in the Ni(dmit)$_2$ units. Such a slipping occurs only in the first cooling–heating cycle, but does not exist anymore in the successive cycles, resulting in the disappearance of the hysteresis loop.[79] The subsequent temperature cycles show that this complex still undergoes a reversible spin transition at 231 K. It also exhibits a LIESST effect (light-induced excited spin trapping). The role of the solvent (here acetonitrile) should be pointed out here, although it is not well understood. Indeed, a similar compound has also been reported, namely [Fe(qsal)$_2$][Ni(dmit)$_2$], which is nonsolvated.[78] This latter was obtained in a mixture of acetone and methanol and [Fe (qsal)$_2$][Ni(dmit)$_2$]·2CH$_3$CN was synthesized in acetonitrile. Structural arrangements of these complexes are not identical: the solvated polymorph is arranged in a packing favoring π–π interactions in chains of [Fe(qsal)$_2$]$^-$ units,[76] with channels of [Fe(qsal)$_2$]$^+$ and solvent units separated from the [Ni(dmit)$_2$]$^-$ anions (Fig. 5.18, left); the structure of the nonsolvated polymorph is built on mixed layer of $[Ni(dmit)_2]^-$ and $[Fe(qsal)_2]^+$ (Fig. 5.18, right).[78] This prevents any cooperativity in this latter complex and $[Fe(qsal)_2][Ni(dmit)_2]$ only exhibits a gradual decrease in the magnetic moment.

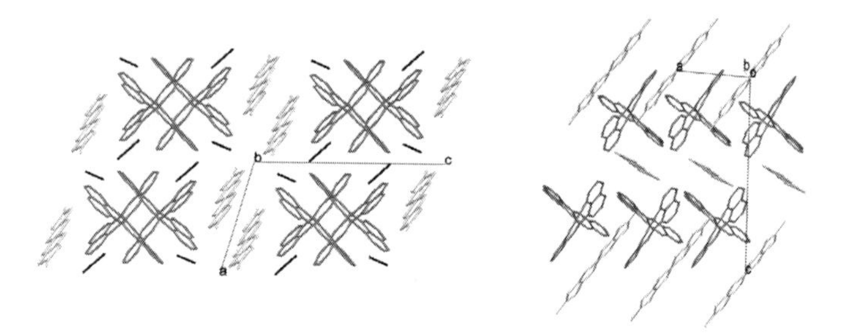

Figure 5.18 Comparison of the structural arrangement of $[Fe(qsal)_2][Ni(dmit)_2]\cdot 2CH_3CN$ (left) and $[Fe(qsal)_2][Ni(dmit)_2]$ (right).

The electrocrystallization of the solvated compound $[Fe(qsal)_2][Ni(dmit)_2]\cdot 2CH_3CN$ in acetonitrile yields the fractional oxidation state complex $[Fe(qsal)_2][Ni(dmit)_2]_3\cdot CH_3CN\cdot H_2O$.[77] In this compound, $Ni(dmit)_2$ molecules are arranged in columns and $Fe(qsal)_2$ cations are dimerized by π–π interactions and construct a 1-D chain (Fig. 5.19, left). $[Fe(qsal)_2][Ni(dmit)_2]_3\cdot CH_3CN\cdot H_2O$ behaves like a semiconductor between 300 and 70 K, with a hysteretic behavior in resistivity in the temperature range of 90 120 K (Fig. 5.19, right). At 300 K, the Fe(III) complexes are mainly in the HS state. When decreasing temperature, a gradual decrease in the $\chi_M T$ value is

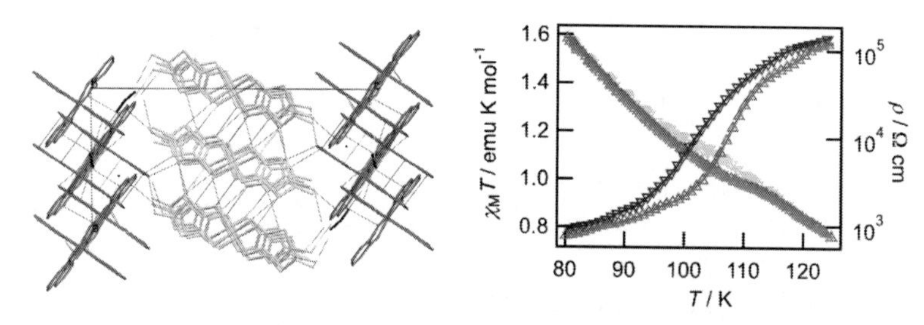

Figure 5.19 Structural arrangement of $[Fe(qsal)_2][Ni(dmit)_2]_3\cdot CH_3CN\cdot H_2O$ plane (dotted lines represent short contacts between atoms [less than the sum of van der Waals radii]) (left) and variation of the $\chi_M T$ product and of the resistivity as a function of the temperature (right).

observed until most of the Fe(III) complexes are in the LS state at 60 K. Meanwhile, at 90 and 120 K, in the cooling and the heating cycles respectively, an abrupt two-step change appears which also coincides with the anomalies observed in the electrical behavior (Fig. 5.19, right).

This complex is the first in which a clear synergy between the SCO phenomenon and electrical conduction is observed. Moreover, it also exhibits a LIESST effect.

$[Fe(qsal)_2]$ has also been used in combination with $[Ni(dmise)_2]^-$[79] to yield $[Fe(qsal)_2][Ni(dmise)_2]\cdot 2CH_3CN$, which is isostructural with $[Fe(qsal)_2][Ni(dmit)_2]\cdot 2CH_3CN$.[76] The dmise-based complex exhibits a cooperative spin transition with a thermal hysteresis loop of 15 K around 250 K. Like the dmit-based complex, $[Fe(qsal)_2][Ni(dmise)_2]\cdot 2CH_3CN$ exhibits π–π interactions between $[Fe(qsal)_2]^+$ units, together with a reversible molecular slipping of the $Ni(dmise)_2$ molecules along the molecular long axis, during the HS-LS transition.

$[Fe(salten)Mepepy](BPh_4)$ (H_2salten = 4-azaheptamethylene-1,7-bis(salicylidene-iminate); Mepepy = 1-(pyridin-4-yl)-2-(*N*-methylpyrrol-2-yl) ethane), previously reported by Sour *et al.*,[95] has been used for combination with $M(dmit)_2$ units (M = Ni, Pd, Pt), since this complex undergoes a gradual, thermally induced SCO. Moreover, the ligand Mepepy is photoisomerizable, which results in a partial photoinduced spin-state change in solution. Such a physical property might also play a role in the electrical properties of the final compound. The complexes $[Fe(salten)Mepepy][M(dmit)_2]$ (M = Ni, Pd, Pt) all exhibit a gradual HS-to-LS crossover of the Fe^{III} ions. Contrary to what has been observed with $[Fe(sal_2\text{-}trien)]^+$ and $[Fe(qsal)_2]^+$, in the present case the $M(dmit)_2$ units did not force the Fe complexes to interact strongly with each other. Preliminary attempts to synthesize fractional oxidation-state complexes have resulted in the synthesis of a conductive powder (σ_{RT} = 0.1 S cm^{-1}) that exhibits a spin conversion with antiferromagnetic interactions.

By using the cationic complex $[Fe(3\text{-}R\text{-}salEen)]^+$ (salEen stands for N-(2-ethylamino)-ethyl)-salicylaldimine, R = H, CH_3O) with $Ni(dmit)_2$, three compounds have been obtained.[81] Two of them are simple 1:1 salts, without any electrical properties, but exhibiting spin transition: $[Ni(dmit)_2][Fe(3\text{-}OMe\text{-}salEen)_2]$. CH_3OH shows an apparent hysteresis loop, due to an irreversible desolvatation process. $[Ni(dmit)_2](NO_3)[Fe(salEen)_2]_2$ exhibits a gradual and incomplete spin transition between 400 and 300 K. $[Ni(dmit)_2]_5[Fe(salEen)_2]_2 \cdot 6CH_3CN$ is a fractional oxidation state complex, which behaves like a semiconductor (σ_{RT} = 0.1 S cm^{-1}, E_a = 80 meV), and exhibits a gradual but complete spin transition between 300 and 4 K. The structural arrangement of $[Fe(salEen)_2]_2[Ni(dmit)_2]_5 \cdot 6CH_3CN$ is built on layers of $[Ni(dmit)_2]^-$ units separated from each other by double layers of $[Fe(salEen)_2]^+$ cations and solvent molecules (Fig. 5.20, left). When the temperature is decreased, the $[Ni(dmit)_2]^-$ units tend to form pentamers, which become isolated from each other, reducing the efficiency of the interactions between the conducting layer. This results in an increase in the resistivity of $[Fe(salEen)_2]_2[Ni(dmit)_2]_5 \cdot 6CH_3CN$ and in its semiconducting behavior[81] (Fig. 5.20, right).

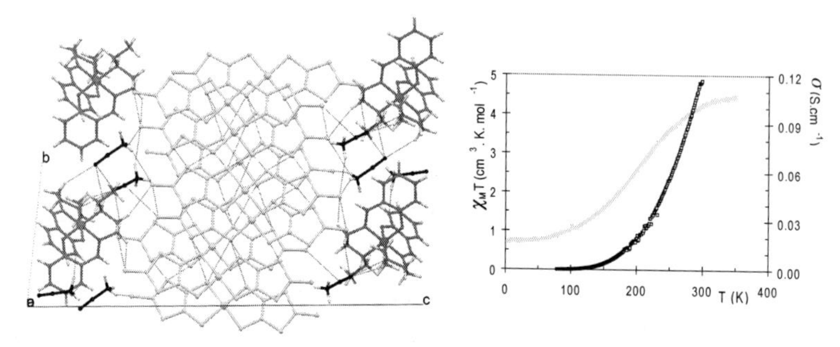

Figure 5.20 Structural arrangement of $[Fe(salEen)_2]_2[Ni(dmit)_2]_5 \cdot 6CH_3CN$ (left) and variation of $\chi_M T$ and σ as a function of the temperature (right).

In order to increase the π–π interactions between the SCO cations, a successful strategy was adopted by Takahashi *et al.* They chose to work with the $[Fe(qnal)_2]^+$ moiety, since H_2qnal contains an additional aromatic six-ring compared to H_2qsal. They combined the $[Fe(qnal)_2]^+$ complex with the $[Pd(dmit)_2]^-$ unit, to obtain $[Fe(qnal)_2][Pd(dmit)_2]$.[82] This latter was then oxidized at a constant potential, to yield the fractional oxidation state complex $[Fe(qnal)_2][Pd(dmit)_2]_5 \cdot (CH_3)_2CO$. The crystal structure indeed reveals the occurrence of π–π interactions between $[Fe(qnal)_2]^+$ moieties. As a consequence, a complete spin transition was observed between 240 and 200 K. $[Fe(qnal)_2][Pd(dmit)_2]_5 \cdot (CH_3)_2CO$ behaves like a semiconductor (σ_{RT} = 1.6 10^{-2} S cm^{-1}), and the temperature dependence of resistivity shows an anomaly at ~220 K, i.e., at the same temperature as the spin transition, with a change of the activation energies below and above the temperature of the anomaly, 0.37 and 0.24 eV, respectively. This is also accompanied at 220 K by a shortening of the *a* parameter of the unit cell. As observed in $[Fe(qsal)_2][Ni(dmit)_2]\cdot 2CH_3CN$[76] and in $[Fe(qsal)_2][Ni(dmit)_2]_3\cdot CH_3CN\cdot H_2O$,[77] $[Fe(qnal)_2][Pd(dmit)_2]_5\cdot (CH_3)_2CO$ also exhibits LIESST effects.

The penultimate compound reported in this chapter is $[Fe(HB(pz)_3)_2]$, which was first studied in 1967.[83] Its magnetic properties were recently revisited.[84] This compound undergoes a gradual spin transition between 450 and 300 K, with an apparent hysteresis after the first cycle of temperature. The origin of this irreversibility might be due to a self-grinding process[96] or it might arise from a structural phase transition.[84] More importantly, Salmon *et al.* noticed a change of the electrical conductivity of ~3–4 orders during the first irreversible transition. As mentioned by the authors, such a feature might be useful for possible applications in read-only memory devices.[84]

Very recently, a new complex was reported by Takahashi *et al.*[85] It concerns, again, the $Fe(qsal)_2$ unit which was combined with a phtalocyanine cobalt complex, namely $[Co(Pc)(CN)_2]$ (Fig. 5.21, left), to yield $[Fe(qsal)_2][Co(Pc)(CN)_2]_3$. As in $[Fe(qsal)_2][Ni(dmit)_2]_3\cdot CH_3CN\cdot H_2O$,[77] the phtalocyanine-based complex exhibits a clear synergy between its magnetic behavior and its electrical properties, as shown in Fig. 5.21, right: a clear hysteresis is observed in the spin transition at ~120 and ~130 K, temperatures at which another hysteresis is observed in the resistivity values.

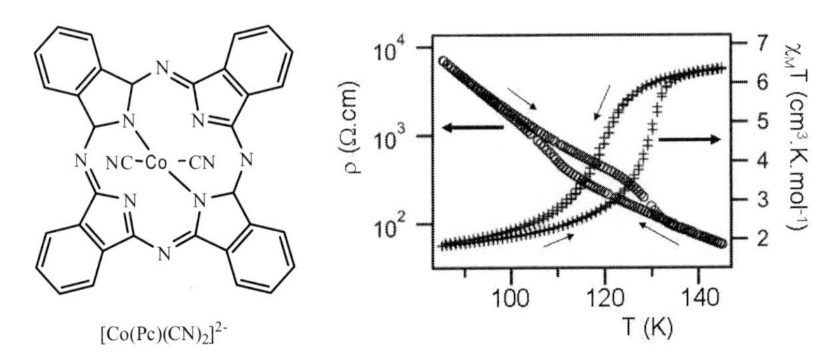

Figure 5.21 The $[Co(Pc)(CN)_2]$ moiety (left) and the variation of the resistivity and of the $\chi_M T$ product as a function of temperature for $[Fe(qsal)_2][Co(Pc)(CN)_2]_3$ (right).

5.4 Conclusion

This chapter considered two aspects of the formation of multi-property materials. In both aspects, conductivity was the common property, and it was studied in combination with either photochromism or SCO properties. The interest for building such materials is multiple. First, it is interesting to analyze the structural organization of the resulting materials, as the presence of an extra, noninnocent building block may induce modifications that have further consequences on the properties of both blocks. Second, it is of importance to evidence interactions between joint properties as interplay is the basis of efficient switchable materials.

Numerous conductive/photochromic systems have been studied. A majority of them are built from donor radical cations and photochromic anionic counter-ions. Only a few examples associate photochromic cationic species with acceptor radical anions. The photochromic properties of these materials have not yet been clearly evidenced, although the recently isolated phase associating the $[Ru(Cl)(py)_4NO]^{2+}$ complex and TCNQ molecules[53] shows the expected shift of the ν_{NO} band upon irradiation.

The combination of conductive building blocks with SCO complexes has led to several phases, and interactions between individual properties have been observed in many cases. For example, in the $[Fe(sal_2\text{-trien})][Ni(dmit)_2]$ complex, the tendency for stacking of the $[Ni(dmit)_2]$ units forces the $[Fe(sal_2\text{-trien})]$ units to be in close contact, affording enhanced SCO properties to the compound.[71] Moreover, clear synergy between properties were reported, for example, in $[Fe(qsal)_2][Ni(dmit)_2]_3{\cdot}CH_3CN{\cdot}H_2O$.[77]

Both fields associating conductivity to photochromic or SCO properties are active because, as they are based on molecular building blocks, the number of compounds that can be studied is almost infinite. Moreover, we have noticed that a very slight change in the molecular building blocks often drastically changes the properties of the materials; conductive properties are highly sensitive to such features. The examples gathered in this chapter encourage the continuation of studies in this field, as enhanced properties may appear from such associations and be applied in future generations of devices.

Acknowledgments

The authors acknowledge all researchers and students who have been part of the work described in this review. They are especially thankful to K. Takahashi from the University of Tokyo for sending information on his last work on $[Fe(qsal)_2][Co(Pc)(CN)_2]_3$.

References

1. Setifi, F., Ouahab, L., Golhen, S., Yoshida, Y., Saito, G. (2003). First radical cation salt of paramagnetic transition metal complex containing TTF as ligand, $[Cu^{II}(hfac)_2(TTF\text{-}py)_2](PF_6)\cdot 2CH_2Cl_2$ (hfac = Hexafluoroacetylacetonate and TTF-py = 4-(2-Tetrathiafulvalenyl-ethenyl)pyridine), *Inorg. Chem.*, **42**, 1791–1793.
2. Hervé, K., Gal, Y. L., Ouahab, L., Golhen, S., Cador, O. (2005). Pure TTF chains in π-d material made of paramagnetic transition metal complex containing TTF as ligand, $[Cu^{II}(hfac)_2(TTF\text{-}py)_2](BF_4)_2\cdot 2CH_2Cl_2$ (hfac = hexafluoroacetylacetonate and TTF-py = 4-(2-tetrathiaful-valenyl-ethenyl)pyridine), *Synth. Met.*, **153**, 461–464.
3. Kubo, K., Nakao, A., Ishii, Y., Yamamoto, T., Tamura, M., Kato, R., Yakushi, K., Matsubayashi, G.-E. (2008). Electrical properties and electronic states of molecular conductors based on unsymmetrical organometallic-dithiolene Gold(III) complexes, *Inorg. Chem.*, **47**, 5495–5502.
4. Itkis, M. E., Chi, X., Cordes, A. W., Haddon, R. C. (2002). Magneto-opto-electronic bistability in a phenalenyl-based neutral radical, *Science*, **296**, 1443–1445.
5. Fraxedas, J. (2006). *Molecular Organic Materials: Synthesis, Characterisation and Physical Properties. From Molecules to Crystalline Solids*, Cambridge University Press, Cambridge, UK.
6. Valade, L., de Caro, D., Malfant, I. (2004). Thin films and nano-objects of molecule-based materials: Processing methods and application to materials exhibiting conductive, magnetic or photochromic properties, *NATO Sci. Series, II: Math. Phys. Chem.*, **139**, 241–268.
7. de Caro, D., Basso-Bert, M., Casellas, H., Elgaddari, M., Savy, J.-P., Lamere, J.-F., Bachelier, A., Faulmann, C., Malfant, I., Etienne, M., Valade, L. (2005). Metal complexes-based molecular materials as thin films on silicon substrates, *Comptes Rendus Chimie*, **8**, 1156–1173.

8. de Caro, D., Basso-Bert, M., Sakah, J., Casellas, H., Legros, J.-P., Valade, L., Cassoux, P. (2000). CVD-grown thin films of molecule-based magnets, *Chem. Mat.*, **12**, 587–589.
9. Shibata, K., Wada, H., Ishikawa, K., Takezoe, H., Mori, T. (2007). (Tetrat hiafulvalene)(tetracyanoquinodimethane) as a low-contact-resistance electrode for organic transistors, *Appl. Phys. Lett.*, **90**, 193509.
10. de Caro, D., Sakah, J., Basso-Bert, M., Faulmann, C., Legros, J.-P., Ondarcuhu, T., Joachim, C., Aries, L., Valade, L., Cassoux, P. (2000). Application of conversion coatings to the growth of TTF-TCNQ thin films by CVD and conducting nanowires by dipping process, *C. R. Acad. Sci. IIc*, **3**, 675–680.
11. Hiraoka, M., Hasegawa, T., Yamada, T., Takahashi, Y., Horiuchi, S., Tokura, Y. (2007). On-substrate synthesis of molecular conductor films and circuits, *Adv. Mat.*, **19**, 3248–3251.
12. de Caro, D., Malfant, I., Savy, J.-P., Valade, L. (2008). A review of molecule-based conductors electrodeposited as thin films on silicon wafers, *J. Phys.: Cond. Matter*, **20**, 184012.
13. Valade, L., de Caro, D., Basso-Bert, M., Malfant, I., Faulmann, C., de Bonneval, B. G., Legros, J.-P. (2005). Thin films of transition metal-containing molecule-based materials: A highlight on electrochemically processed systems, *Coord. Chem. Rev.*, **249**, 1986–1996.
14. Mas-Torrent, M., Ribera, E., Tkatcheva, V., Mata, I., Molins, E., Vidal-Gancedo, J., Khasanov, S., Zorina, L., Shibaeva, R., Wojciechowski, R., Ulanski, J., Wurst, K., Vaciana, J., Laukhin, V., Canadell, E., Laukhina, E., Rovira, C. (2002). New molecular conductors based on ETEDT-TTF trihalides: from single crystals to conducting layers of nanocrystals, *Chem. Mater.*, **14**, 3295–3304.
15. Savy, J.-P., Caro, D. D., Faulmann, C., Valade, L., Almeida, M., Koike, T., Fujiwara, H., Sugimoto, T., Fraxedas, J., Ondarçuhu, T., Pasquier, C. (2007). Nanowires of molecule-based charge-transfer salts, *New J. Chem.*, **31**, 519–527.
16. Sugimoto, T., Tanaka, H., de Caro, D., Valade, L. (2010). New development in the preparation of micro/nano-wires of molecular (magnetic) conductors, *Materials*, **3**, 1640–1673.
17. Clavel, G., Larionova, J., Guari, Y., Guerin, C. (2006). Synthesis of cyano-bridged magnetic nanoparticles using room temperature ionic liquids, *Chem. Eur. J.*, **12**, 3798–3804.
18. Catala, L., Gloter, A., Stephan, O., Rogez, G., Mallah, T. (2006). Superparamagnetic bimetallic cyanide-bridged coordination nano-particles with T-B = 9 K, *Chem. Commun.*, 1018–1020.

19. Coronado, E., Galán-Mascarós, J. R., Monrabal-Capilla, M., García-Martínez, J., Pardo-Ibáñez, P. (2007). Bistable spin-crossover nanoparticles showing magnetic thermal hysteresis near room temperature, *Adv. Mater.*, **17**, 1359–1361.

20. de Caro, D., Jacob, K., Faulmann, C., Legros, J.-P., Senocq, F., Fraxedas, J., Valade, L. (2010). Ionic liquid-stabilized nanoparticles of charge transfer-based conductors, *Synth. Met.*, **160**, 1223–1227.

21. Cormary, B., Malfant, I., Valade, L. (2009). New photochromic xerogels composites based on nitrosyl complexes, *J. Sol-Gel Sci. Technol.*, **52**, 19–23.

22. Cormary, B., Malfant, I., Valade, L., Kawai, S., Kawai, T. (2009). Photochromic nanocomposite films of mononitrosyl complexes embedded in polymer matrices, *Inorg. Chem. Commun.*, **12**, 769–772.

23. Gütlich, P., Garcia, Y., Woike, T. (2001). Photoswitchable coordination compounds, *Coord. Chem. Rev.*, **219–221**, 839–879.

24. Kawata, S., Kawata, Y. (2000). Three-dimensional optical data storage using photochromic materials, *Chem. Rev.*, **100**, 1777–1788.

25. Bousseksou, A., Vieu, C., Létard, J.-F., Demont, P., Tuchagues, J. P., Malaquin, L., Menegotto, J., Salmon, L. (2005). Molecular memory and method for making same, US2005161728.

26. Letard, J.-F., Guionneau, P., Goux-Capes, L. (2004). Towards spin crossover applications, *Top. Curr. Chem.*, **235**, 221–249.

27. Kahn, O., Martinez-Jay, C. (1998). Spin-transition polymers: From molecular materials toward memory devices, *Science*, **279**, 44–48.

28. Gaspar, A. B., Ksenofontov, V., Seredyuk, M., Gütlich, P. (2005). Multifunctionality in spin crossover materials, *Coord. Chem. Rev.*, **249**, 2661–2676.

29. Saito, G., Yoshida, Y. (2005). Development of conductive organic molecular assemblies: Organic metals, superconductors, and exotic functional materials, *Bull. Chem. Soc. Jpn.*, **80**, 1–137.

30. Williams, J. M., Schultz, A. J., Geiser, U., Carlson, K. D., Kini, A. M., Wang, H. H., Kwok, W.-K., Whangbo, M. H., Schirber, J. E. (1991). Organic superconductors–New benchmarks, *Science*, **252**, 1501–1508.

31. Gütlich, P., Garcia, Y., Woike, T. (2001). Photoswitchable coordination compounds, *Coord. Chem. Rev.*, **219–221**, 839–879.

32. Woike, T., Krasser, W., Bechthold, P., Haussühl, S. (1984). Extremely long-living metastable state of $Na_2[Fe(CN)_5NO]\cdot 2H_2O$ single crystals: Optical properties, *Phys. Rev. Lett.*, **53**, 1767–1770.

33. Zöllner, H., Woike, T., Krasser, W., Haussühl, S. (1989). Thermal decay of laser-induced long living metastable electronic states in $Na_2[Fe(CN)_5NO]\cdot 2H_2O$, *Z. Kristallogr.*, **188**, 139–153.

34. Woike, T., Haussühl, S. (1993). Infrared-spectroscopic and differential scanning calorimetric studies of the two light-induced metastable states in $K_2[Ru(NO_2)_4(OH)(NO)]$, *Solid State Commun.*, **86**, 333–337.

35. Pressprich, M. R., White, M. A., Vekhter, V., Coppens, P. (1994). Analysis of a metastable electronic excited state of sodium nitroprusside by X-ray crystallography, *J. Am. Chem. Soc.*, **116**, 5233–5238.

36. Carducci, M. D., Pressprich, M. R., Coppens, P. (1997). Diffraction studies of photoexcited crystals: Metastable nitrosyl-linkage isomers of sodium nitroprusside, *J. Am. Chem. Soc.*, **119**, 2669–2678.

37. Khasanov, S. S., Zorina, L. V., Shibaeva, R. P. (2001). Structure of (BEDT-TTF)-based organic metals with photochromic nitroprusside anion $(BEDT\text{-}TTF)_4M[FeNO(CN)_5]_2$; where M = Na^+; K^+; NH_4^+; Tl^+; and Cs^+, *Russ. J. Coordinat. Chem.*, **27**, 259–269.

38. Gener, M., Canadell, E., Khasanov, S. S., Zorina, L. V., Shibaeva, R. P., Kushch, L. A., Yagubskii, E. B. (1999). Band structure and Fermi surface of the $(BEDT\text{-}TTF)_4M[FeNO(CN)_5]_2$ (M = Na; K; Rb;...) molecular metals containing the photochromic nitroprusside anion, *Solid State Commun.*, **111**, 329–333.

39. Simonov, S. V., Shevyakova, I. Y., Zorina, L. V., Khasanov, S. S., Buravov, L. L., Emel'yanov, V. A., Canadell, E., Shibaeva, R. P., Yagubskii, E. B. (2005). Variety of molecular conducting layers in the family of radical cation salts based on BEDT-TTF with the metal mononitrosyl complex $[OsNOCl_5]_2$, *J. Mater. Chem.*, **15**, 2476–2488.

40. Zorina, L. V., Khasanov, S. S., Shibaeva, R. P., Gener, M., Rousseau, R., Canadell, E., Kushch, L. A., Yagubskii, E. B., Drozdova, O. O., Yakushi, K. (2000). A new stable organic based on the BEDO-TTF donor and the doubly charged nitroprusside anion; $(BEDO\text{-}TTF)_4[FeNO(CN)_5]$, *J. Mater. Chem.*, **10**, 2017–2023.

41. Drozdova, O., Yamochi, H., Yakushi, K., Uruichi, M., Horiuchi, S., Saito, G. (2000). Determination of the charge on BEDO-TTF in its complexes by raman spectroscopy, *J. Am. Chem. Soc.*, **122**, 4436–4442.

42. Clemente-Leon, M., Coronado, E., Galan-Mascaros, J. R., Gomez-Garcia, C. J., Canadell, E. (2000). Hybrid molecular materials based upon the photochromic nitroprusside complex; $[Fe(CN)_5NO]^{2-}$; and organic π-electron donors. Synthesis; structure; and properties of the radical salt $(TTF)_7[Fe(CN)_5NO]_2$ (TTF = tetrathiafulvalene), *Inorg. Chem.*, **39**, 5394–5397.

43. Clemente-Leon, M., Coronado, E., Galan-Mascaros, J. R., Gimenez-Saiz, C., Gomes-Garcia, C. J., Fabre, J. M. (1999). Molecular conductors based upon TTF-type donors and octahedral magnetic complexes, *Synth. Met.*, **103**, 2279–2282.

44. Clemente-Leon, M., Coronado, E., Galan-Mascaros, J. R., Gimenez-Saiz, C., Gomes-Garcia, C. J., Ribera, E., Vidal-Gancedo, J., Rovira, C., Canadell, E., Lauthkhin, V. (2001). Hybrid molecular materials based upon organic π-electron donors and metal complexes; radical salts of bis (ethylenethia)tetrathiafulvalene (BET-TTF) with the octahedral anions hexacyanoferrate(III) and nitroprusside. The first kappa phase in the BET-TTF family, *Inorg. Chem*, **40**, 3526–3533.

45. Sanchez, M. E., Doublet, M. L., Faulmann, C., Malfant, I., Cassoux, P., Kushch, L. A., Yagubskii, E. B. (2001). Anion conformation and physical properties in BETS salts with the nitroprusside anion and its related ruthenium halide (X = Cl; Br) mononitrosyl complexes: θ-$(BETS)_4[Fe(CN)_5NO]$, $(BETS)_2[RuBr_5NO]$, and $(BETS)_2[RuCl_5NO]$, *Eur. J. Inorg. Chem.*, 2797–2804.

46. Valade, L., Malfant, I., Glaria, A., Lamère, J. F., Bonneval, B. G. d., Caro, D. d., Vincendeau, S., Jacob, K., Doublet, M. L., Zwick, A. (2006). $(BETS)_2[RuX_5NO]$ (X = Cl; Br): an explanation of different conductive properties through structural and spectroscopic studies, *J. Low Temp. Phys.*, **142**, 445–448.

47. Zorina, L. V., Gener, M., Khasanov, S. S., Shibaeva, R. P., Canadell, E., Kushch, L. A., Yagubskii, E. B. (2002). Crystal and electronic structures of the radical cation salt based on EDT-TTF and the photochromic nitroprusside anion; $(EDT\text{-}TTF)_3[Fe(CN)_5NO]$, *Synth. Met.*, **128**, 325–332.

48. Kushch, L., Buravov, L., Tkacheva, V., Yagubskii, E., Zorina, L., Khasanov, S., Shibaeva, R. (1999). Molecular metals based on radical cation salts of ET and some of its analogues with the photochromic nitroprusside anion; $[Fe(CN)_5NO]_2$, *Synth. Met.*, **102**, 1646–1649.

49. Malfant, I., Faulmann, C., Kushch, L. A., Sanchez, M. E., Pilia, L., Bonneval, B. G. d., Cassoux, P., Doublet, M. L. (2003). New salts derived from organic donor molecules with long-living excited states counter-ions, *Synth. Met.*, **133–134**, 377–380.

50. Ueda, K., Sugimoto, T., Faulmann, C., Cassoux, P. (2003). Synthesis, crystal structure, and electrical properties of novel radical cation salts of Iodine-substituted TTF-derived donors with the Nitroprusside anion exhibiting strong I···NC–interactions, *Eur. J. Inorg. Chem.*, 2333–2338.

51. Ueda, K., Takahashi, M., Tomizawa, H., Miki, E., Faulmann, C. (2005). Crystal structure and electrical conductivity of {*trans*-[Ru(ox)(en)$_2$NO]}$_2$(TCNQ)$_3$ (ox = oxalate; en = ethylenediamine; TCNQ=7; 7; 8; 8-tetracyanoquinodimethane) at ground state, *J. Mol. Str.*, **751**, 12–16.

52. Cassoux, P., Valade, L., Fabre, P. L. (2004). Electrochemical methods. Electrocrystallization, *Comp. Coord. Chem.*, **1**, 761–773.

53. Salado, D., Cormary, B., Malfant, I. Synthesis of photochromic and conductive salts based on $[Ru(py)_4ClNO]^{2+}$ complex and TCNQ organic acceptors, *unpublished results*.

54. Alcacer, L., Maki, A. H. (1976). Magnetic properties of some electrically conducting perylene-metal dithiolate complexes, *J. Phys. Chem.*, **80**, 1912–1916.

55. Graham, A. W., Kurmoo, M., Day, P. (1995). β''-(BEDT-TTF)$_4$[(H$_2$O)Fe(C$_2$O$_4$)$_3$]·PhCN: the first molecular superconductor containing paramagnetic metal ions, *J. C. S., Chem. Soc.*, 2061–2062.

56. Kobayashi, H., Tomita, H., Naito, T., Kobayashi, A., Sakai, F., Watanabe, T., Cassoux, P. (1996). New BETS conductors with magnetic anions (BETS = bis(ethylenedithio)tetraselenafulvalene), *J. Am. Chem. Soc.*, **118**, 368–377.

57. Goze, F., Laukhin, V. N., Brossard, L., Audouard, A., Ulmet, J. P., Askenazy, S., Naito, T., Kobayashi, H., Kobayashi, A., Tokumoto, M., Cassoux, P. (1994). Magnetotransport measurements on the lambda-phase of the organic conductors (BETS)$_2$MCl$_4$ (M = Ga, Fe). Magnetic-field-restored highly conducting state in λ-(BETS)$_2$FeCl$_4$, *Europhys. Lett.*, **28**, 427–431.

58. Kobayashi, H., Udagawa, T., Tomita, H., Bun, K., Naito, T., Kobayashi, A. (1993). A new organic superconductor, λ-(BEDT-TSF)$_2$GaCl$_4$, *Chem. Lett.*, 1559–1562.

59. Coronado, E., Galán-Mascarós, J. R., Gómez-García, C. J., Laukhin, V. (2000). Coexistence of ferromagnetism and metallic conductivity in a molecule-based layered compound, *Nature*, **408**, 447–449.

60. Kobayashi, H., Kobayashi, A., Cassoux, P. (2000). BETS as a source of molecular magnetic superconductors (BETS = bis(ethylenedithio)tetraselenafulvalene), *Chem. Soc. Rev.*, **29**, 325–333.

61. Fujiwara, H., Fujiwara, E., Nakazawa, Y., Narymbetov, B. Z., Kato, K., Kobayashi, H., Kobayashi, A., Tokumoto, M., Cassoux, P. (2001). A novel antiferromagnetic organic superconductor κ-(BETS)$_2$FeBr$_4$ [where BETS = Bis(ethylenedithio)tetraselenafulvalene], *J. Am. Chem. Soc.*, **123**, 306–314.

62. Uji, S., Shinagawa, H., Terashima, T., Yakabe, T., Terai, Y., Tokumoto, M., Kobayashi, A., Tanaka, H., Kobayashi, H. (2001). Magnetic-field-induced superconductivity in a two-dimensional organic conductor, *Nature*, **410**, 908–910.

63. Zhang, B., Tanaka, H., Fujiwara, H., Kobayashi, H., Fujiwara, E., Kobayashi, A. (2002). Dual-action molecular superconductors with magnetic anions, *J. Am. Chem. Soc.*, **124**, 9982–9983.

64. Real, J. A., Gaspar, A. B., Munoz, M. C. (2005). Thermal, pressure and light switchable spin-crossover materials, *Dalton Trans.*, 2062–2079.

65. Kahn, O. (1993) *Molecular Magnetism*, Wiley VCH, New York.

66. Muller, R. N., Vander Elst, L., Laurent, S. (2003). Spin transition molecular materials: intelligent contrast agents for magnetic resonance imaging, *J. Am. Chem. Soc.*, **125**, 8405–8407.

67. Bousseksou, A., Molnar, G., Demont, P., Menegotto, J. (2003). Observation of a thermal hysteresis loop in the dielectric constant of spin crossover complexes: Towards molecular memory devices, *J. Mater. Chem.*, **13**, 2069–2071.

68. Kunkeler, P. J., van Koningsbruggen, P. J., Cornelissen, J. P., van der Horst, A. N., van der Kraan, A. M., Spek, A. L., Haasnoot, J. G., Reedijk, J. (1996). Novel hybrid spin systems of 7,7′,8,8′-tetracyanoquinodimethane (TCNQ) radical anions and 4-amino-3,5-bis(pyridin-2-yl)-1,2,4-triazole (abpt). Crystal structure of $[Fe(abpt)_2(TCNQ)_2]$ at 298 and 100 K, Moessbauer spectroscopy, magnetic properties, and infrared spectroscopy of the series $[MII(abpt)_2(TCNQ)_2]$ (M = Mn, Fe, Co, Ni, Cu, Zn), *J. Am. Chem. Soc.*, **118**, 2190–2197.

69. Nakano, M., Fujita, N., Matsubayashi, G.-E., Mori, W. (2002). Modified chesnut model for spin-crossover semiconductors $[Fe(acpa)_2](TCNQ)_n$, *Mol. Cryst. Liq. Cryst.*, **379**, 365–370.

70. Dorbes, S., (2005) Thèse d'Université, *Conducteurs moléculaires magnétiques associant des complexes $M(dmit)_2$ à des complexes à transition de spin*, Université Paul Sabatier, Toulouse, France, p. 225.

71. Dorbes, S., Valade, L., Real, J. A., Faulmann, C. (2005). $[Fe(sal_2\text{-}trien)][Ni(dmit)_2]$: Towards switchable spin crossover molecular conductors, *Chem. Commun.*, 69–71.

72. Szilágyi, P. Á., Dorbes, S., Molnár, G., Real, J. A., Homonnay, Z., Faulmann, C., Bousseksou, A. (2008). Temperature and pressure effects on the spin state of ferric ions in the $[Fe(sal_2\text{-}trien)][Ni(dmit)_2]$ spin crossover complex, *J. Phys. Chem. Solids*, **69**, 2681–2686.

73. Faulmann, C., Szilagyi, P. A., Jacob, K., Chahine, J., Valade, L. (2009). Polymorphism and its effects on the magnetic behaviour of the [Fe(sal_2-trien)][Ni($dmit)_2$] spin-crossover complex, *N. J. Chem.*, **33**, 1268–1276.

74. Faulmann, C., Dorbes, S., Real, J. A., Valade, L. (2006). Electrical conductivity and spin crossover: Towards the first achievement with a metal Bis Dithiolene complex, *J. Low Temp. Phys.*, **142**, 265–270.

75. Pereira, L. C. J., Gulamhussen, A. M., Dias, J. C., Santos, I. C., Almeida, M. (2007). Searching for switchable molecular conductors: Salts of [M($dcbdt)_2$] (M = Ni, Au) anions with [Fe(sal_2-trien)]$^+$ and [Fe($phen)_3$]$_2^+$, *Inorg. Chim. Acta*, **360**, 3887–3895.

76. Takahashi, K., Cui, H., Kobayashi, H., Yasuaki, E., Sato, O. (2005). The light-induced excited spin state trapping effect on Ni($dmit)_2$ salt with an Fe(III) spin-crossover cation: [Fe($qsal)_2$][Ni($dmit)_2$]·2CH_3CN, *Chem. Lett.*, **34**, 1240–1241.

77. Takahashi, K., Cui, H.-B., Okano, Y., Kobayashi, H., Einaga, Y., Sato, O. (2006). Electrical conductivity modulation coupled to a high-spin-low-spin conversion in the molecular system [Fe^{III}($qsal)_2$][Ni($dmit)_2$]$_3$·CH_3CN·H_2O, *Inorg. Chem.*, **45**, 5739–5741.

78. Faulmann, C., Dorbes, S., Lampert, S., Jacob, K., Garreau de Bonneval, B., Molnar, G., Bousseksou, A., Real, J. A., Valade, L. (2007). Crystal structure, magnetic properties and Mossbauer studies of [Fe($qsal)_2$][Ni($dmit)_2$], *Inorg. Chim. Acta*, **360**, 3870–3878.

79. Takahashi, K., Mori, H., Kobayashi, H., Sato, O. (2009). Mechanism of reversible spin transition with a thermal hysteresis loop in [Fe^{III} ($qsal)_2$][Ni($dmise)_2$]·2CH_3CN: Selenium analogue of the precursor of an Fe(III) spin-crossover molecular conducting system, *Polyhedron*, **28**, 1776–1781.

80. Faulmann, C., Dorbes, S., Garreau de Bonneval, B., Molnar, G., Bousseksou, A., Gomez-Garcia, C. J., Coronado, E., Valade, L. (2005). Towards molecular conductors with a spin-crossover phenomenon: Crystal structures, magnetic properties and mossbauer spectra of [Fe(salten)Mepepy][M($dmit)_2$] complexes, *Eur. J. Inorg. Chem.*, 3261–3270.

81. Faulmann, C., Jacob, K., Dorbes, S., Lampert, S., Malfant, I., Doublet, M.-L., Valade, L., Real, J. A. (2007). Electrical conductivity and spin crossover: A new achievement with a metal bis dithiolene complex, *Inorg. Chem.*, **46**, 8548–8559.

82. Takahashi, K., Cui, H.-B., Okano, Y., Kobayashi, H., Mori, H., Tajima, H., Einaga, Y., Sato, O. (2008). Evidence of the chemical uniaxial strain

effect on electrical conductivity in the spin-crossover conducting molecular system: $[Fe^{III}(qnal)_2][Pd(dmit)_2]_5$. Acetone, *J. Am. Chem. Soc.*, **130**, 6688–6689.

83. Jesson, J. P., Trofimenko, S., Eaton, D. R. (1967). Spin equilibria in octahedral iron(II) poly(1-pyrazolyl)-borates, *J. Am. Chem. Soc.*, **89**, 3158–3164.

84. Salmon, L., Molnar, G., Cobo, S., Oulie, P., Etienne, M., Mahfoud, T., Demont, P., Eguchi, A., Watanabe, H., Tanaka, K., Bousseksou, A. (2009). Re-investigation of the spin crossover phenomenon in the ferrous complex $[Fe(HB(pz)_3)_2]$, *New J. Chem.*, **33**, 1283–1289.

85. Takahashi, K. (2010). ICSM'2010, Kyoto, invited talk, July 8, 2010.

86. Ferraris, A. J., Cowan, D. O., Walatka, V. V., Perlstein, J. H. (1973). Electron transfer in a new highly conducting donor-acceptor complex, *J. Am. Chem. Soc.*, **95**, 948.

87. Coleman, L. B., Cohen, M. J., Sandman, D. J., Yamagishi, F. G., Garito, A. F., Heeger, A. J. (1973). Superconducting fluctuations and the peierls instability in an organic solid, *Solid State Commun.*, **12**, 1125.

88. Chi, X., Itkis, M. E., Kirschbaum, K., Pinkerton, A. A., Oakley, R. T., Cordes, A. W., Haddon, R. C. (2001). Dimeric phenalenyl-based neutral radical molecular conductors, *J. Am. Chem. Soc.*, **123**, 4041–4048.

89. Huang, J., Kertesz, M. (2003). Spin crossover of spiro-biphenalenyl neutral radical molecular conductors, *J. Am. Chem. Soc.*, **125**, 13334–13335.

90. van Koningsbruggen, P. J., Maeda, Y., Oshio, H. (2004). Iron(III) spin crossover compounds, *Top. Curr. Chem.*, **233**, 259–324.

91. Tweedle, M. F., Wilson, L. J. (1976). Variable spin iron(III) chelates with hexadentate ligands derived from triethylenetetramine and various salicylaldehydes. Synthesis, characterization, and solution state studies of a new ^2T-^6A spin equilibrium system, *J. Am. Chem. Soc.*, **98**, 4824–4834.

92. Sinn, E., Sim, G., Dose, E. V., Tweedle, M. F., Wilson, L. J. (1978). Iron(III) chelates with hexadentate ligands from triethylenetetramine and β-diketones or salicylaldehyde. Spin state dependent crystal and molecular structures of $[Fe(acac)_2trien]PF_6$(S = 5/2), $[Fe(acacCl)_2 trien]PF_6$ (S = 5/2), $[Fe(sal)_2trien]Cl{\cdot}2H_2O$ (S = 1/2), and $[Fe(sal)_2 trien]NO_3{\cdot}H_2O$ (S = 1/2), *J. Am. Chem. Soc.*, **100**, 3375–3390.

93. Maeda, Y., Oshio, H., Tanigawa, Y., Oniki, T., Takashima, Y. (1991). Physical characteristic and molecular structure of spin-crossover iron(III) complexes of monoclinic form with hexadentate ligands derived

from triethylenetetramine and salicylaldehyde [Fe(sal$_2$trien)]BPh$_4$· acetone, *Bull. Chem. Soc. Jap.*, **64**, 1522–1527.

94. Maeda, Y., Oshio, H., Tanigawa, Y., Oniki, T., Takashima, Y. (1992). Moessbauer spectra and molecular structure of spin-crossover iron (III) complexes of monoclinic form with hexadentate ligands derived from triethylenetetramine and salicylaldehyde [Fe(sal$_2$trien)]BPh$_4$. acetone, *Hyperfine Interact.*, **68**, 157–160.
95. Sour, A., Boillot, M.-L., Riviere, E., Lesot, P. (1999). First evidence of a photoinduced spin change in an Fe(III) complex using visible light at room temperature, *Eur. J. Inorg. Chem.*, 2117–2119.
96. Grandjean, F., Long, G. J., Hutchinson, B. B., Ohlhausen, L., Neill, P., Holcomb, J. D. (1989). Study of the high-temperature spin-state crossover in the iron(II) pyrazolylborate complex Fe[HB(pz)$_3$]$_2$, *Inorg. Chem.*, **28**, 4406–4414.

Chapter 6

Electroactive 4*f* Lanthanides Complexes Involving Tetrathiafulvalene Derivatives as Ligands: Magnetism and Luminescence

Fabrice Pointillart, Stéphane Golhen, Olivier Cador, and Lahcène Ouahab

Institut des Sciences Chimiques de Rennes, UMR 6226 CNRS, Université de Rennes 1, 263 Avenue du General Leclerc, 35042 Rennes Cedex, France

fabrice.pointillart@univ-rennes1.fr

On the one hand, the tetrathiafulvalene (TTF) moiety and its analogues are well known to present very good redox activity leading to electric conductivity. On the other, the lanthanide ions are intensively studied for their peculiar magnetic and luminescent properties. In this chapter, we present new TTF ligands suitable for the coordination of 4*f* elements. Thus, original coordinating complexes involving redox-active TTF-based ligands and 4*f* lanthanide ions have been obtained and studied because they may give rise to new types of multifunctional materials — conducting, magnetic, and luminescent. X-ray structures and physical properties of seven coordinating complexes and one salt are presented in these lines.

Multifunctional Molecular Materials
Edited by Lahcène Ouahab

ISBN 978-981-4364-29-4 (Hardcover), 978-981-4364-30-0 (eBook)
www.panstanford.com

6.1 Introduction

Intense investigations are devoted to multifunctional molecular materials. In particular, the chemists and physicists are attracted to design new molecules and materials, which possess synergy or interplay between electrical conductivity with magnetism.[1-9] The objective of this combination is to establish a coupling between mobile and localized electrons. In molecular-based materials, conduction electrons mainly arise from organic moieties assembled in networks while the spins mainly arise from transition metal ions. The strategy is to combine these moieties with supramolecular chemistry tools. Two approaches are developed: (i) a "through-space" approach, in which the interactions between mobile and localized electrons take place through short contacts between chemical units, which usually lead weak interactions,[10-17] (ii) a "through-bond" approach, in which the magnetic and mobile electrons are covalently linked. The second approach appears as a promising alternative to obtain strong interactions and a large number of coordination complexes employing this approach have been studied.[18-44] However, only few of them have been successfully oxidized as radical cation complexes.[45-50] So far, the two approaches concerned almost exclusively π–d systems: interactions between π and d electrons. Nevertheless, such approaches can be applied to the elaboration of π–f systems.[51-59]

In this context, trivalent lanthanides present many advantages. Indeed, their large spins and pronounced spin-orbit coupling, in particular for Dy(III) and Tb(III) ions, result in strong Ising-type magnetic anisotropy[60] leading to good candidates for obtaining single-molecule magnets (SMM) and single-chain magnets (SCM).[61-68] In addition, the lanthanides are widely studied for their specific luminescence properties (emission lines ranging from the visible to the near-infra red spectra, s-ms luminescence lifetime allowing time-gated detection and large pseudo-Stokes shifts considering a ligand excitation).[69-70] Owing to the weak absorption of the forbidden f–f transitions, the rare-earth emission is usually sensitized by organic antenna chromophores, which strongly absorb the UV-visible light and whose triplet state matches the accepting level of the lanthanide ion so that an efficient energy transfer occurs.[71-74] According to this classical strategy, highly stable complexes associating a chromophore to a tris-β-diketonates lanthanides platform have been designed leading to potential applications in chelate lasers,[75] or efficient organic light-emitting diodes (OLEDs),[76,77] and polymer light-emitting diodes (PLEDs).[78] Nowadays, a great research endeavor is devoted to the sensitization of near-infrared (NIR) luminescent lanthanides (Nd(III), Yb(III), Er(III) and to a less extend Ho(III) and Pr(III)) for potential applications in telecommunication or *in vivo* bio-imaging.[79]

The aim of the works presented in this chapter is the elaboration of lanthanide-based coordination complexes involving redox-active TTF derivatives. High-spin [Gd(III)] and anisotropic [Tb(III)] lanthanides are employed for their magnetic response. Near-infrared emitters [Pr(III), Nd(III), and Yb(III)] are incorporated, and in these cases, the functionalized redox-active TTFs play the role of organic antenna for the sensitization of the lanthanide.[80,81] In this line, we present the tetrathiafulvalene-2-pyrimidine-1-oxide (TTF-CONH-2-Pym-1-oxide), tetrathiafulvalene-2-pyridine-*N*-oxide (TTF-CONH-2-Py-*N*-oxide), tetrathiafulvalenecarboxylic acid (TTF–COOH), and 4-(2-tetrathiafulvalenyl-ethenyl)pyridine (TTF–CH=CH–Py) ligands (Fig. 6.1), which have been successfully associated and coordinated to 4*f* elements.

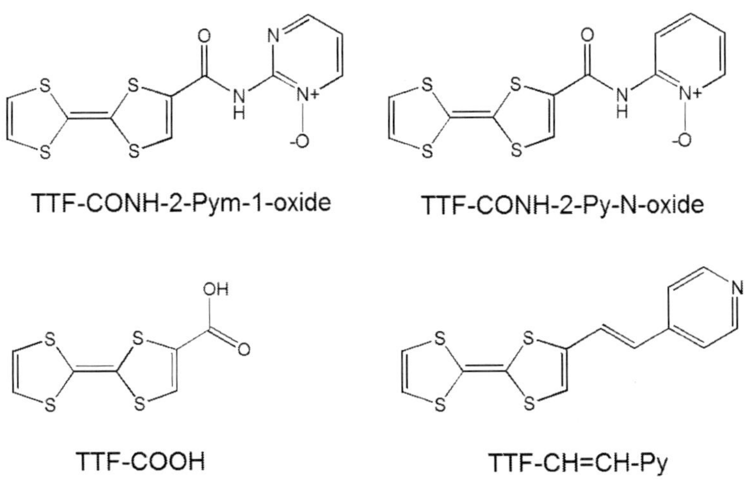

TTF-CONH-2-Pym-1-oxide

TTF-CONH-2-Py-N-oxide

TTF-COOH

TTF-CH=CH-Py

Figure 6.1 Chemical structures of TTF-CONH-2-Pym-1-oxide, TTF-CONH-2-Py-*N*-oxide, TTF–COOH, and TTF–CH=CH–Py electroactive donor molecules.

6.2 Neutral Coordination Complexes: Molecular Building Blocks for π–*f* Magnetic Conducting Materials

6.2.1 Mononuclear Complex [Gd(hfac)$_3$(TTF-CONH-2-Pym-1-oxide)$_3$](CH_2Cl_2)0.5 (C_6H_{14})

TTF-CONH-2-Pym-1-oxide ligand was synthesized by condensation between 2-aminopyrimidine-1-oxide and 4-chlorocarbonyl-tetrathiafulvalene.[59] The reaction of coordination between this ligand and the Gd(hfac)$_3$·2H_2O precursor[82] in dichloromethane leads to the mononuclear complex of formula [Gd(hfac)$_3$(TTF-CONH-2-Pym-1-oxide)$_3$](CH_2Cl_2)0.5 (C_6H_{14}), which is crystallized by layering a *n*-hexane solution.

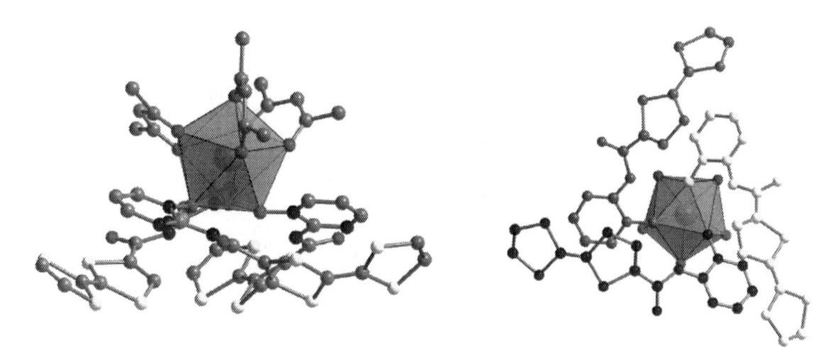

Figure 6.2 Ball-and-stick representation of the structure of [Gd(hfac)$_3$ (TTF-CONH-2-Pym-1-oxide)$_3$] complex, highlighting the coordination polyhedron, the *fac* conformation, and pseudo-helicoidal arrangement of the ligands. Hydrogen and fluorine atoms have been removed for clarity.

The asymmetric unit is composed of a metallic Gd(III) ion, three bischelate hfac$^-$ anions, and three TTF-CONH-2-Pym-1-oxide, acting as monochelating ligands. One-half *n*-hexane and one dichloromethane molecules of crystallization complete the asymmetric unit. The TTF derivatives are coordinated to the Gd(III) ions through the oxygen atoms of the N–O groups. Six additional oxygen atoms from three hfac$^-$ ligands complete the coordination sphere of the Gd(III) ion. The gadolinium surroundings can be described as a 4,4,4-tricapped trigonal prism where all the faces are triangular (Fig. 6.2). The coordination polyhedron has a D_{3h}

symmetry. The coordinated oxygen atoms from three TTF derivatives form one of the triangular faces of the coordinating polyhedron leading to a *fac*-conformation of the three ligands TTF-CONH-2-Pym-1-oxide. The arrangement of the ligands TTF-CONH-2-Pym-1-oxide can be described as a pseudo-propeller. It is noteworthy that the three ligands TTF-CONH-2-Pym-1-oxide are on the same side of the Gd(III) coordination sphere and the three $hfac^-$ anions occupy the opposite site (Fig. 6.2). This segregation permits optimization of the intermolecular interactions between the donors and between the perfluorated ligands. The two organic and inorganic networks are formed by the three coordinated TTF ligands in *fac* conformation and the $Gd(hfac)_3$ entities, respectively. In the organic network, the crystalline organization of the donors is built on unusual $S–O_{carbonyl}$ and classic S–S short contacts.[83] The inorganic network is composed of a double layer of $Gd(hfac)_3$ entities along the *c* axis. The $Gd(hfac)_3$ species interact together through the perfluorated groups.

Free ligand and coordination complex have similar electrochemical properties with two reversible single-electron oxidation waves corresponding successively to the formation of the radical cations and the dicationic species.

The thermal variation of the magnetic susceptibility and the first magnetization show that the $[Gd(hfac)_3(TTF\text{-}CONH\text{-}2\text{-}Pym\text{-}1\text{-}oxide)_3]$ is a paramagnetic system with very small antiferromagnetic interactions between the Gd(III) centers.

The $[Gd(hfac)_3(TTF\text{-}CONH\text{-}2\text{-}Pym\text{-}1\text{-}oxide)_3]$ complex is the first example of 4*f* lanthanide-based coordination complex involving TTF-derivative as ligand, which has been characterized by X-ray diffraction on single crystal. It is the first step in the elaboration of π–f system, which should lead to new combinations of physical properties.

6.2.2 Binuclear Complexes $[Ln_2(hfac)_5(O_2CPhCl)(TTF\text{-}CONH\text{-}2\text{-}Py\text{-}N\text{-}oxide)_3]\cdot 2H_2O$ with Ln = Pr and Gd

The 2:3 binuclear complexes have been obtained for Pr(III) and Gd(III). Both compounds are isotructural.[84] The asymmetric unit is composed of one $Ln(hfac)_3$ and one $Ln(hfac)_2$ units linked by two $bridging_2$-TTF-CONH-2-Py-*N*-oxide ligands and one bridging $_2(_{1,2})$ 3-chloro-benzoate anion (Fig. 6.3). The presence of an anionic bridge explains the loss of one $hfac^-$ ligand in the coordination sphere of one of the two lanthanides. Two water solvent molecules of crystallization are found. The coordination sphere of the Ln(hfac)$_2$ fragment is filled by a terminal TTF-CONH-2-Py-*N*-oxide ligand. The metal centers are nine-coordinated and the arrangements of the ligands lead to coordination polyhedra, which can be described by distorted 4,4,4-tricapped trigonal prisms, like in [Gd(hfac)$_3$(TTF-CONH-2-Pym-1-oxide)$_3$], with 14 triangular faces.[85] The symmetry of the coordination polyhedra is close to D_{3h}.

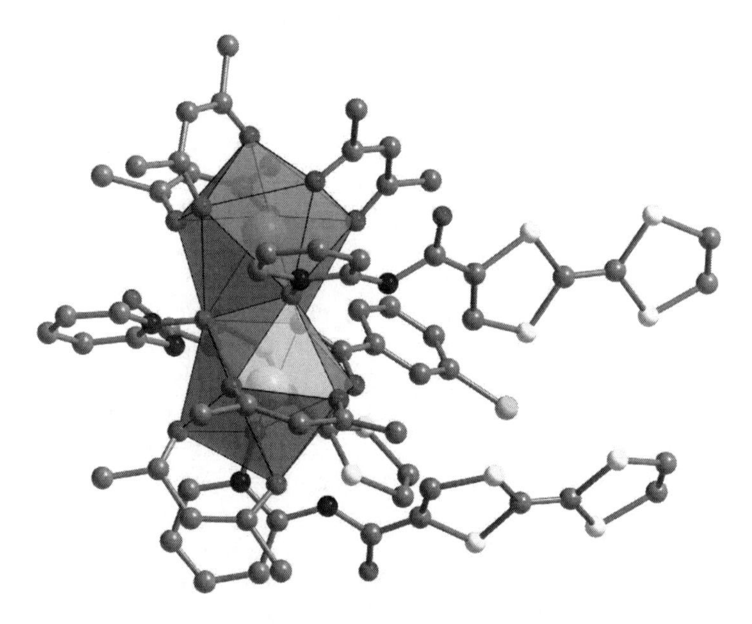

Figure 6.3 Ball-and-stick representation of the structure of [Gd$_2$(hfac)$_5$(O$_2$CPhCl)(TTF-CONH-2-Py-*N*-oxide)$_3$]·2H$_2$O with coordination polyhedra around the Gd(III). Hydrogen and fluorine atoms have been removed for clarity.

The flexibility and the disposition of the donors (all the donors are on the same side of the molecule, see Fig. 6.3), optimize the formation of two segregated inorganic and organic networks with good – overlapping between the TTF fragments. The inorganic network is formed by the Ln(hfac)$_3$ and Ln(hfac)$_2$ moieties while the organic network is formed by the TTF-CONH-2-Py-*N*-oxide donors and the 3-chloro-benzoate anions. The latter is composed of dimers and tetramers of TTF-CONH-2-Py-*N*-oxide. The tetramers are sandwiched by 3-chloro-benzoate anions that performed intermolecular – interactions with both extremities of the tetramer of TTF-CONH-2-Py-*N*-oxide. The TTF cores are almost planar, and the central C=C bond lengths [mean value = 1.339(13) Å] attest that the donors are neutral. Several intermolecular S–S contacts in the range of the sum of the van der Waals radii are observed forming a one-dimensional reminiscent packing of organic donors along the *b*-axis.

The determination of the nature of the exchange interaction in the lanthanides-based complexes is more difficult than in the case of the 3*d* elements. In the case of 4*f* ions, the first excited multiplet is separated by more than 1000 cm^{-1} from the ground multiplet.[86] The latter is split in Stark sublevels under the influence of a crystal field.[87] The crystal field effects are of the order of 100 cm^{-1} for lanthanides. When the temperature decreases, the depopulation of these sublevels leads to a deviation from the Curie law in the absence of any exchange interaction. But any deviation from Curie law observed is not enough to testify the existence of superexchange interactions in polynuclear complexes containing at least one lanthanide ion. One must, at first, be able to estimate the influence of crystal field effects to, at second, determine the nature of the magnetic exchange interaction with the 4*f* elements. It has been shown that the energies of stark sublevels depends on Ln(III) site symmetry.[88]

The thermal variation of the $\chi_M T$ product for the $[Ln_2(hfac)_5(O_2CPhCl)$(TTF-CONH-2-Py-*N*-oxide)$_3]\cdot 2H_2O$ compounds are given in Fig. 6.4. The product $\chi_M T$ for the Pr(III) derivative shows a monotonic decrease in this range of temperature 2–300 K taking respectively the values of 0.17 and 3.18 cm^3 K mol^{-1}. The experimental room temperature value of $\chi_M T$ is in agreement with the theoretical value of 3.20 cm^3 K mol^{-1} expected for two magnetically isolated Pr(III) ions.[86]

The short metal–metal distance [4.017(1) Å] makes possible the existence of an exchange interaction between the paramagnetic Pr(III) (J_{Pr-Pr}). Fortunately, Pr(III) in $Pr(hfac)_3\cdot 3H_2O$ is surrounded by nine oxygen atoms and its coordination sphere can be described as a distorted 4,4,4-tricapped trigonal prism with a D_{3h} symmetry. Similar coordination sphere and symmetry is observed in the Pr(III)-based dinuclear complex. Thus, the nature of the exchange interaction J_{Pr-Pr} in $[Pr_2(hfac)_5(O_2CPhCl)$(TTF-CONH-2-Py-*N*-oxide)$_3]\cdot 2H_2O$ can be determined from the comparison of the magnetic properties of the coordination compound and $Pr(hfac)_3\cdot 3H_2O$ as defined by the following equation: $\chi_M T = \chi_M T_{([Pr_2(hfac)_5(O_2CPhCl)(TTF\text{-}CONH\text{-}2\text{-}Py\text{-}N\text{-}oxide)_2])} - 2^*\chi_M T_{Pr}$, where $\chi_M T_{([Pr_2(hfac)_5(O_2CPhCl)(TTF\text{-}CONH\text{-}2\text{-}Py\text{-}N\text{-}oxide)_2])}$ and $\chi_M T_{Pr}$ are the $\chi_M T$ product for the complex and for the precursor $Pr(hfac)_3 \cdot 3H_2O$ respectively, and $\chi_M T$ can be considered free of single-ion effects and thus representative of the nature of the exchange interaction between the Pr(III) ions (inset in the Fig. 6.4a). $\chi_M T(T)$ takes almost constant values from 300 to 40 K and decreases for lower temperatures. The shape of the $\chi_M T(T)$ curve is in agreement with weak antiferromagnetic exchange interaction ($J_{Pr-Pr} < 0$) between the Pr(III) ions through the nitroxide group of TTF-CONH-2-Py-*N*-oxide and 3-chloro-benzoate bridges.

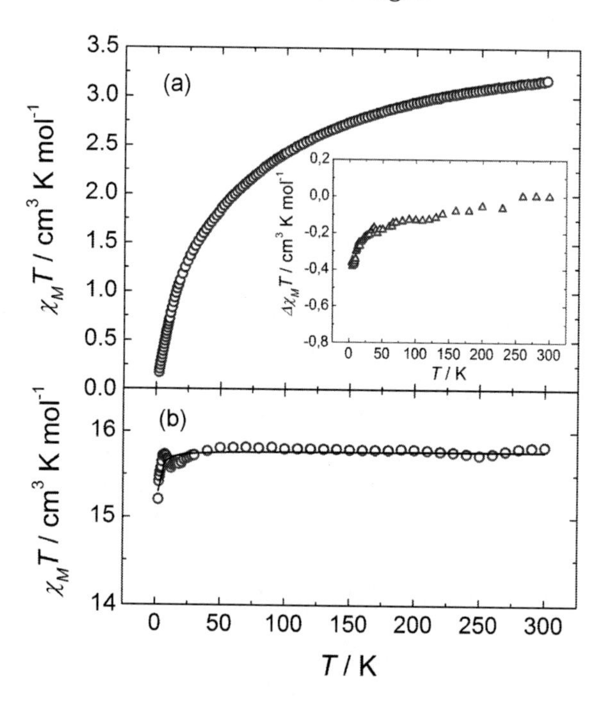

Figure 6.4 (a) Thermal variation of the $\chi_M T$ product (gray circles) and $\chi_M T = \chi_M T_{complex} - 2^*\chi_M T_{Pr}$ (gray triangles) (in inset) for $[Gd_2(hfac)_5(O_2CPhCl)(TTF\text{-}CONH\text{-}2\text{-}Py\text{-}N\text{-}oxide)_3] \cdot 2H_2O$. (b) Thermal variation of the $\chi_M T$ product (gray lozenges) with the best fit (full black line) for $[Gd_2(hfac)_5(O_2CPhCl)(TTF\text{-}CONH\text{-}2\text{-}Py\text{-}N\text{-}oxide)_3] \cdot 2H_2O$.

The $\chi_M T$ for $[Gd_2(hfac)_5(O_2CPhCl)(TTF\text{-}CONH\text{-}2\text{-}Py\text{-}N\text{-}oxide)_3] \cdot 2H_2O$ is constant (15.83 cm^3 K mol^{-1}) in the temperature range

40–300 K (Fig. 6.3b). The experimental value of $\chi_M T$ at room temperature is close to that expected for two noninteracting Gd(III) (S = 7/2) ions (15.75 cm^3 K mol^{-1}). Gd(III) is the prototype of spin-only contribution to magnetism. The nature of the exchange interaction between the Gd(III) can be determined from the shape of the $\chi_M T(T)$ curve. $\chi_M T$ for Gd(III) complex is constant till 40 K while for lower temperatures, weak antiferromagnetic interactions can be observed between the Gd(III) through the TTF-CONH-2-Py-*N*-oxide and 3-chloro-benzoate bridges. A quantitative determination of the exchange interaction can be performed using Eq. 6.1 derived from the isotropic spin Hamiltonian $H = -J\, \mathbf{S}_{Gd1} \cdot \mathbf{S}_{Gd2}$, where J is the exchange interaction constant and $\mathbf{S}_{Gd1}$ and $\mathbf{S}_{Gd2}$ are the spin operators for the interacting spin centers ($S_{Gd1} = S_{Gd2} = 7/2$).[89]

$$\chi_M T = \frac{2Ng^2\beta^2}{k} \frac{140 + 91e^{7x} + 55e^{13x} + 30e^{18x} + 14e^{22x} + 5e^{25x} + e^{27x}}{15 + 13e^{7x} + 11e^{13x} + 9e^{18x} + 7e^{22x} + 5e^{25x} + 3e^{27x} + e^{28x}}$$

$$\text{with } x = -\frac{J}{kT} \qquad (6.1)$$

The best fit is shown on Fig. 6.3b with J_{Gd-Gd} = −0.01 0.0008 cm^{-1} and g_{Gd} = 2. The amplitude of the antiferromagnetic exchange interaction is in agreement with J values in $[Gd(hfac)_3(4\text{-cpyNO})]_2$ (4-cpyNO = 4-cyanopyridine-*N*-oxide) magnetic systems[89] but smaller than the antiferromagnetic value found in some other compounds.[90,91] The differences should be due to the change of symmetry around the Gd(III) centers.

6.2.3 Binuclear Complex $[Tb_2(hfac)_4(O_2CPhCl)_2$ (TTF-CONH-2-Py-*N*-oxide)$_2]$

When the Pr(III) or Gd(III) are replaced by the smaller Tb(III) ions, a 2:2 dinuclear coordination complex is obtained instead of the 2:3 ones.[84] The structure of $[Tb_2(hfac)_4(O_2CPhCl)_2$(TTF-CONH-2-Py-*N*-oxide)$_2]$ consists of a centrosymmetric dimetallic unit made of two Terbium ions in a distorted dodecahedral oxygenated coordination sphere (Fig. 6.5), the coordination sphere is made of eight oxygen atoms arising from two bischelate $hfac^-$ anions, one terminal ligand TTF-CONH-2-Py-*N*-oxide, one water molecule, and two 3-chloro-benzoate anions. The coordination polyhedron has almost a D_{2d} symmetry. The two $Tb(hfac)_2(H_2O)$(TTF-CONH-2-Py-*N*-oxide) units are linked by two bridging 2(1,1) 3-chloro-benzoate anions (Fig. 6.5).

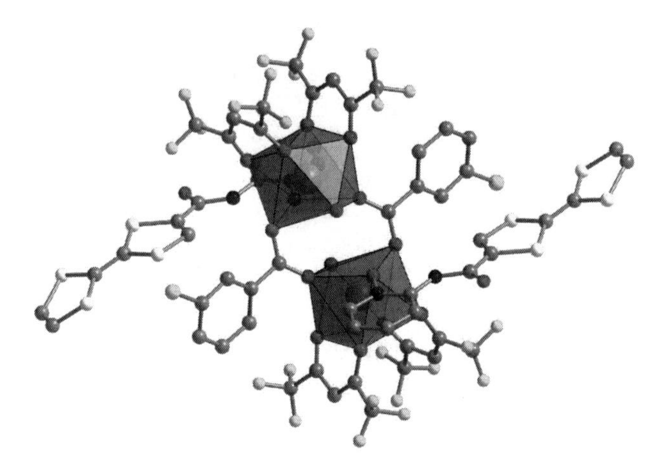

Figure 6.5 Ball-and-stick representation of the structure of $[Tb_2(hfac)_4(O_2CPhCl)_2(TTF\text{-}CONH\text{-}2\text{-}Py\text{-}N\text{-}oxide)_2]$ with coordination polyhedra around the Tb(III). Hydrogen atoms have been removed for clarity.

As observed in the structures involving the Pr(III) and Gd(III) ions, the presence of an anionic bridge explains the loss of one $hfac^-$ ligand in the coordination sphere of Tb(III) ions. The central C=C bond length of the TTF fragment [1.327(9) Å] confirms the neutral form of the donor TTF-CONH-2-Py-*N*-oxide. As described previously, the coordination sphere of Tb(III) is completed by a water molecule, which plays an important role in the cohesion of the dinuclear complex and in the crystal packing. In fact, this water molecule participates in intramolecular and intermolecular hydrogen bonds. Each ligand TTF-CONH-2-Py-*N*-oxide of the dinuclear complex of Tb(III), is "head-to-tail" stacked with the neighbor donor. The two "head-to-tail" stacked donors formed dimers with another TTF derivative through S–S contacts and one-dimensional organic chains run along the *c* axis. The inorganic network includes the $[Tb(hfac)_2(\text{-3-chloro-benzoate})]_2$ units.

$\chi_M T$ takes constant value of 23.47 cm^3 K mol^{-1} in the temperature range 300–100 K, this value is close to the expected value for two magnetically isolated Tb(III) ions (23.64 cm^3 K mol^{-1}). The $\chi_M T(T)$ curve shows a monotonic decrease between 100 and 15 K mainly due to the crystal field effect on Tb(III) ions and decreases more rapidly on cooling further due to some antiferromagnetic exchange interactions in the system.

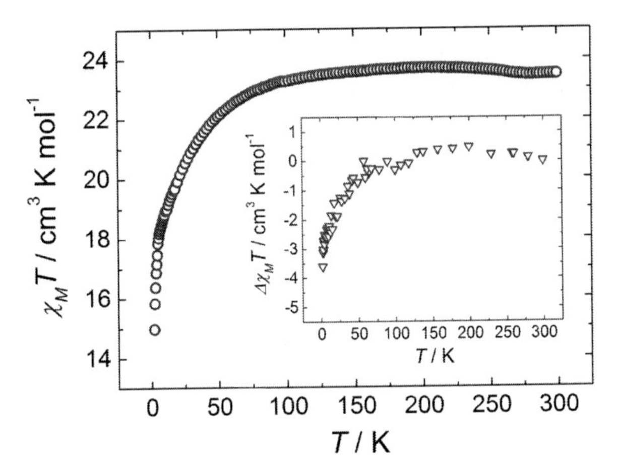

Figure 6.6 Thermal variation of the $\chi_M T$ product for $[Tb_2(hfac)_4 (O_2CPhCl)_2 (TTF\text{-}CONH\text{-}2\text{-}Py\text{-}N\text{-}oxide)_2]$ (gray circles). In inset, thermal variation of the $\chi_M T = \chi_M T_{([Tb_2(hfac)_4(O_2CPhCl)_2(TTF\text{-}CONH\text{-}2\text{-}Py\text{-}N\text{-}oxide)_2])}$ $-2{*}_M T_{Tb}$ (gray triangles).

The method used for the determination of the exchange interaction in the Pr(III) complex can be adapted in the case of $[Tb_2(hfac)_4(O_2CPhCl)_2(TTF\text{-}CONH\text{-}2\text{-}Py\text{-}N\text{-}oxide)_2]$. In fact, the structural analysis has revealed a dodecahedral coordination sphere for the Tb(III) centers as observed in the crystal structure of the precursors $Ln(hfac)_3{\cdot}2H_2O$ (Ln = Gd-Lu).[92] The expression of the magnetic coupling constant between the Tb(III) ions through the bridging 3-chloro-benzoate anions can be written as $\chi_M T = \chi_M T_{([Tb_2(hfac)_4(O_2CPhCl)_2(TTF\text{-}CONH\text{-}2\text{-}Py\text{-}N\text{-}oxide)_2])} - 2{*}_M T_{Tb}$ where the parameters are defined as done previously for Terbium instead of Praseodymium. The $\chi_M T(T)$ is presented in inset of Fig. 6.6. The value of $\chi_M T(T)$ remains almost equal to 0 down to 50 K and decreases for lower temperatures. The shape and the sign of the $\chi_M T(T)$ curve attests that weak antiferromagnetic exchange interaction ($J_{Tb\text{-}Tb} < 0$) takes place between the Tb(III) ions through the 3-chloro-benzoate bridges.

The three complexes of Pr(III), Gd(III), and Tb(III) are the first reported polynuclear coordination complexes involving lanthanides and TTF ligands.

6.3 Oxidized Coordination Complex {[Gd(hfac)$_3$ (-TTF$^+$COO$^-$)]$_2$}: A Coordinating π–f System

The electroactive donors must be oxidized to induce electronic conductivity but only a few examples of such compounds, in which metal–radical and radical–radical couplings take place, have been reported.[45–50] The oxidation of the TTF ligand implies the insertion of additional anions such as PF_6^- or BF_4^- to maintain the electroneutrality of the structure. Their incorporation in the crystal lattice modifies the crystal packing of the neutral equivalent complexes.[34,46,47] One approach to overcome these problems is the elaboration of TTF derivatives, which are substituted with potential anionic groups. One possibility is the use of the TTF carboxylate, which reacts with the Gd(hfac)$_3$·2H$_2$O precursor under aerobic conditions to give the binuclear complex of formula [Gd(hfac)$_3$ (-TTF$^+$COO$^-$)]$_2$ (Fig. 6.7).[58]

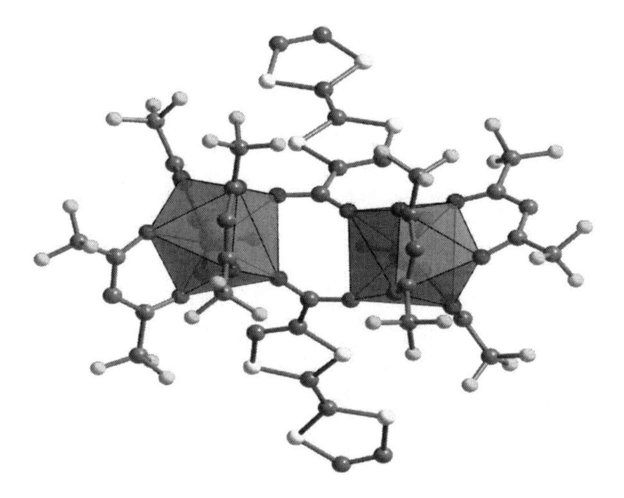

Figure 6.7 Molecular structure of [Gd(hfac)$_3$(–TTF$^+$COO$^-$)]$_2$ with coordination polyhedra around the Gd(III).

Gd(III) ions lie in a distorted square antiprism environment of oxygen atoms, six belonging to three hfac$^-$ anions and two belonging to two TTF$^{\bullet+}$COO$^-$ ligands. In the asymmetric unit, both TTF$^{\bullet+}$COO$^-$ moieties are 2(1,1) bridging ligands. These donors are planar and the mean central C=C and C–S bond lengths are equal to 1.402(10) Å and 1.714(7) Å, respectively, indicating that the donors are radical cations (charge + 1). These values are in agreement with those found in the complex [Cu(hfac)$_2$(TTF–CH=CH–Py$^+$)$_2$](BF$_4$)$_2$2CH$_2$Cl$_2$[44] containing oxidized donors. In contrast, the mean central C=C and C–S distances are equal to 1.338(2) Å and 1.761(2) Å, 1.334(9) Å and 1.759(7) Å, 1.334(2) Å and 1.753(2) Å, respectively, for the compounds [Rh$_2$(ButCO$_2$)(TTFCO$_2$)(NEt$_3$)$_2$], *cis*-, and *trans*-[Rh$_2$(Bu tCO$_2$)(TTFCO$_2$)$_2$(NEt$_3$)$_2$],[93] in which the TTF cores of the 2(1,1) TTF–COO$^-$ anions are neutral. Since the Gd(hfac)$_3$ is neutral and the TTF fragments are radical cations, the acid functions are deprotonated and no additional counter anions are needed in the lattice. Figure 6.8 depicts the crystal packing of [Gd(hfac)$_3$(–TTF$^+$COO$^-$)]$_2$, which shows that the dinuclear complexes interact to form pseudo one-dimensional zigzag chains.

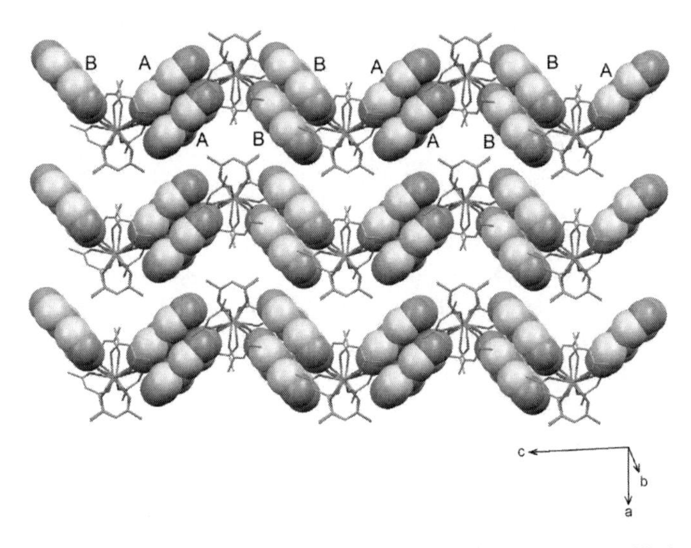

Figure 6.8 Packing view of the compound [Gd(hfac)$_3$(–TTF$^+$COO$^-$)]$_2$ highlighting the zigzag chains and the formation of the two different A–A and B–B dimers of TTF$^{\bullet+}$COO$^-$ radicals. Hydrogen atoms and fluorine atoms have been removed for clarity.

The intermolecular interactions in one of these zigzag chains take place between two crystallographically equivalent donors to form two different centrosymmetric dimers (A–A and B–B). The interplane distance, defined by the central six-member atoms of the TTF core, is close to 3.50(3) Å in both cases. The carboxylate groups are located close to the TTF fragments of neighboring TTF$^{\bullet+}$COO$^-$.

This structural arrangement is in agreement with the relative charge carried by the carboxylate (−1) and TTF cores (+1). The shortest intradimer S–S contacts (3.600(4), 3.674(4) Å and 3.947(3), 4.203 (4) Å) are found in the range (A–A) or longer (B–B) than the sum of the van der Waals radii (3.65 Å). These distances are relatively long for radical cation dimers because of the electrostatic interaction between the carboxylate group and TTF fragment. In other words, the competition between electrostatic and S–S interactions takes place, leading to a compromise.

A solid-state electron paramagnetic resonance (EPR) spectrum has been measured at 290 K. It shows a very large signal centered at $g = 2.00$ corresponding to the paramagnetic Gd(III) ions ($S = 7/2$) and an isotropic one centered at $g = 2.008$, which is characteristic of an organic radical coming from the $TTF^{\bullet+}COO^-$. The A donors are strongly dimerized and should not give an EPR signal, and the weak organic signal should come from the B donors. Single crystals are then dissolved in DMF but no radical signal has been observed due to the instability of free radical of TTFCOOH in solution.[93]

$\chi_M T$ for $[Gd(hfac)_3(-TTF^+COO^-)]_2$ remains quasi constant at 16.00 cm^3 K mol^{-1} in the range 5–300 K (Fig. 6.9). A small decrease of $\chi_M T$ is observed down to 5 K. The experimental Curie constant is slightly higher than the expected value for two magnetically isolated Gd(III) ions ($S = 7/2$) (15.76 cm^3 K mol^{-1})[86] but lower than the expected one for two isolated $S = 7/2$ and two $S = ½$ coming from the radical cations (16.51 cm^3 K mol^{-1}). This result clearly means that the magnetism arises principally from Gd(III) spins. The structural analysis suggest that the *A* donors are strongly dimerized and, therefore, do not contribute to magnetic susceptibility of $[Gd(hfac)_3(-TTF^+ COO^-)]_2$.[45–50] In B–B dimers, the S–S contacts are long and magnetic contribution is expected. In first approximation, the experimental $\chi_M T$ value of 16.00 cm^3 K mol^{-1} at 300 K can be attributed to both isolated Gd(III) ions ($S = 7/2$) and to an additional magnetic contribution of the antiferromagnetically coupled B donors ($S = ½$). The first magnetization curve of $[Gd(hfac)_3(-TTF^+COO^-)]_2$ at 2 K is shown in the inset of the Fig. 6.9. It can be fitted with the Brillouin function for two uncoupled Gd(III) spins.[86] Negligible intramolecular exchange interactions are expected between the radical cations as observed in electrochemistry and ESR measurements for oxidized *cis*-$[Rh_2(ButCO_2)(TTF-CO_2)_2(NEt_3)_2]$.[93] The small value of the Curie–Weiss constant = −0.43 K suggests that the exchange interactions are very weak in $[Gd(hfac)_3(-TTF^+COO^-)]_2$. The exchange interactions between the 4*f* Gd(III) and the electrons of the radical cations are undeterminable due to the strong antiferromagnetic interactions between the donors.

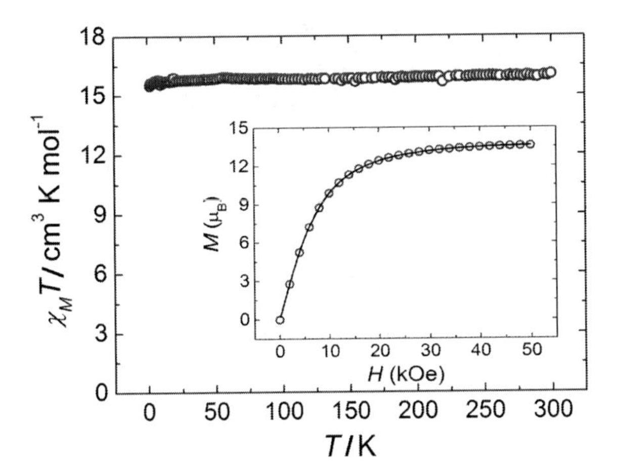

Figure 6.9 Thermal variation of the magnetic susceptibility of $[Gd(hfac)_3(-TTF^+COO^-)]_2$ in an applied field. Inset: First magnetization at 2 K.

First we have presented three examples of molecular precursors for the design of π–*f* systems, the dinuclear $[Gd(hfac)_3(-TTF^+COO^-)]_2$ compound is an unprecedented coordination π–*f* system.

Infrared Luminescent Coordination Complexes ($\{[Nd(hfac)_4(H_2O)][(TTF–CH=CH–Py^{•+})]\}_2$): A Luminescent π–*f* System

From selection of other lanthanide ions may emerge new physical properties such as infrared luminescence. Thus, the $Nd(hfac)_3 \cdot 3H_2O$ precursors have been associated with the 4-(2-tetrathiafulvalenyl-ethenyl)pyridine ligand (TTF–CH=CH–Py) at 60°C to lead after spontaneous oxidation of the redox-active ligand to the $\{[Nd(hfac)_4(H_2O)][(TTF–CH=CH–Py^{•+})]\}_2$ compound.[57] The crystal-lographic structure is composed of two monoanionic $[Nd(hfac)_4(H_2O)]^-$ complexes and two TTF–CH=CH–Py$^{•+}$ derivatives (Fig. 6.10). To ensure the electroneutrality of the structure, both TTF–CH=CH–Py species must be radical cations. Each Nd(III) ion is surrounded by nine oxygen atoms from the four bischelating hfac$^-$ ligands and one water molecule. The coordination geometry of the Nd(III) ions can be described as a distorted capped square antiprism.[94] The $[Nd(hfac)_4(H_2O)]^-$ complexes related through the inversion center form pseudo dimeric units with strong hydrogen bonds between water molecules and oxygen atoms of hfac$^-$ anions. The nitrogen atom of TTF–CH=CH–Py$^{\bullet+}$ radical cations interact weakly with Nd(III), leading to small distortions of the coordination sphere of lanthanide. The NdN interactions pass through the center of the square face of the coordination polyhedron of Nd(III) ions.

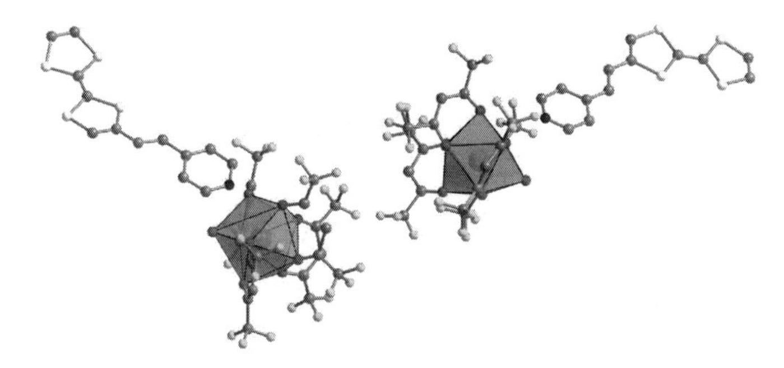

Figure 6.10 Representation of the asymmetric unit of $\{[Nd(hfac)_4(H_2O)][(TTF–CH=CH–Py^{\bullet+})]\}_2$ highlighting the coordination polyhedra around the Nd(III). Hydrogen atoms have been removed for clarity.

The crystal packing in $\{[Nd(hfac)_4(H_2O)][(TTF–CH=CH–Py^{\bullet+})]\}_2$ is composed of an inorganic network of pseudo dimeric unit $[(Nd(hfac)_4(H_2O)]_2^{2-}$ and an organic network, which is composed of stacked TTF–CH=CH–Py$^{\bullet+}$ radical cations. In the organic network, the TTF–CH=CH–Py$^{\bullet+}$ radical cations form columns along the[100] direction (Fig. 6.10). The columns consist of pseudo-tetrameric units of TTF–CH=CH–Py$^{\bullet+}$ radical cations with short S–S contacts slightly longer than the sum of the van der Waals radii while in the previous reported materials involving lanthanide ions the S–S contacts are shorter than 3.6 Å.[51–56] Two reasons can explain the absence of short S–S contacts in $\{[Nd(hfac)_4(H_2O)][(TTF–CH=CH–Py^{\bullet+})]\}_2$: (i) the nitrate and thiocyanate anions in the previously reported structures are smaller than the hexafluoroacetylacetonate, (ii) the TTF–CH=CH–Py$^{\bullet+}$ radicals are not free to autoassemble to build short S–S contacts because the potentially coordinating pyridyl group of each radical interacts with neodymium. The crystal packing then results from the competition between Nd–Py and S–S interactions. The pseudo-tetrameric units are separated by long S–S distance [5.518(5) Å]. The aromatic systems containing the radical (mainly the TTF cores) almost fit in parallel planes but are shifted one with respect to the other. In spite of the absence of short S–S interatomic contacts, the systems overlap. The shortest distance between the systems of the TTF core is equal to 3.537 Å. Such a short distance can lead to strong exchange interactions.

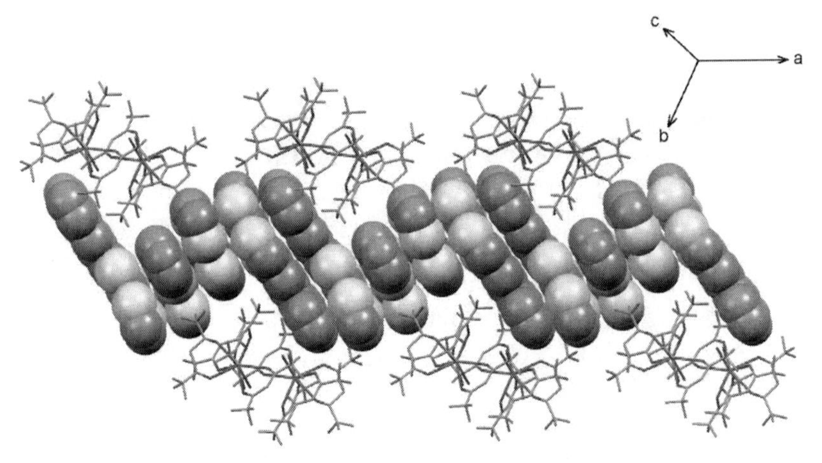

Figure 6.11 View of the packing in {[Nd(hfac)$_4$(H$_2$O)][(TTF–CH=CH–Py$^{\bullet+}$)]}$_2$ showing the inorganic and organic networks along the *a*-axis and highlighting the overlap between the systems.

The electrical resistivity of single crystals has been measured between 300 and 70 K by the standard two-probe method. Gold wires were attached to the crystal with carbon paste. The values of resistivity are close to 10 kΩ·cm^{-1}, and it can be described as insulators in this temperature range. These experimental observations are in agreement with the crystal structure (tetramers) and the oxidized state +1 of all donors.

The EPR spectrum of {[Nd(hfac)$_4$(H$_2$O)][(TTF–CH=CH–Py$^{\bullet+}$)]}$_2$ shows a weak and broad signal centered at the isotropic g value (g = 2.008) typical of a radical cation TTF derivative.[43,95] The signal of the paramagnetic Nd(III) ion is not observed for higher temperature than 70 K. The presence of a weak signal at 70 K indicates that the TTF–CH=CH–Py$^{\bullet+}$ radicals are almost completely antiferromagnetically coupled at this temperature. The intensity of the signal is in agreement with a tetramerization of the TTF derivatives in the crystal structure leading to strong antiferromagnetic interactions between the radical cations.

The $\chi_M T(T)$ curve shows a monotonic decrease from 300 to 2 K. The room temperature value of $\chi_M T$ is equal to 3.21 cm^3 K mol^{-1} while it is equal to 1.43 cm^3 K mol^{-1} at 2 K. The electronic configuration $4f^3$ is split by interelectronic repulsions and gives the ^{4}I ground state spectroscopic term. Each term $^{2S+1}$L is split by spin-orbit coupling and give the $^4I_{9/2}$ ground state spectroscopic level with a Zeeman factor $g_{9/2}$ equal to 8/11.[96] Finally, each $^{2S+1}L_J$ term is split by crystal field in Stark sublevels separated by about 100 cm^{-1}.[87] This splitting is smaller than that for first transition metallic ions because the $4f$ orbitals are not involved in first approximation in the bonds with neighboring atoms. In addition, the energy splitting between the ground state $^4I_{9/2}$ and the first exited state $^4I_{11/2}$ is equal to 1670 cm^{-1}. So at room temperature, only the Stark sublevels from the $^4I_{9/2}$ ground state are populated. When the temperature decreases, the depopulation of these sublevels leads to a deviation from the Curie law observed by a variation of the $\chi_M T$ even in the absence of any exchange interaction. The room temperature value of $\chi_M T$ is in agreement with the expected value for two isolated Nd(III) ions in a O_9 crystal field (the theoretical value is 3.20 cm^3 K mol^{-1}).[86,97] No contribution of the radical cations (S = 1/2) is observed and the decrease of the $\chi_M T$ is only due to the crystal field effect. The absence of contribution of the spin of the radical cations supports the observation of weak signals of the organic radicals on EPR spectra.

The spectrum of TTF–CH=CH–Py shows four strong absorption bands centered around 38,200, 30,000, 22,900, and 19,900 cm^{-1} (Fig. 6.12a). The two lowest-energy transitions correspond to intramolecular charge transfers HOMO LUMO+n from the donor TTF core to the acceptor 4-pyridine moiety.[98,99] The two highest-energy transitions correspond to intramolecular π–π* transitions of the TTF–CH=CH–Py ligand. The absorption spectrum of the complex shows three new absorption bands compared to the spectrum of the neutral TTF–CH=CH–Py ligand (Fig. 6.12b). The first one at 33,000 cm^{-1} (red deconvolution) is principally due to intramolecular π–π* transition of the hfac$^-$ anions.[94] The two other ones at 16,000 cm^{-1} (CT_1) and 26,500 cm^{-1} (CT_2) (blue deconvolutions) are characteristic of oxidized TTF derivatives[99] and confirm to the radical form of the TTF–

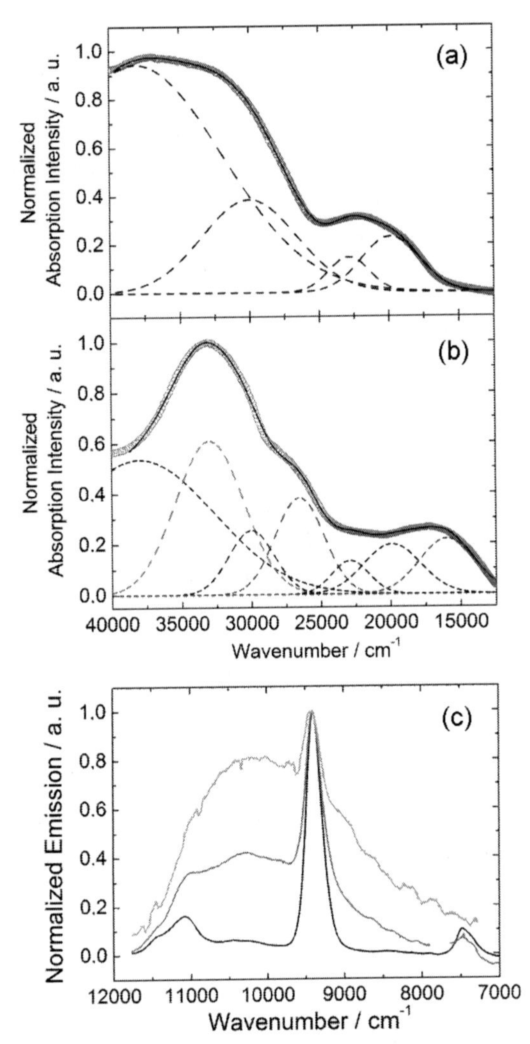

Figure 6.12 Experimental solid state (KBr pellets) UV-visible absorption spectra of TTF–CH=CH–Py (a) and {[Nd(hfac)$_4$(H$_2$O)][(TTF–CH=CH–Py$^{\bullet+}$)]}$_2$ (open circles) (b), deconvolution of the experimental curve (dashed lines) and the best fit R = 0.9999 (for TTF–CH=CH–Py) (a) and R = 0.9996 (for {[Nd(hfac)$_4$(H$_2$O)][(TTF–CH=CH–Py$^{\bullet+}$)]}$_2$) (b). Emission spectra of {[Nd(hfac)$_4$(H$_2$O)][(TTF–CH=CH–Py$^{\bullet+}$)]}$_2$ in the near-IR (curve in black for λ_{ex} = 25,000 cm^{-1}, in gray for λ_{ex} = 15,625 cm^{-1} and in light gray for λ_{ex} = 14,300 cm^{-1}) at room temperature (293 K) in the solid state (c). See also Color Insert.

CH=CH–Py in {[Nd(hfac)$_4$(H$_2$O)][(TTF–CH=CH–Py$^{\bullet+}$)]}$_2$. By analogy with previous works,[99] the band centered at 16,000 cm^{-1} can be attributed to an intramolecular charge transfer of one electron from a SOMO-*n* localized on the 4-pyridine group to the SOMO localized on the TTF$^{\bullet+}$ fragment. The absorption band centered at 26,500 cm^{-1} is attributed to the SOMO LUMO transition of the TTF–CH=CH–Py$^{\bullet+}$ radical cation. The intermolecular electronic interactions with the anionic lanthanide complexes have been considered playing a negligible role in the interpretation of the UV-visible absorption spectroscopy.

The Nd^{3+} luminescence is observed upon excitation in the entire spectrum up to 14,300 cm^{-1} and an additional broad organic fluorescence band is observed mainly by irradiation in the CT$_1$ transition (Fig. 6.12c). The Nd(III) luminescence is composed of the classic three emission bands localized at 11,070 ($^4F_{3/2}$ $^4I_{9/2}$), 9400 ($^4F_{3/2}$ $^4I_{11/2}$), and 7470 ($^4F_{3/2}$ $^4I_{13/2}$) cm^{-1} while the fluorescence band of the radical cation is centered at 10,200 cm^{-1} (relaxed CT$_1$ ground state). Higher energy excitation (25,000 cm^{-1}) (corresponding to the CT$_2$ and intramolecular transition of the 4-pyridine substituent) leads to a significant decrease of the fluorescence band of the organic radical cation (Fig. 6.12c). It is noteworthy that similar Nd(III) luminescence is observed by excitation of [NEt$_4$][Nd(hfac)$_4$] anion due to the presence of a large number of *f–f* transitions in the 25,000–14,300 cm^{-1} spectral range.[94]

These transitions are forbidden and feature a very low extinction coefficient but in the solid state, the concentration is high enough to allow this direct luminescence sensitization process. Therefore, it is not possible to conclude that in {[Nd(hfac)$_4$(H$_2$O)][(TTF–CH =CH–Py$^{\bullet+}$)]}$_2$, the neodymium luminescence is sensitized through antenna effect from the radical cation-centered charge transfer transition[100,101] or by direct process through the *f–f* transitions.

Infrared Luminescent Coordination Complexes ([Ln(hfac)$_3$(TTF-CONH-2-Py-*N*-Oxide)$_2$]) with Ln = Y and Yb): Efficient Tetra-thiafulvalene-based Antenna for the Sensitization of Yb(III) Luminescence

To overcome the incertitude about the process involved in the sensitization of the luminescence of the lanthanide, the Nd(III) ion is replaced by the Yb(III) to cancel the *f-f* transitions. In addition,

TTF-CONH-2-Py-*N*-oxide ligand substitutes TTF–CH=CH–Py to obtain coordination complexes (see Sections 6.2.2 and 6.2.3).

The Y(III) analog was also synthesized to perform density function theory (DFT) and time-dependent density function theory (TD-DFT) calculations. The reaction between Ln(hfac)$_3$·2H$_2$O (Ln = Y and Yb) and the TTF-CONH-2-Py-*N*-oxide ligand in the same condition than previously described in Sections 6.2.2 and 6.2.3 gives the coordination complex of formula [Ln(hfac)$_3$(TTF-CONH-2-Py-*N*-oxide)$_2$] (Ln = Y and Yb) (Fig. 6.13).[84] These two compounds are isostructural. The asymmetric unit is composed of one Ln(III) metal center linked to eight oxygen atoms coming from three bischelate hfac$^-$ anions and two terminal donors TTF-CONH-2-Py-*N*-oxide. The coordination sphere around the Ln(III) can be described as distorted square antiprism polyhedra. The donors are coordinated through the N–O group in a *cis* conformation. The central C=C bond length of the TTF fragment [C–C = 1.338(7) Å] confirms the neutral form of the donor.

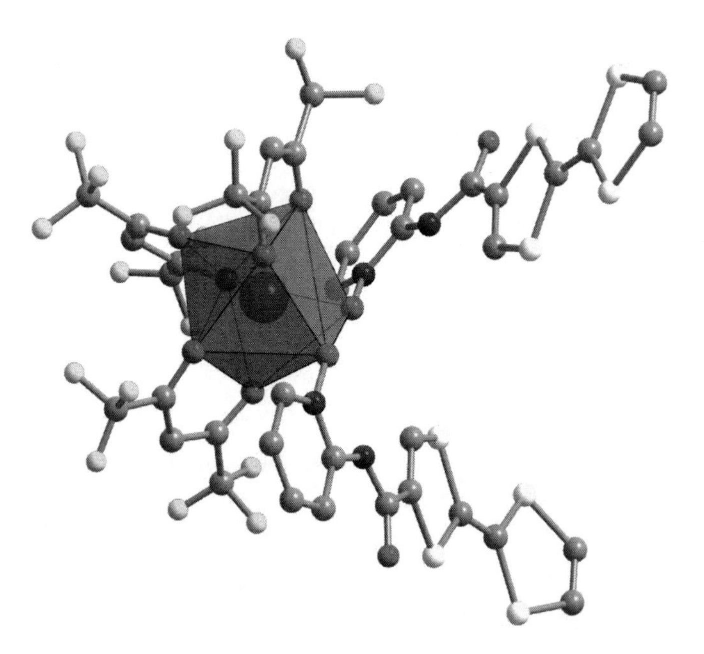

Figure 6.13 Ball-and-stick representation of the asymmetric unit of [Yb(hfac)$_3$(TTF-CONH-2-Py-*N*-oxide)$_2$] highlighting the coordination polyhedra around the Yb(III). Hydrogen atoms have been removed for clarity.

In the crystal packing, the inversion center generates two dimers of TTF-CONH-2-Py-*N*-oxide in which the shortest contacts between sulfur atoms are longer than the sum of the van der Waals radii. Nevertheless, lateral short S–S contacts are found [3.464(2) Å]. These interactions are shorter than those observed in the dinuclear complexes (Sections 6.2.2 and 6.2.3). The intra-dimers and lateral S–S contacts lead to the formation of a bidimensional organic layer while the $Ln(hfac)_3$ unit, constituting the inorganic network, is localized between the organic network.

At room temperature, $\chi_M T$ is equal to 2.49 cm^3 K mol^{-1} and shows a monotonic decrease on cooling down to 2 K. The value at 300 K is close to the expected value for one isolated Yb(III) ion (2.57 cm^3 K mol^{-1}) while the decrease of the $\chi_M T(T)$ curve is only attributed to crystal field effects. It reveals a classic behavior for such lanthanide ion.

The experimental solid state absorption spectra of the free TTF-CONH-2-Py-*N*-oxide ligand and $Ln(hfac)_3$(TTF-CONH-2-Py-*N*-oxide)$_2$] complexes are composed of 4 and 5 Gaussian deconvolutions, respectively (Fig. 6.14).

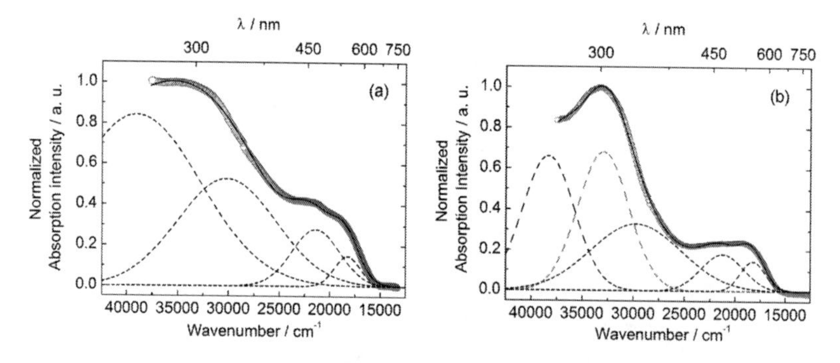

Figure 6.14 Experimental solid state (KBr pellets) UV-visible absorption spectra (open circles), respective Gaussian deconvolutions (dashed lines) and best fit (full black lines), $R = 0.9994$ (for TTF-CONH-2-Py-*N*-oxide) and R = 0.9992 [for $Yb(hfac)_3$(TTF-CONH-2-Py-*N*-oxide)$_2$]. See also Color Insert.

The molecular orbital diagram (Fig. 6.15) and the simulation of the absorption spectrum (Fig. 6.16) for the diamagnetic Y(III) analogue have been obtained by DFT and TD-DFT calculations, respectively.

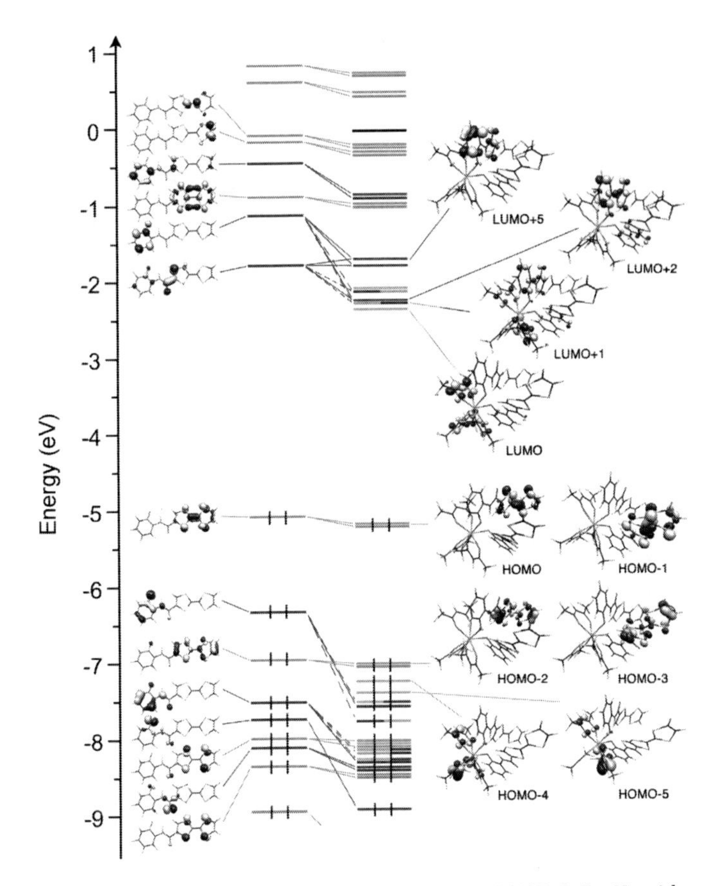

Figure 6.15 Molecular orbital diagram of TTF-CONH-2-Py-*N*-oxide and Y(hfac)$_3$(TTF-CONH-2-Py-*N*-oxide)$_2$]. Energy level of the TTF, pyridine-*N*-oxide, hfac$^-$-, and Y(III)-centered orbitals are represented in orange, blue, green, and black, respectively. See also Color Insert.

For the complex, the experimental lowest-energy deconvolution centered at the mean value of 19,600 cm^{-1} (calculated 18,611 cm^{-1}), is identified as intramolecular charge transfers HOMO LUMO+1/+2. The HOMO is centered on the TTF core while the LUMO+1 and LUMO +2 are centered on the amidopyridine-*N*-oxide acceptor (called A). The experimental deconvolution centered at 29,500 cm^{-1} is attributed to several intra-TTF moiety transitions. The most intense excitations are calculated at the following energies: 34,739, 35,802, 37,274, and 37,475 cm^{-1}. The TD–DFT calculations show that these excitations are attributed to the following transitions: HOMO-3 LUMO+7,

HOMO-1 LUMO+13/+18, and HOMO LUMO+14/+16/+17/+19. The third experimental deconvolution centered to 32,800 cm^{-1} is attributed to π–π* intra-hfac⁻ (Ihfac⁻) excitations that is mainly composed of the three HOMO-4 LUMO+2 (calculated at 34,900 cm^{-1}) and HOMO-5 LUMO+3/+4 (calculated at 36,959 and 37,826 cm^{-1}, respectively) transitions. Finally, the highest-energy experimental deconvolution (3,8600 cm^{-1}) is identified as π–π* intra-amidopyridine-*N*-oxide acceptor (IA) excitation with a participation of the π–π* intra-hfac⁻ ones. Theses excitations are attributed to the HOMO-8 LUMO+1 and HOMO-4 LUMO+5 transitions with a calculated absorption maximum at 39,743 cm^{-1}.

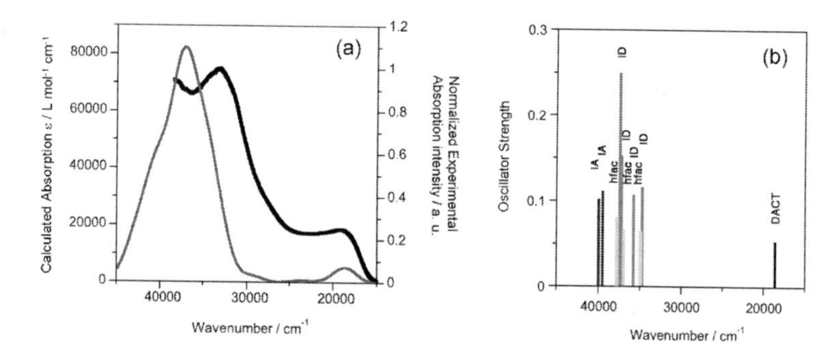

Figure 6.16 (a) Theoretical and experimental absorption spectra of Y(hfac)$_3$(TTF-CONH-2-Py-*N*-oxide)$_2$]. Solid state and solution (CH_3CN) measurements correspond respectively to black and blue lines while the red lines correspond to calculated spectra. (b) Most pertinent low-lying electronic transitions and assignment of Y(hfac)$_3$(TTF-CONH-2-Py-*N*-oxide)$_2$] where DACT, ID, hfac, and IA represent respectively the donor–acceptor charge transfers, intramolecular donor (TTF), hfac, and acceptor (amido-pyridine-*N*-oxide) transitions. See also Color Insert.

The molecular orbital diagrams of TTF-CONH-2-Py-*N*-oxide and [Y(hfac)$_3$(TTF-CONH-2-Py-*N*-oxide)$_2$] highlight the coordination effect of the Y(III) ions on the energy level of the orbitals (Fig. 6.14). The coordination stabilizes the acceptor-centered orbitals (blue levels) as expected because the coordination takes place through the N–O group of the 2-pyridine-*N*-oxide moiety, whereas the TTF-centered orbitals (orange levels) are only weakly stabilized due to the long pathway between the TTF core and the N–O group. Nevertheless, a significant electronic effect of the Y(III) coordination is observed through the amido bridge. The most important consequence of the stabilization of the acceptor-centered orbitals is the red-shift of the donor–acceptor charge transfer (DACT) in Y(hfac)$_3$(TTF-CONH-2-Py-*N*-oxide)$_2$] compared to TTF-CONH-2-Py-*N*-oxide (expected: 2260 cm^{-1}; calculated: 3500 cm^{-1}).

Upon excitation in the lower-energy DACT transitions HOMOLUMO+1/+2 (λ_{ex} = 510 nm, 19,600 cm^{-1}), both free ligand and yttrium complex exhibit a broad fluorescence band centered at 715 and 737 nm (13,986 and 13,568 cm^{-1}), respectively (Fig. 6.17).

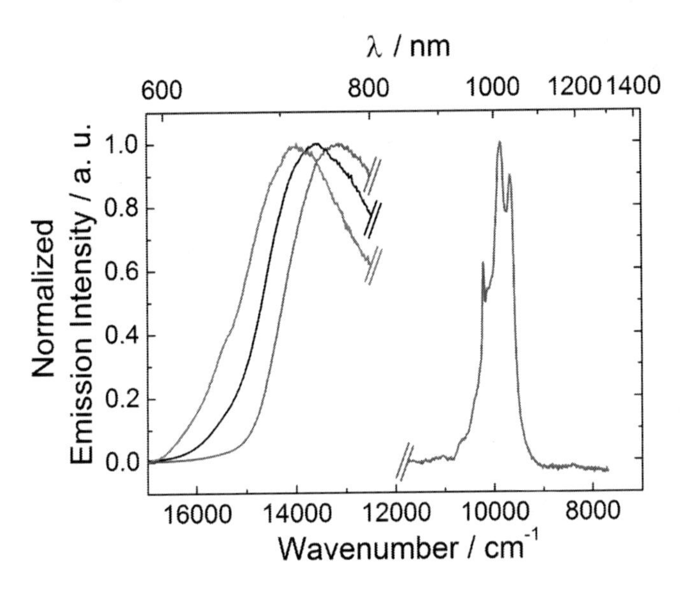

Figure 6.17 Emission spectra of TTF-CONH-2-Py-N-oxide, Y(hfac)$_3$(TTF-CONH-2-Py-*N*-oxide)$_2$, and Yb(hfac)$_3$(TTF-CONH-2-Py-*N*-oxide)$_2$ in the near-IR for λ_{ex} = 19,600 cm^{-1} at room temperature (293 K) in the solid state. See also Color Insert.

The bathochromic shift observed in the absorption spectra upon complexation is amplified in the emission spectra confirming the CT character of these transitions. TTF-fused phenazine ligand featuring strong CT transition are known to exhibit emission band in similar wavelength range and to present analogous bathochromic shift upon complexation with transition metal ions.[99,102–105] In the case of the ytterbium complex, excitation in the same CT transition (λ_{ex} = 510 nm, 19,600 cm^{-1}) induces the line shape emissions in the NIR spectral range assigned to the $^2F_{5/2} \rightarrow {}^2F_{7/2}$ (980 nm) transition centered on the ytterbium ion. In addition, the residual CT emission is observed around 760 nm (13,157 cm^{-1}) (Fig. 6.16). It is noteworthy that this residual CT emission is slightly red-shifted with respect to that of the yttrium complex. Similar bathochromic shift in the ligand CT emission along the 4*f* element raw has already been described and is generally explained by the enhanced sensitivity of luminescence transitions to the central ion Lewis acidity compared to the absorption ones. Yb(III) is smaller than Y(III), and presents a more pronounced Lewis acid character and consequently a larger red-shift in the CT emission.[101] It is noteworthy that the presence of these low-energy CT transitions allow the sensitization of Yb(III) with long wavelength up to 500 nm localized in the visible part of the spectra. The energy of the donor-excited state can be estimated from the zero-phonon transition wavelength λ_{0-0} estimated to 635 nm (15,750 cm^{-1}) from the intersection of the absorption and emission spectra. This result is in the range of the longest wavelengths reported for the Yb sensitization using antenna chromophores.

In the particular case of Yb(III), the sensitization process remains a matter of debate due to the presence of a single excited state responsible for the absorption and emission. The sensitization can proceed either through energy transfer from the triplet or the singlet excited state of a chromophore antenna or via a stepwise photo-induced electron transfer in the case of electroactive ligands. This latter process constrains to use Ln(III) ions that can be reduced at a rather modest potential. It involves the transient formation of an oxidized donor/Yb(II) species that will generate the emissive Yb(III) excited state after back electron transfer. Due to the presence of electro-active TTF containing ligand[80–81] in the $Yb(hfac)_3$(TTF-CONH-2-Py-*N*-oxide$)_2$] complex, this sensitization process must be considered. According to the above-mentioned studies,[80–81,106–110] it is possible to estimate the feasibility of the photo-induced electron transfer (ΔG_{ET}) using the extended Raehm–Weller[111] Eq. 6.2.

$$\Delta G_{ET} \geq e_0(E_{ox} - E_{red}) - E_s - w \tag{6.2}$$

where e_0 represents the elementary electron charge, E_{ox}, the oxidation potential of the electron-donating ligand, E_{red}, the reduction potential of the acceptor, and E_s the excited-state energy. In addition, w represents the stabilization energy between the different components of the ion pair (0.15 eV for a closely associated ion pair). In the present case, the oxidation potential of the TTF moieties in [Yb(hfac)$_3$(TTF-CONH-2-Py-*N*-oxide)$_2$] is estimated to 0.512 V vs. SCE from cyclic voltametry. On the other hand, the reduction potential of the couple Yb(III)(hfac)$_3$(H$_2$O)$_2$/Yb(II)(hfac)$_3$(H$_2$O)$_2$ is found to be −1.65 V vs. SCE. The energy of the excited state corresponds to the zero phonon transition, $E_s = E_{0-0} = 1.95$ eV (Eq. 6.2) becomes $\Delta G_{ET} \geq 1.(0.512 + 1.65) - 1.95 - 0.15$, hence $\Delta G_{ET} \geq 0.112$ eV. Since ΔG_{ET} is a positive value, the electron-transfer process is not thermodynamically favored suggesting that the sensitization occurs via antenna effect.

This antenna-mediated sensitization process generally involves a resonant energy-transfer process (Forster or Dexter mechanism) and, therefore, requires an overlap between the emission of the donating state and the absorption of the central metal ion.[71–74] As a general rule, the energy level of the donating state must lie at about 2–3000 cm^{-1} above that of the accepting state to optimize the energy transfer and to avoid the thermally activated back-energy transfer. In the case of ytterbium, the $^2F_{7/2} \rightarrow {}^2F_{5/2}$ absorption is located at 10,240 cm^{-1} so a good antenna should present a donating state around 12,000–13,000 cm^{-1}. Since the triplet state associated to the hfac$^-$ ligand lies around 21,370 cm^{-1} (in the analogous Gd complex),[89] the lowest excited state in [Yb(hfac)$_3$(TTF-CONH-2-Py-*N*-oxide)$_2$] is the charge transfer state. It is worth noting that CT excited state are known to present significant relaxation in their excited state, so the energy of the donating state is better described by the CT emission band taking into account its broadness.[101] In the present case (Fig. 6.17), the emission band starts with the zero-phonon transition (15,750 cm^{-1}), presents a maximum around 13,160 cm^{-1}, and its red tail can be estimated to 11,360 cm^{-1} using a rough deconvolution into Gaussian function. (The detection set-up is blind in the 800–950 nm region). The energy of the CT donating, comprised in the energy range 15,750–11,360 cm^{-1}, is located at the optimal position to sensitize efficiently the Yb(III) luminescence, and finally, the most-probable sensitization process in Yb(hfac)$_3$(TTF-CONH-2-Py-*N*-oxide)$_2$] involves a direct sensitization via the singlet CT state.[101,112–116]

6.4 Conclusion

We have demonstrated throughout this chapter that 4*f* lanthanides can be coordinated by redox-active ligands derived from TTF involving oxygenated site. Mononuclear and dinuclear coordination complexes have been obtained with different metal:ligand ratio from 1:1 to 1:3. We have also shown that the lanthanide-based compounds can involve neutral and radical cation forms of the TTF donors. In most cases, inorganic and organic moieties are segregated in two networks in the crystal lattice. The organic network can be constituted by a one-dimensional arrangement, tetramers or dimers of TTF derivatives. Lanthanides complexes are extensively studied for two main reasons — magnetism and luminescence.

Both isotropic [Gd(III) ions] and anisotropic [Tb(III) ions] are incorporated in the coordination complexes. The possibility to synthesize polynuclear complexes involving both coordinating functionalized TTFs and anisotropic lanthanides is a very exciting way to elaborate redox-active single molecule magnets.

We have explored luminescent properties of infrared emissive lanthanides, such as Nd(III) and Yb(III) in salt and coordination complex. The infra-red luminescence of these ions has been observed not only when the TTF ligands are neutral but also in their radical cation form. The active participation of the TTF-based ligand as organic chromophore has been demonstrated in the case of the sensitization of the Yb(III) luminescence.

The studies of molecular-based materials containing oxidized or mixed valence TTF-type organic donor ligand should open new perspectives in the field of multifunctional materials with the potentiality to obtain π–f luminescent conducting magnets.

References

1. Kobayashi, H., Tomita, H., Naito, T., Kobayashi, A., Sakai, F., Watanabe, T., Cassoux, P. (1996). *J. Am. Chem. Soc.*, **118**, 368.
2. Kobayashi, H., Kobayashi, A., Cassoux, P. (2000). *Chem. Soc. Rev.*, **29**, 325.
3. Kobayashi, A., Fijiwara, E., Kobayashi, H. (2004). *Chem. Rev.*, **104**, 5243 and references therein.
4. Enoki, T., Miyasaki, A. (2004). *Chem. Rev.*, **104**, 5449.
5. Kurmoo, M., Graham, A. W., Day, P., Coles, S. J., Hurtsthouse, M. B., Caufield, J. M., Singleton, J., Ducasse, L., Guionneau, P. (1995). *J. Am. Chem. Soc.*, **117**, 12209.
6. Coronado, E., Day, P. (2004). *Chem. Rev.*, **104**, 5419 and the references therein.

7. Ouahab, L., Enoki, T. (2004). *Eur. J. Inorg. Chem.*, 933.
8. Lorcy, D., Bellec, N., Fourmigué, M., Avarvari, N. (2009). *Coord. Chem. Rev.*, **253**, 1398 and references therein.
9. Fourmigué, M., Ouahab, L. (2009). Conducting and Magnetic Organometallic Molecular Materials, Springer.
10. Ojima, E., Fujiwara, H., Kato, K., Kobayashi, H., Tanaka, H., Kobayashi, A., Tokumoto, M., Cassoux, P. (1999). *J. Am. Chem. Soc.*, **121**, 5581.
11. Fujiwara, H., Ojima, E., Nakazawa, Y., Narymbetov, B., Kato, K., Kobayashi, H., Kobayashi, A., Tokumoto, M., Cassoux, P. (2001). *J. Am. Chem. Soc.*, **123**, 306.
12. Otsuka, T., Kobayashi, A., Miyamoto, Y., Kiuchi, J., Wada, N., Ojima, E., Fujiwara, F., Kobayashi, H. (2000). *Chem. Lett.*, **29**, 732.
13. Otsuka, T., Kobayashi, A., Miyamoto, Y., Kiuchi, J., Nakamura, S., Wada, N., Fujiwara, E., Fujiwara, H., Kobayashi, H. (2001). *J. Solid State Chem.*, **159**, 407.
14. Kobayashi, H., Sato, A., Arai, E., Akutsu, H., Kobayashi, A., Cassoux, P. (1997). *J. Am. Chem. Soc.*, **119**, 12392.
15. Sato, A., Ojima, E., Akutsu, H., Nakazawa, Y., Kobayashi, H., Tanaka, H., Kobayashi, A., Cassoux, P. (1998). *Chem. Lett.*, **27**, 673.
16. Tanaka, H., Kobayashi, H., Kobayashi, A., Cassoux, P. (2000). *Adv. Mater.*, **12**, 1685.
17. Uji, S., Shinagawa, H., Terashima, T., Terakura, C., Yakabe, T., Terai, Y., Tokumoto, M., Kobayashi, A., Tanaka, H., Kobayashi, H. (2001). *Nature*, **410**, 908.
18. Bouguessa, S., Gouasmia, A. K., Golhen, S., Ouahab, L., Fabre, J.-M. (2003). *Tetrahedron Lett.*, **44**, 9275.
19. Liu, S.-X., Dolder, S., Rusanov, E. B., Stoecki-Evans, H., Decurtins, S. (2003). *C. R. Chim.*, **6**, 657.
20. Liu, S.-X., Dolder, S., Pilkington, M., Decurtins, S. (2002). *J. Org. Chem.*, **67**, 3160.
21. Jia, C., Zhang, D., Xu, Y., Xu, W., Hu, U., Zhu, D. (2003). *Synth. Met.*, **132**, 249.
22. Becher, J., Hazell, A., McKenzie, C. J., Vestergaard, C. (2000). *Polyhedron*, **19**, 665.
23. Devic, T., Avarvari, N., Batail, P. (2004). *Chem-Eur. J.*, **10**, 3697.
24. Ota, A., Ouahab, L., Golhen, S., Cador, O., Yoshida, Y., Saito, G. (2005). *New. J. Chem.*, **29**, 1135.
25. Liu, S.-X., Dolder, S., Franz, P., Neels, A., Stoeckli-Evans, H., Decurtins, S. (2003). *Inorg. Chem.*, **42**, 4801.

26. Xue, H., Tang, X.-J., Wu, L.-Z., Zhang, L.-P., Tung, C.-H. (2005). *J. Org. Chem.*, **70**, 9727.
27. Benbellat, N., Le Gal, Y., Golhen, S., Gouasmia, A., Ouahab, L., Fabre, J.-M. (2006). *Eur. J. Org. Chem.*, **18**, 4237.
28. Hervé, K., Liu, S.-X., Cador, O., Golhen, S., Le Gal, Y., Bousseksou, A., Stoeckli-Evans, H., Decurtins, S., Ouahab, L. (2006). *Eur. J. Inorg. Chem.*, **17**, 3498.
29. Massue, J., Bellec, N., Chopin, S., Levillain, E., Roisnel, T., Clérac, R., Lorcy, D. (2005). *Inorg. Chem.*, **44**, 8740.
30. Pellon, P., Gachot, G., Le Bris, J., Marchin, S., Carlier, R., Lorcy, D. (2003). *Inorg. Chem.*, **42**, 2056.
31. Smucker, B. W., Dunbar, K. R. J. (2000). *J. Chem. Soc., Dalton Trans.*, **29**, 1309.
32. Devic, T., Batail, P., Fourmigué, M., Avarvari, N. (2004). *Inorg. Chem.*, **43**, 3136.
33. Avarvari, N., Fourmigué, M. (2004). *Chem. Commun.*, **40**, 1300.
34. Iwahori, F., Golhen, S., Ouahab, L., Carlier, R., Sutter, J.-P. (2001). *Inorg. Chem.*, **40**, 6541.
35. Ouahab, L., Iwahori, F., Golhen, S., Carlier, R., Sutter, J.-P. (2003). *Synth. Met.*, **133–134**, 505.
36. Jia, C., Liu, S.-X., Ambrus, C., Neels, A., Labat, G., Decurtins, S. (2006). *Inorg. Chem.*, **45**, 3152.
37. Ichikawa, S., Kimura, S., Mori, H., Yoshida, G., Tajima, H. (2006). *Inorg. Chem.*, **45**, 7575.
38. Wang, L., Zhang, B., Zhang, J. (2006). *Inorg. Chem.*, **45**, 6860.
39. Gavrilenko, K. S., Le Gal, Y., Cador, O., Golhen, S., Ouahab, L. (2007). *Chem. Commun.*, **43**, 280.
40. Cosquer, G., Pointillart, F., Le Gal, Y., Golhen, S., Cador, O., Ouahab, L. (2009). *Dalton Trans.*, **38**, 3495.
41. Pointillart, F., Le Gal, Y., Golhen, S., Cador, O., Ouahab, L. (2008). *Inorg. Chem.*, **47**, 9730.
42. Umezono, Y., Fujita, W., Awaga, K. (2006). *J. Am. Chem. Soc.*, **128**, 1084.
43. Keniley, L. K., Ray, L., Kovnir, K., Dellinger, L. A., Hoyt, J. M., Shatruk, M. (2010). *Inorg. Chem.*, **49**, 1307.
44. Zhu, Q.-Y., Liu, Y., Lu, Z.-J., Wang, J.-P., Huo, J.-B., Qin, Y.-R., Dai, J. (2010). *Synth. Met.*, **160**, 713.
45. Liu, S.-X., Ambrus, C., Dolder, S., Neels, A., Decurtins, S. (2006). *Inorg. Chem.*, **45**, 9622.

46. Hervé, K., Le Gal, Y., Ouahab, L., Golhen, S., Cador, O. (2005). *Synth. Met.*, **153**, 461.

47. Setifi, F., Ouahab, L., Golhen, S., Yoshida, Y., Saito, G. (2003). *Inorg. Chem.*, **42**, 1791.

48. Ichikawa, S., Kimura, S., Takahashi, K., Mori, H., Yoshida, G., Manabe, Y., Matsuda, M., Tajima, H., Yamaura, J. I. (2008). *Inorg. Chem.*, **47**, 4140.

49. Lu, W., Zhang, Y., Dai, J., Zhu, Q.-Y., Bian, G.-Q., Zhang, D.-Q. (2006). *Eur. J. Inorg. Chem.*, **8**, 1629.

50. Kubo, K., Nakao, A., Ishii, Y., Yamamoto, T., Tamura, M., Kato, R., Yakushi, K., Matsubayashi, G. (2008). *Inorg. Chem.*, **47**, 5495.

51. Imakubo, T., Sawa, H., Tajima, H., Kato, R. (1997). *Synth. Met.*, **86**, 2047.

52. Tamura, M., Matsuzaki, F., Nishio, Y., Kajita, K., Kitazawa, T., Mori, H., Tanaka, S. (1999). *Synth. Met.*, **102**, 1716.

53. Dyachenko, O. A., Kazheva, O. N., Gritsenko, V. V., Kushch, N. D. (2001). *Synth. Met.*, **120**, 1017.

54. Otsuka, T., Cui, H., Kobayashi, A., Misaki, Y., Kobayashi, H. (2002). *J. Solid State Chem.*, **168**, 444.

55. Kushch, N. D., Kazheva, O. N., Gritsenko, V. V., Buravov, L. I., Van, K. V., Dyachenko, O. A. (2001). *Synth. Met.*, **123**, 171.

56. Tamura, M., Yamanaka, K., Mori, Y., Nishio, Y., Kajita, K., Mori, H., Tanaka, S., Yamaura, J. I., Imakubo, T., Kato, R., Misaki, Y., Tanaka, K. (2001). *Synth. Met.*, **120**, 1041.

57. Pointillart, F., Maury, O., Le Gal, Y., Golhen, S., Cador, O., Ouahab, L. (2009). *Inorg. Chem.*, **48**, 7421.

58. Pointillart, F., Le Gal, Y., Golhen, S., Cador, O., Ouahab, L. (2009). *Chem. Commun.*, **45**, 3777.

59. Pointillart, F., Le Gal, Y., Golhen, S., Cador, O., Ouahab, L. (2009). *Inorg. Chem.*, **48**, 4631.

60. Carlin, R. L. (1986). *Magnetochemistry*, Springer, Berlin.

61. Ishikawa, N., Sugita, M., Ishikawa, T., Koshihara, S., Kaizu, Y. (2003). *J. Am. Chem. Soc.*, **125**, 8694.

62. Ishikawa, N., Sugita, M., Wernsdorfer, W. (2005). *J. Am. Chem. Soc.*, **127**, 3650.

63. Tang, J., Hewitt, I., Madhu, N. T., Chastanet, G., Wernsdorfer, W., Anson, C. E., Benelli, C., Sessoli, R., Powell, A. K. (2006). *Angew. Chem. Int. Ed.*, **45**, 1729.

64. Lin, P.-H., Burchell, T.-J., Clérac, R., Murugesu, M. (2008). *Angew. Chem. Int. Ed.*, **47**, 8848.

65. Ishikawa, N. (2007). *Polyhedron*, **26**, 2147 and references herein.
66. Ishikawa, N., Sugita, M., Wernsdorfer, W. (2005). *Angew. Chem. Int. Ed.*, **44**, 2931.
67. Cinti, F., Rettori, A., Pini, M. G., Mariani, M., Micotti, E., Lascialfari, A., Papinutto, N., Amato, A., Canechi, A., Gatteschi, D., Affronte, M. (2008). *Phys. Rev. Lett.*, **100**, 057203.
68. Chen, Z., Zhao, B., Cheng, P., Zhao, X.-Q., Shi, W., Song, Y. (2009). *Inorg. Chem.*, **48**, 3493.
69. Sabbatini, N., Guardigli, M., Manet, I. (1996). *Handbook of the Physics and Chemistry of Rare Earths*, Elsevier, Amsterdam, Vol. **23**, p. 69.
70. Comby, S., Bünzli, J.-C. G. (2007). *Handbook on the Physics and Chemistry of Rare Earths*, Elsevier BV, Amsterdam, Chapter 235, p. 37.
71. Parker, D. (2000). *Coord. Chem. Rev.*, **205**, 109.
72. Parker, D. (2004). *Chem. Soc. Rev.*, **33**, 156.
73. Bünzli, J.-C. G., Piguet, C. (2005). *Coord. Chem. Rev.*, **34**, 1048.
74. Eliseeva, S. V., Bünzli, J.-C. G. (2010). *Coord. Chem. Rev.*, **39**, 189 and references therein.
75. Lempicki, A., Samelson, H. (1936). *Phys. Lett.*, **4**, 133.
76. De Bettencourt-Diaz (2007). *Dalton Trans.*, **36**, 2229–2241.
77. Kido, J., Okamota, Y. (2002). *Chem. Rev.*, **102**, 2357.
78. Burroughes, J. H., Bradley, D. D. C., Holmes, A. B. (1990). *Nature*, **347**, 539.
79. Faulkner, S., Pope, S. J. A., Burton-Pye, B. P. (2004). *Appl. Spec. Rev.*, **39**, 1.
80. Faulkner, S., Burton-Pye, B. P., Khan, T., Martin, L. R., Wray, S. D., Skabara, P. J. (2002). *Chem. Commun.*, **16**, 1668.
81. Pope, S. J. A., Burton-Pye, B. P., Berridge, R., Khan, T., Skabara, P., Faulkner, S. (2006). *Dalton Trans.*, **35**, 2907.
82. Richardson, M. F., Wagner, W. F., Sands, D. E. (1968). *J. Inorg. Nucl. Chem.*, **30**, 1275.
83. Bondi, A. (1964). *J. Phys. Chem.*, **68**, 441.
84. Pointillart, F., Cauchy, T., Maury, O., Le Gal, Y., Golhen, S., Cador, O., Ouahab, L. (2010). *Chem.-Eur. J.*, **16**, 11926.
85. King, R. B. (1969). *J. Am. Chem. Soc.*, **91**, 7211.
86. Kahn, O. (1993). *Molecular Magnetism*, VCH, Weinhem.
87. Sutter, J.-P., Kahn, M. L. (2005). *Magnetism: Molecules to Materials V*, VCH.

88. Bünzli, J. C. in Bünzli, J. C. G., Chopin, G. R. (eds.) (1989). *Lanthanide Probes in Life, Chemicals and Earth Sciences: Theory and Practice*. Elsevier, Amsterdam, p. 219.

89. Eliseeva, S. V., Ryazanov, M., Gumy, F., Troyanov, S. I., Lepnev, L. S., Bünzli, J.-C. G., Kuzmina, N. P. (2006). *Eur. J. Inorg. Chem.*, **23**, 4809.

90. Panagiotopoulos, A., Zafiropoulos, T. F., Perlepes, S. P., Bakalbassis, E., Masson-Ramade, I., Kahn, O., Terzis, A., Raptopoulou, C. P. (1995). *Inorg. Chem.*, **34**, 4918.

91. Avecilla, F., Platas-Iglesias, C., Rodriguez-Cortinas, R., Guillemot, G., Bünzli, J.-C. G., Brondino, C. D., Geraldes, C. F. G. C., De Blas, A., Rodriguez-Blas, T. (2002). *Dalton Trans.*, **31**, 4658.

92. Rinehart, J. D., Harris, T. D., Kozimor, S. A., Bartlett, B. M., Long, J. R. (2009). *Inorg. Chem.*, **48**, 3382.

93. Ebihara, M., Nomura, M., Sakai, S., Kawamura, T. (2007). *Inorg. Chim. Acta*, **360**, 2345.

94. Mech, A. (2008). *Polyhedron*, **27**, 393.

95. Scout, J. S. (1988). In: Highly conducting quasi-one-dimensional organic crystals, Conwell, E. (ed.), *Semiconductors and Semimetals*, Academic Press, New York, Vol. **27**, pp. 385–436.

96. Benelli, C., Gatteschi, D. (2002). *Chem. Rev.*, **102**, 2369.

97. Pointillart, F. (2005). Magnetism and Chirality: From the Tridimensional Oxalate Based Magnets to Verdazyl Based Chains, Thesis of the Marie Curie University.

98. Andreu, R., Malfant, I., Lacroix, P. G., Cassoux, P. (2000). *J. Eur. Org. Chem.*, **7**, 737.

99. Jia, C., Liu, S.-X., Tanner, C., Leiggener, C., Neels, A., Sanguinet, L., Levillain, E., Leutwyler, S., Hauser, A., Decurtins, S. (2007). *Chem. Eur. J.*, **13**, 3804.

100. Yang, C., Fu, L.-M., Wang, Y., Zhang, J.-P., Wong, W.-T., Ai, X.-C., Qiao, X.-F., Zou, B.-S., Gui, L.-L. (2004). *Angew. Chem. Int. Ed.*, **43**, 5010.

101. D'Aléo, A., Picot, A., Beeby, A., Williams, J. A. G., Le Guennic, B., Andraud, C., Maury, O. (2008). *Inorg. Chem.*, **47**, 10258.

102. Sénéchal, K., Hemeryck, A., Tancrez, N., Toupet, L., Williams, J. A. G., Ledoux, I., Zyss, J., Boucekkine, A., Guégan, J.-P., Le Bozec, H., Maury, O. (2006). *J. Am. Chem. Soc.*, **128**, 12243.

103. Wang, B., Wasielewski, M. R. (1997). *J. Am. Chem. Soc.*, **119**, 12.

104. Sénéchal, K., Toupet, L., Ledoux, I., Zyss, J., Le Bozec, H., Maury, O. (2004). *Chem. Commun.*, **40**, 2180.

105. Goze, C., Dupont, N., Beitler, E., Leiggener, C., Jia, H., Monbaron, P., Liu, S.-X., Neels, A., Hauser, A., Decurtins, S. (2008). *Inorg. Chem.*, **47**, 11010.

106. Horrocks De Jr., W. W., Bolender, J. P., Smith, W. D., Supkowski, R. M. (1997). *J. Am. Chem. Soc.*, **119**, 5972.

107. Supkowski, R. M., Bolender, J. P., Smith, W. D., Reynolds, L. E. L., Horrocks De Jr., W. W. (1999). *Coord. Chem. Rev.*, **185–186**, 307.

108. Beeby, A., Faulkner, S., Williams, J. A. G. (2002). *Dalton Trans.*, **31**, 1918.

109. Lazarides, T., Alamiry, M. A. H., Adams, H., Pope, S. J. A., Faulkner, S., Weinstein, J. A., Ward, M. D. (2007). *Dalton Trans.*, **36**, 1484.

110. Lazarides, T., Tart, N. M., Sykes, D., Faulkner, S., Barbieri, A., Ward, M. D. (2009). *Dalton Trans.*, **38**, 3971.

111. Rehm, D., Weller, A. (1970). *Isr. J. Chem.*, **8**, 259.

112. Huang, W., Wu, D., Guo, D., Zhu, X., He, C., Meng, Q., Duan, C. (2009). *Dalton Trans.*, **38**, 2081.

113. Hebbink, G. A., Grave, L., Woldering, L. A., Reinhoudt, D. N., van Veggel, F. C. J. M. (2003). *J. Phys. Chem. A.*, **107**, 2483.

114. Werts, M. H. V., Hofstraat, J. W., Geurts, F. A. J., Verhoeven, J. W. (1997). *Chem. Phys. Lett.*, **276**, 196.

115. Ziessel, R., Ulrich, G., Charbonnière, L., Imbert, D., Scopelliti, R., Bünzli, J.-C. G. (2006). *Chem. Eur. J.*, **12**, 5060.

116. Shavaleev, N. M., Scopelliti, R., Gumy, F., Bünzli, J.-C. G. (2008). *Eur. J. Inorg. Chem.*, **9**, 1523.

Chapter 7

Multifunctional Materials of Interest in Molecular Electronics

M. Laura Mercuri, Paola Deplano, Angela Serpe, and Flavia Artizzu

Dipartimento di Chimica Inorganica e Analitica, Università di Cagliari, S.S. 554-Bivio per Sestu, I09042 Monserrato-Cagliari, Italy

mercuri@unica.it

7.1 Introduction

As the physical limits of conventional silicon-based electronics are approached, the search for new types of materials that can deliver smaller devices grows up. Molecular electronics, which uses assemblies of individual molecules to reproduce conventional structures, such as switches or semiconductors, represents the new frontier. In this context, molecule-based materials, namely, materials built from predesigned molecular building blocks, play a key role since they are known to exhibit many technologically important properties (e.g., magnetic ordering, conductivity, superconductivity), traditionally considered to be solely available for classic atom-based inorganic solids such as metals, alloys, or oxides. Their relevance in material science is mainly due to the tunability of their

Multifunctional Molecular Materials
Edited by Lahcène Ouahab

ISBN 978-981-4364-29-4 (Hardcover), 978-981-4364-30-0 (eBook)
www.panstanford.com

physical properties by conventional synthetic methods; molecular materials in fact are obtained through soft routes, traditionally from organic chemistry, coordination chemistry, and supramolecular chemistry, and this opens unprecedented possibilities to the design of molecules with the desired size, shape, charge, polarity, and electronic properties, in response to the changing demands of technology. The area of molecular materials with interesting technological properties started almost 75 years ago with the discovery of the first complexes showing spin crossover transitions $[Fe(S_2NCR_2)_3]$.[1] Since then molecular materials have given rise to complexes with spin crossover transition[2] semiconductors, metals and superconductors,[3] ferrimagnets, ferromagnets and weak ferromagnets,[4] chromophores,[5] including those for nonlinear optics (NLO) and Visible-NIR (NearInfrared) emitters based on lanthanide complexes with polyconjugated ligands.[6] Because of their versatility and peculiar optical, magnetic, and conducting properties, molecule-based materials are appealing candidates for practical applications in post-silicon molecular electronics and spintronics (a new paradigm of electronics based on the spin degree of freedom of the electron).[7] A summary of their most appealing applications in technologies such as in electronic, magnetic, and photonic devices, ranging from data storage to telecommunication/information technology (switches, sensors, etc.)[8] is reported in Scheme 7.1.

Scheme 7.1

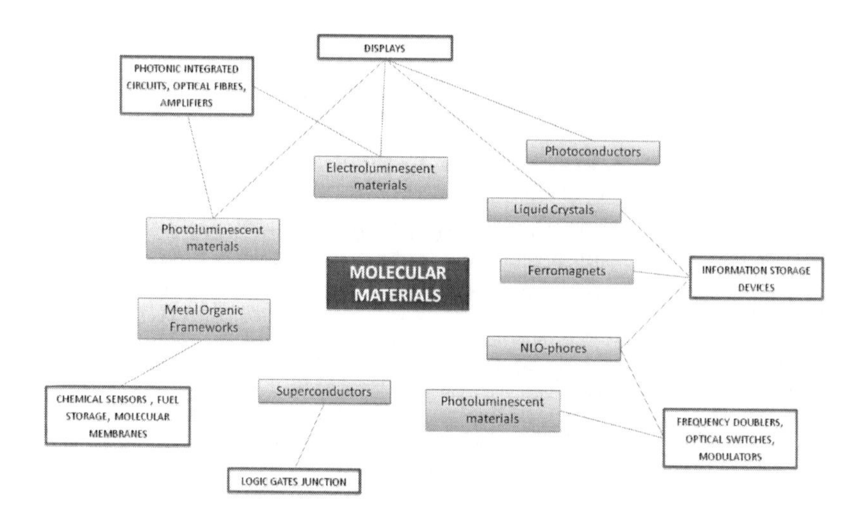

In addition to monofunctional materials, a more recent and appealing goal in the field of molecule-based materials is the search for multifunctionality in these systems — proper selection of molecular building blocks yields the combination (or even the interplay) in the same crystal lattice of two or more physical properties, such as magnetism and conductivity or optical properties, which are difficult or impossible to achieve in conventional inorganic solids. In this aspect the possibilities offered by the molecular bottom-up approach are unprecedented and many interesting and challenging combinations of physical properties can be envisaged.

Because of the great diversity of metals, ligands, structures, and physical properties, it is difficult to present a unified picture of what has been accomplished so far and several reviews have been reported. Cassoux and Faulmann[9] and Robertson,[10] separately, reported on $M(dmit)_2$ and isologs, $M(mnt)_2$, $M(dddt)_2$ systems; Kato[11] analyzed the electronic structure of some of these systems, especially $M(dmit)_2$, by discussing geometrical aspects, such as dimensionality, dimerization, and frustration as well as the pressure effect on crystal and electronic structures and also on the metallic and superconducting states. Kobayashi and coworkers[12] reported on single-component molecular metals (SCMMs) with extended-TTF dithiolate ligands. Coronado and Day[13] discussed on the transport properties of magnetic molecular conductors formed by TTF derivatives (especially BEDT-TTF) and tetrahalometalate, trisoxalatometalate (III) anions, polyoxometalate clusters, and chain anions such as $M(mnt)_2$ dithiolenes. Recently, Coronado and Galán-Mascarós[14] reviewed all aspects of the molecular ferromagnetic conductors built from the combination, in a single compound of organic cationic radicals that are able to give rise to conducting architectures, with polymeric anionic metal complexes, that are able to give rise to ferromagnetism. The correlation between the intermolecular interactions modes, in particular, van der Waals interactions (S···S, S···Se, ... shorter than the van der Waals radii), π–π, and À–d interactions, H-bonding, and the physical properties of these materials have been recently highlighted by some of us.[15] A summary of the molecular building blocks of the above-mentioned molecular materials with technologically important physical properties is presented in Chart 7.1.

Chart 7.1

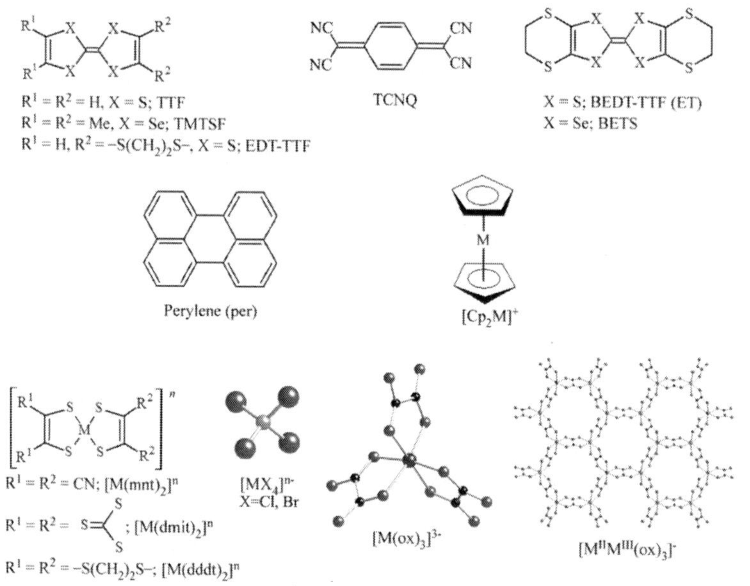

$R^1 = R^2 = H$, X = S; TTF
$R^1 = R^2 = Me$, X = Se; TMTSF
$R^1 = H$, $R^2 = -S(CH_2)_2S-$, X = S; EDT-TTF

TCNQ

X = S; BEDT-TTF (ET)
X = Se; BETS

Perylene (per)

$[Cp_2M]^+$

$R^1 = R^2 = CN$; $[M(mnt)_2]^n$
$R^1 = R^2 =$ S=C(S)(S); $[M(dmit)_2]^n$
$R^1 = R^2 = -S(CH_2)_2S-$; $[M(dddt)_2]^n$

$[MX_4]^{n-}$
X=Cl, Br

$[M(ox)_3]^{3-}$

$[M^{II}M^{III}(ox)_3]^-$

The complexity of these materials makes the design of the molecular building blocks, the control of the intermolecular interactions, and the crystallization techniques crucial for obtaining functional materials exhibiting technologically useful physical properties. In this chapter, we will report on relevant examples based on chalcogenolene transition metal complexes showing magnetic and/or conducting properties whose structural features or physical properties are unusual with respect to analogous compounds reported in the literature up to now.

7.2 Magnetic Molecular Conductors

From a physical point of view, molecular materials which combine conducting (π-electrons) and magnetic (localized d electrons) properties have attracted major interest since they can exhibit a coexistence of two distinct physical properties furnished by the two networks or novel and improved properties with respect to those of the two networks, due to the interactions established between them. The development of these π–d systems as multifunctional materials represents one of the main targets in current material science for their potential applications for future molecular electronics. Important milestones in the area of molecular magnetic conductors have been achieved using molecular building blocks as the BEDT-TTF organic donor or its selenium derivatives and charge-compensating anions ranging from simple mononuclear complexes $[MX_4]^{n-}$ (M = Fe^{III}, Cu^{II}; X = Cl, Br) and $[M(ox)_3]^{3-}$ (ox = oxalate = $C_2O_4^{2-}$) with tetrahedral and octahedral geometry, to layered structures such as the bimetallic oxalate complexes $[M^{II}M^{III}(ox)_3]^-$ (M^{II} = Mn, Co, Ni, Fe, Cu; M^{III} = Fe, Cr). Three representative hybrid materials have been synthesized up to now: (i) paramagnetic/superconductor $(BEDT\text{-}TTF)_4[(H_3O)M^{III}(ox)_3]S$ (M^{III} = Cr, Fe, and Ga; S = C_6H_5CN, $C_6H_5NO_2$, PhF, PhBr, etc.);[16a–f] (ii) antiferromagnetic/superconductor κ-$(BETS)_2[FeBr_4]$;[17] and (iii) ferromagnetic/metal, the $(BEDT\text{-}TTF)_3[MnCr(ox)_3]$[18] being the most recent and significant advance in this field. The most successful synthetic procedure for obtaining these materials, the electrocrystallization technique, involves the slow oxidation of the organic donors to form radical cations that crystallize with the charge-compensating inorganic counterions. Typically, the structure of these materials is formed by segregated stacks of the self-assembled organic donors and the inorganic counterions, which add the second functionality to the conducting material. The intermolecular interactions, in particular, van der Waals interactions (S···S, S···Se, shorter than the van der Waals radii), π–π, and π–d interactions, H-bonding, play a crucial role in self-assembling these predesigned molecular units and may provide a powerful way to afford layered mono- or multifunctional molecular materials with new or unknown physical properties. The first paramagnetic superconductor $(BEDT\text{-}TTF)_4[H_3OFe^{III}(ox)_3]\cdot C_6H_5CN$[16a] and the first ferromagnetic conductor $(BEDT\text{-}TTF)_3[CrMn(ox)_3]$[18] were successfully obtained by reacting, via electrocrystallization, the mononuclear chiral $[Fe(ox)_3]^{3-}$ and the $[Cr^{III}Mn^{II}(ox)_3]_n$ (2-D honeycomb with oxalate bridges) anions with the BEDT-TTF organic donor and carriers, of magnetism and conductivity, respectively. Furthermore, by reacting the BETS molecule with the zero-dimensional $FeCl_4^-$ anion, a field-induced superconductivity with π–d interaction was observed, which may be mediated through S···Cl interactions between the BETS molecule and the anion.[19]

In this section, the engineering skills to control the packing of the organic network and therefore the conducting properties by playing with the size, shape, symmetry, and charge of the charge-compensating inorganic counterions will be presented. As far as we know, tris-chelated metal complexes with octahedral geometry are the most successful counterions for favoring the metallic state because of their capability to segregate the stacks compared to square-planar complexes such as d^8 metal dithiolenes, which instead favor alternate stacks and thus the semiconducting or insulating state. In addition, they have the possibility of a specific assembly order of Λ and Δ chirality that may influence the packing and thus the physical properties of the material, as well as to introduce functionalities such as magnetic properties through both the metal and a suitably tailored ligand. The role of intermolecular interactions in determining the packing pattern and therefore the physical properties in these materials will be highlighted as well.

7.2.1 Paramagnetic Conductors

By combining TTF and BETS molecules with the one-dimensional (1-D) $[Fe^{III}(ox)Cl_2^-]_n$ polymeric magnetic chain anions, dual-function materials were recently obtained by Zhang and coworkers,[20a] containing both the oxalate ligand and the Cl^- anion with the aim to introduce both the magnetic interaction by the oxalate bridges between moment carriers and the π–d interaction through S···Cl contacts. When using the BETS molecule, the first weak-ferromagnetic organic conductor κ-$(BETS)_2[Fe^{III}$-$(ox)Cl_2]$ **(1)** was obtained. The structure consists of BETS molecules stacks, arranged in a 2-D κ-phase, slightly different from that of the κ- $(BETS)_2[FeBr_4]$;[17] there are Se···Se contacts within the BETS dimers and Se···S, S···S between dimers in one layer where the 1-D $[Fe^{III}(ox)Cl_2^-]_n$ chain anions fits in the channel between the two BETS layers (Fig. 7.1). κ-$(BETS)_2[Fe^{III}(ox)Cl_2]$ is a molecular conductor with σ_{300K} = 10^2 S cm^{-1} and it retains the metallic state down to 2 K, in good agreement with the calculated band structures.[21] The susceptibility data in the 400–425 K range were fitted to a model of a 1-D antiferromagnetic S = 5/2 chain plus a contribution from a Curie impurity as well as a temperature independent term. The long-range magnetic ordering below 5 K can only be ascribed to the presence of π–d interaction mediated by S···Cl interaction between BETS molecules and anions, because in principle, no long-range order is possible at any temperature for a 1-D magnetic system.

κ-$(BETS)_2[Fe^{III}$-$(ox)Cl_2]$ is the first organic conductor showing weak ferromagnetism deriving from the two sublattices.

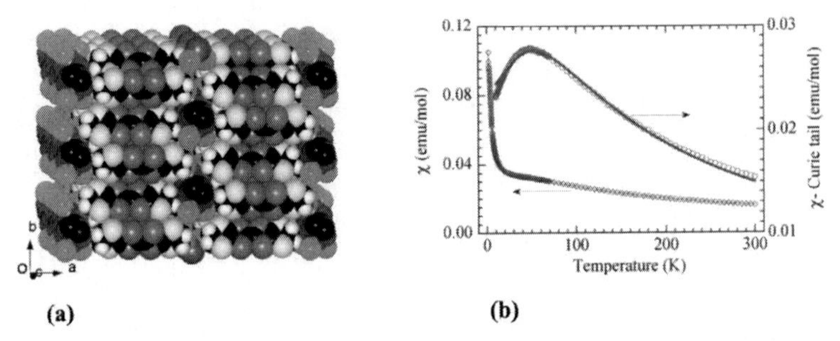

Figure 7.1 Structure of κ-$(BETS)_2[Fe^{III}(ox)Cl_2]$**(1)**: a space-filling view of the packing of BETS layers and $[Fe^{III}(ox)Cl_2^-]_n$ chains viewed along the *c*-axis. Color scheme: C, black; H, white; O, red; S, yellow; Cl, green; Fe, brown-green; Se, orange (left). Experimentally observed susceptibility (green diamonds), after correction for the Curie impurity contribution (red circles) and the expected susceptibility for a 1-D Heisenberg, $S = 5/2$ antiferromagnet (blue line) (right). Reprinted from Ref. 21. Copyright 2006, American Chemical Society. See also Color Insert.

When using the peculiar $[Fe(tdas)_2]^-$ dithiolene anion which exhibits unusual magnetic properties, co-existence of delocalized and localized unpaired *d* electrons has been also found in the $(TTF)_2[Fe(tdas)_2]$ **(2)** (tdas is 1,2,5-thiadiazole-3,4-dithiolate).[22] Its tetrabutylammonium salt, TBA$[Fe(tdas)_2]$,[23] which exhibits a dimeric structure and magnetic properties typical of an antiferromagnetic dimer, shows two unusual phase transitions in the 190–240 K range. These transitions have been explained as a re-entrant phase behavior, where the low- and high-temperature phases are identical while a second phase with a different value of the antiferromagnetic exchange coupling constant exists at intermediate temperatures. Structural data and magnetic measurements show that this compound consists of antiferromagnetic $[Fe(tdas)_2]_2^{2-}$ dimers separated by organic TTF layers and behaves as a semiconductor. An increase in the number of *S* atoms in the TTF molecule should increase the intermolecular interactions leading to metal-like behavior. Thus, when the BEDT-TTF molecule, which contains eight sulfur atoms (twice the number of TTF) and an extended π-system, is used, the obtained (BEDT-TTF)$_2[Fe(tdas)_2]$ **(3)**[24] still shows semiconducting behavior over the whole temperature range, but the extended sulfur framework of BEDT-TTF, when compared to TTF, allows for increased side-to-side sulfur–sulfur interactions, leading to improved electrical conductivity: room temperature conductivity for the BEDT-TTF salt is about 1 S cm^{-1} as compared to 0.03 S cm^{-1} for the TTF analogue. Substitution of the *S* atoms by Se is expected to result in increased polarizability and orbital overlap between molecules, leading to stable metallic states. In fact, by using BETS, where four of the eight atoms of ET are replaced by Se, metal-like behavior should be achieved. $(BETS)_2[Fe(tdas)_2]_2$**(4)**,[25] which consists of segregated columns of dimers of BETS and columns of dimers of $[Fe(tdas)_2]$, shows metal-like behavior. Numerous chalcogen–chalcogen contacts are observed within and between the columns, responsible for the high room temperature conductivity (0.2 S cm^{-1}) and the metallic character, which is observed down to *ca.* 200 K.

One-dimensional lateral interactions via S···N contacts were observed in the (BEDT-TTF)$[Ni(tdas)_2]$**(5)**,[26] where monomeric $[Ni(tdas)_2]^-$ monoanions are present. The magnetic coupling, in fact, can be tuned by controlling the anisotropic intermolecular interactions. As shown in Fig. 7.2, for example, the 1-D, 2-D, and 3-D magnetic exchange pathways within the crystals are defined by the following arrangements between the $[Ni(dmit)_2]^-$ anions: (a) π-stacking, (b, c) lateral S···S contacts along the short or long axes, and (d) the orthogonal π-overlap.[27] The π-stacking and uniform lateral S···S chain modes along the long axes (Fig. 7.2a, c) of the anions yield antiferromagnetic spin coupling ($J_{AF} < 0$), while the uniform anion arrangement along the short axis (Fig. 7.2b) can generate ferromagnetic spin coupling ($J_{AF} > 0$) within the crystals.

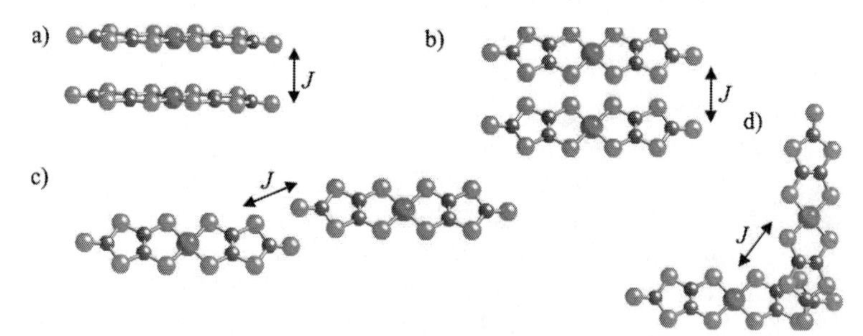

Figure 7.2 Intermolecular interaction modes between the $[Ni(dmit)_2]^-$ anions: (a) π-dimer, (b) lateral S···S interactions along the short axis of the $[Ni(dmit)_2]^-$ anions, (c) lateral S···S interactions along the long axis of the $[Ni(dmit)_2]^-$ anions, and (d) orthogonal π-overlap. Reprinted from Ref. 27. Copyright 2009, American Chemical Society.

The (BEDT-TTF)[Ni(tdas)$_2$] salt forms a layered structure: the first layer contains dimerized BEDT-TTF molecules and isolated [Ni(tdas)$_2$]$^-$ monoanions (A-type anions); the second layer contains chains of [Ni(tdas)$_2$]$^-$ monoanions (B-type anions). According to the structure, the susceptibility data were fitted by a regular $S = 1/2$ antiferromagnetic chain (formed by the B-type anions) and a monomeric contribution coming from the A-type anions. The antiferromagnetic intermolecular coupling between the B-type [Ni(tdas)$_2$]$^-$ units suggests that, although small, there is a non-negligible interaction between these anions, in agreement with the presence of short interanion S–N distances (S23···N21 = 3.265(5)) observed in the crystal structure (Fig. 7.3). The BEDT-TTF sublattice does not contribute to the magnetic properties as it consists of strongly antiferromagnetically coupled (BEDT-TTF)$_2^{2+}$ dimers as shown by the EPR spectra, magnetic susceptibility measurements, diffuse reflectance, and vibrational spectroscopy. This salt shows a semiconductor–semiconductor transition at about 200 K that may be attributed to an ordering of the disordered terminal ethylene group of the BEDT-TTF molecule (Fig. 7.4).

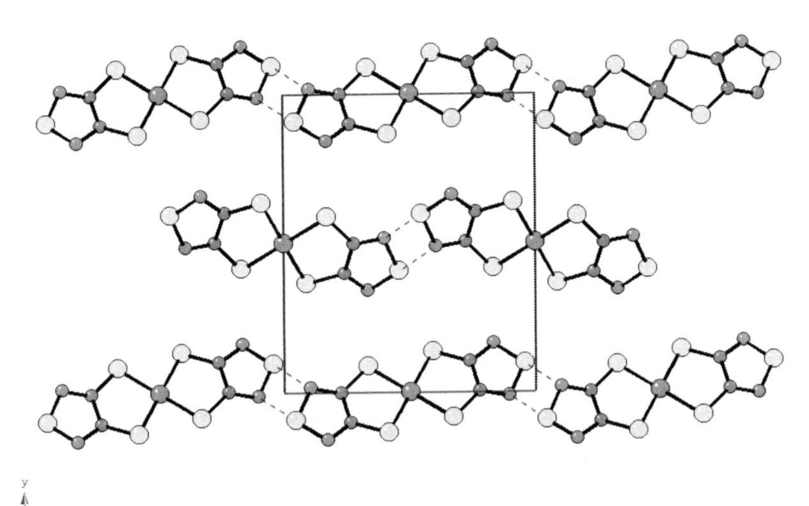

Figure 7.3 View of the [Ni(tdas)$_2$]$^-$ B anionic layer down the *a*-axis. Thermal ellipsoids are drawn at the 20% probability level. Dashed lines indicate the intermolecular contacts shorter than the sum of the van der Waals radii. Adapted from Ref. 26. Copyright 2004, American Chemical Society.

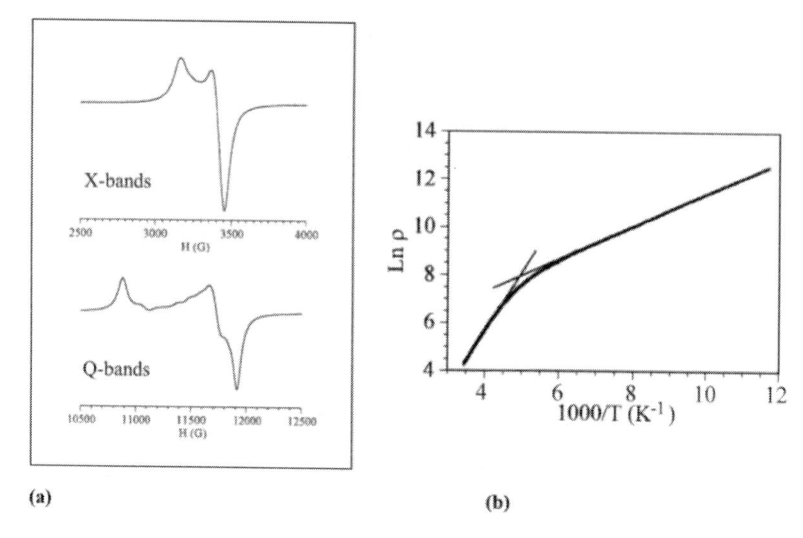

Figure 7.4 (a) EPR spectra at room temperature of a polycrystalline sample of **(5)**. (b) Plot of the logarithm of the dc electrical resistivity of **(5)** as a function of reciprocal temperature showing the two semiconducting regimes and the transition between them at approximately 200 K (right). Adapted from Ref. 26. Copyright 2004, American Chemical Society.

It is noteworthy how the charge-compensating molecular cation controls the network formation and determines the conducting properties of the resulting molecular material.

When using *N*-methyl-3,5-diiodopyridinium(Me-3,5-DIP), the introduction of supramolecular halogen I···S interactions into the $Ni(dmit)_2$ complex has yielded a peculiar magnetic supramolecular conductor, (Me-3,5-DIP)$[Ni(dmit)_2]_2$**(6)**,[28] without localized *d* moments. Systems without localized 3*d* moments have been scarcely reported. The only example is the –$Per_2M(mnt)_2$ series (Per = perylene; mnt = maleonitriledithiolate; M = Ni, Pd, Pt, Fe, Cu, Au, and Co), exhibiting 1-D metallic behavior and paramagnetism with antiferromagnetic (AFM) interactions, which are derived from molecular π-electrons of Per_2^+ and $M(mnt)_2$, respectively.[29] These compounds undergo a metal–insulator (MI) transition in the low-temperature region (T = 8.2 K). Two kinds of layers of $Ni(dmit)_2$ anions are present in (Me-3,5-DIP)$[Ni(dmit)_2]_2$ as shown in Fig. 7.5.

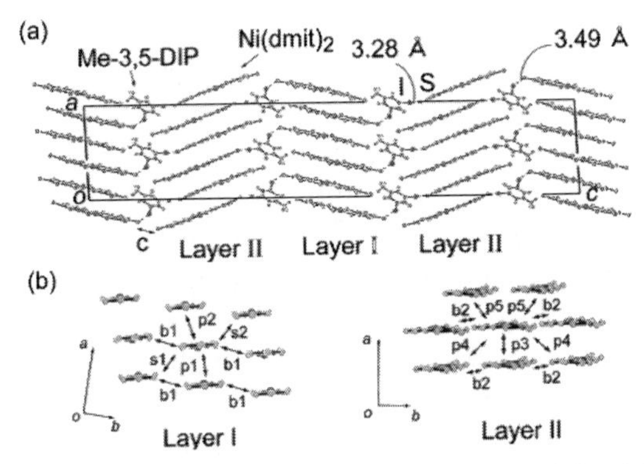

Figure 7.5 Crystal structure of (Me-3,5-DIP)[Ni(dmit)$_2$]$_2$: (a) side view and (b) end-on projection of Ni(dmit)$_2$ anions. Supramolecular I···S interactions are indicated by dotted lines. Reprinted from Ref. 28. Copyright 2007, American Chemical Society.

The considerable difference in the molecular packing of Ni(dmit)$_2$ anions in Layers I and II is due to the "nonequivalence" between the two sides of the cation layer: every methyl group of the cation is projected toward Ni(dmit)$_2$ anions in Layer I, while the opposite side faces those in Layer II. The supramolecular I···S interactions fix the cation orientations with their strong associations, strongly leading to the two kinds of layers. These findings can play a crucial role in the strategies to design multifunctional molecular systems. Two-dimensional metallic conductivity and paramagnetism with AFM interactions down to 4.2 K (Fig. 7.6) are both derived from molecular π-electrons of Ni(dmit)$_2$ anions contained separately in each layer.

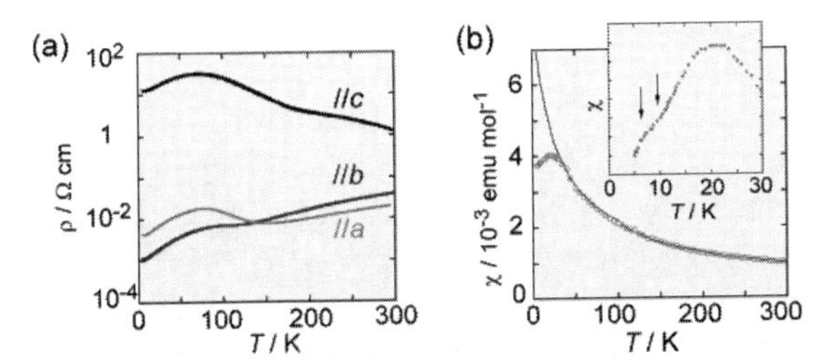

Figure 7.6 (a) Anisotropic temperature dependence of electrical resistivity for (Me-3,5-DIP)[Ni(dmit)$_2$]$_2$ and (b) magnetic susceptibility for (Me-3,5-DIP)[Ni(dmit)$_2$]$_2$. The solid line represents the Curie–Weiss fitting to experimental data. The data below 40 K is expanded in the inset, where the arrows mark anomalies at 6 and 10 K. Reprinted from Ref. 28. Copyright 2007, American Chemical Society.

According to band calculations, a 2-D metallic behavior is observed in the *ab*-plane. The resistivity along the *b*-axis decreases monotonically with decreasing temperature while the one along the *a*-axis also shows an essentially metallic behavior down to 4.2 K. The plot of the magnetic susceptibility (χ) against temperature is shown in Fig. 7.6b. The paramagnetic Curie–Weiss behavior presumably arises from the $Ni(dmit)_2$ anions because the cation $(Me\text{-}3,5\text{-}DIP)^+$ is diamagnetic (closed shell). The anomalies observed at 6 and 10 K (Fig. 7.5b, inset) in the χ vs. T curve presumably arise from magnetic transitions. This is the first example of "one" kind of molecule playing "two" contrasting roles where conducting and magnetic π-electrons coexist down to 4.2 K.

7.2.2 Paramagnetic Chiral Conductors

Introduction of chirality into conducting systems allows the preparation of multifunctional materials in which the chirality may modulate the structural disorder and influence the solid-state conducting properties. Structural chirality, in fact, is a property that has not been exploited in this series of materials up to now. Different families of chiral tetrathiafulvalene (TTF) derivatives have been recently reviewed by Avarvari and Wallis[30] together with the corresponding synthetic strategies. We discuss herein other systems where the use of achiral TTFs with chiral anions has provided molecular conductors, although in most of them, to date, the anion is present in the racemic form. Starting from some of these building blocks (shown in Chart 7.2), several chiral conductors have been prepared and, in two cases, involving either chiral TTF-oxazoline salts or BEDT-TTF salts with metal-oxalate anions and chiral solvent, differences between the conductivity of the racemic and enantiopure forms have been found as a consequence of the structural disorder. Further developments in this field are expected to be addressed especially toward helical architectures, possibly based on supramolecular chirality and systems combining conductivity, magnetism, and chirality in both organic and inorganic lattices. In all examples reported so far with anions such as $[M(ox)_3]^{3-}$ (M = Fe^{3+}, Cr^{3+}, Ga^{3+}), in the extensive series of metals and superconductors $[BEDT\text{-}TTF]_4[(A)M(ox)_3]\cdot S$ (A = H_3O^+, NH_4^+; S = neutral guest molecule);[13,31] $[Cr(2,20\text{-}bipy)(ox)_2]^-$ in the semiconducting salt $[BEDT\text{-}TTF]_2[Cr(2,20\text{-}bipy)(ox)_2]$;[32] and TRISPHAT in 1:1 insulating salts,[33] the anions were present in the crystal as a racemic mixture of Δ and Λ enantiomers.

Chart 7.2

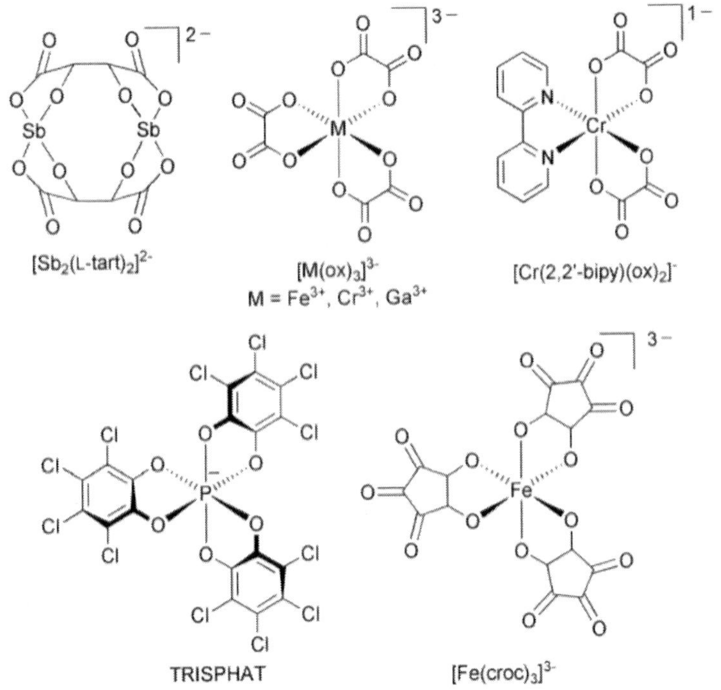

$[Sb_2(L\text{-tart})_2]^{2-}$

$[M(ox)_3]^{3-}$
M = Fe^{3+}, Cr^{3+}, Ga^{3+}

$[Cr(2,2'\text{-bipy})(ox)_2]^-$

TRISPHAT

$[Fe(croc)_3]^{3-}$

Clues for designing the molecular packing of the organic network, carrier of conducting properties were provided with the new paramagnetic and chiral anion $[Fe(croc)_3]^{3-}$ (croc = croconate, $C_5O_5^{2-}$), the magnetic component of the novel dual-function materials α-(BEDT-TTF)$_5$[Fe(croc)$_3$]·5H$_2$O **(7a)** and β-(BEDT-TTF)$_5$[Fe(croc)$_3$]·C$_6$H$_5$CN **(7b)**.[34] This anion contains the croconate ligand whose coordination modes and ability to mediate magnetic interactions[35] are very similar to the oxalate one. The structure of both compounds consists of alternating layers of BEDT-TTF molecules separated by layers of $[Fe(croc)_3]^{3-}$ anions and solvent molecules (water for **7a**, C_6H_5CN for **7b**) molecules where a shortest interanion O···O distance (O4–O5) of 3.122 Å is found. The BEDT-TTF molecules in **7a** are arranged in a herring-bone packing motif, typical of the α-phase, which is induced by the chirality of the anions. The

$[Fe(croc)_3]^{3-}$ anions form zigzag rows, separated by water molecules and along the *c* direction with Δ enantiomers alternating with the Λ ones, giving rise to an achiral structure, are shown in Fig. 7.7. Thus, each Δ enantiomer induces a "right-turned" column of BEDT-TTF molecules in the organic layer above and a "left-turned" column in the layer below. Of course, the Λ enantiomer induces the opposite packing scheme. This effect is possible, thanks to the presence of a supramolecular interlayer S···O cation–anion interaction (S15–O5 = 3.01 Å), much shorter than the sum of the corresponding van der Waals radii (3.32 Å) together with the steric effects. Therefore, the alternation of Δ and Λ enantiomers along the *b*-axis is the driving force that induces the same alternation of "right" and "left" turned columns along this direction, giving rise to the α-packing observed in the organic sublattice in **7a**. As far as we know, **7a** represents the first example of chirality-induced α-phase and the only known pentamerized (θ_{51}) phase. No influence of the chirality of the anions instead has been found in **7b**.

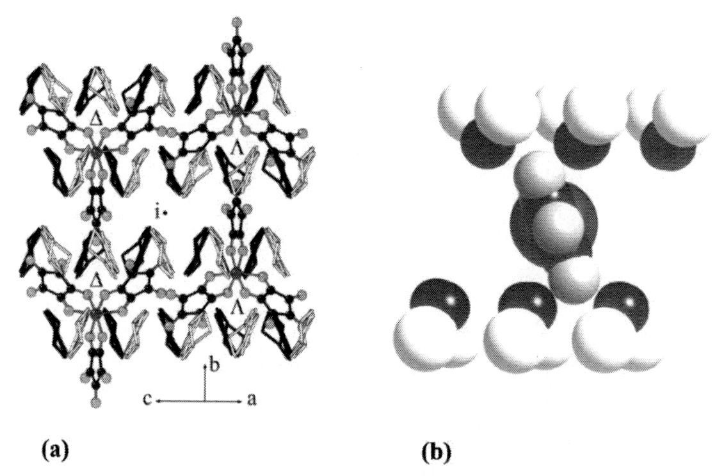

Figure 7.7 (a) View of the anionic layer of **7a** showing the zigzag rows of anions with alternating chirality and the organic layers above (white) and below (black) the anionic layer. The point indicates the location of the inversion center in the anionic layer. (b) View of the croconate ligand located over the BEDT-TTF chains and the different orientation of the BEDT-TTF chains induced above and below it (only the terminal $-S-CH_2-CH_2-S-$ groups of the BEDT-TTF molecules are shown for clarity). Reprinted from Ref. 34. Copyright 2007, American Chemical Society.

Several intermolecular interaction are found in **7b** between the organic and inorganic layers and the solvent molecules. Besides the short S···N contact [3.21(2) Å] between one BEDT-TTF molecule and the N atom of the C_6H_5CN solvent molecule, there are up to five S···O cation–anion interactions in the range 2.887–3.208 Å, shorter than the sum of the van der Waals radii (3.32 Å). Although the room temperature conductivity of **7a** (*ca.* 6 S cm^{-1}) is quite high, the thermal variation of the resistivity shows a continuous decrease with a decrease in temperature and shows a classical semiconducting regime with an activation energy of 116 meV. The thermal variation of the resistivity, reported in Fig. 7.8, shows that **7b** is metallic down to *ca.* 120 K. Below this temperature, the conductivity becomes thermally activated, although it does not follow the Arrhenius law ($\ln\sigma \propto T^{-1}$) nor a hopping regime where $\ln\sigma \propto T^{-\alpha}$ (with α = 1/3 or 1/4). Furthermore, the re-entrance to the metallic state observed below *ca.* 20 K supports the idea that the minimum in the resistivity at *ca.* 120 K is not a true metal–semiconductor transition as supported by the current dependence of this re-entrance.

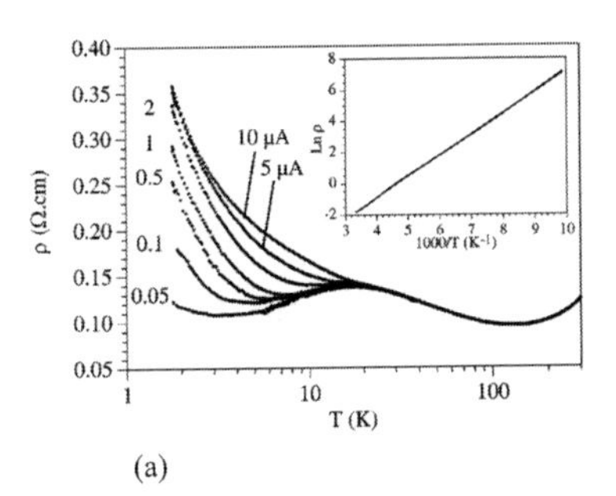

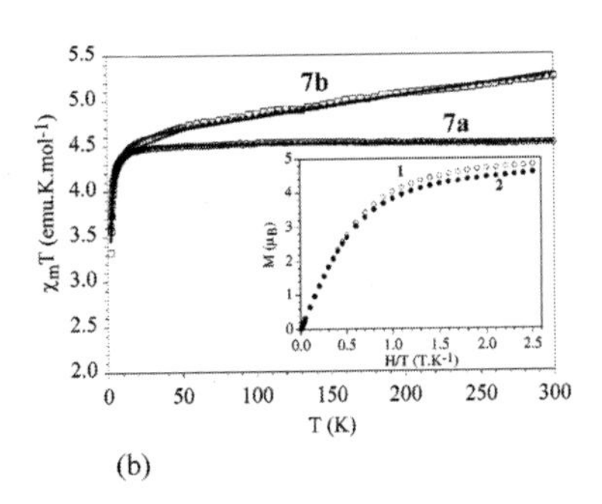

Figure 7.8 (a) Thermal variation of the dc resistivity at different current intensities (in μA) for **7b**; inset: Arrhenius plot of the thermal variation of the resistivity of **7a**. (b) Thermal variation of the $\chi_m T$ product for **7a** and **7b**. Solid lines represent the best fit to a model of isolated S = 5/2 Fe(III) ions with a ZFS. Dashed lines represent the best fit to a simple Curie–Weiss law. For **7b,** both models include a Pauli paramagnetism. The inset shows the isothermal magnetization at 2 K for **7a** and **7b**. Reprinted from Ref. 34. Copyright 2007, American Chemical Society.

This kind of behavior, although unusual, has already been observed in other related BEDT-TTF salts with the chiral anions $[M^{III}(ox)_3]^{3-}$ (M^{III} = Ga, Fe, and Cr).[13,16] The susceptibility data indicates that **7a** and **7b** are $S = 5/2$ paramagnets but **7b** shows an extra temperature-independent paramagnetic contribution (Pauli-type paramagnetism) typical of metallic salts. **7b** is one of the very few examples of paramagnetic molecular metals and the only known example out of the $[M(ox)_3]^{3-}$ and $[MX_4]^{n-}$ series.

These findings suggest the use of enantiomeric pure forms of the chiral anions to induce novel (chiral) packings in the organic layers leading to different conducting behaviors. In the two polymorphs of the racemic salt $[BEDT\text{-}TTF]_4[(H_3O)Cr(ox)_3]\cdot C_6H_5CN$ **(8)**,[36] obtained, in fact, by taking advantage of the chirality of the $[Cr(ox)_3]^{3-}$ anion, superconducting or semiconducting behavior has been observed, depending on the spatial distribution of the Δ and Λ enantiomers. In the superconducting salt, each anion layer contains only a single enantiomer, while the adjacent anion layers contain only the opposite enantiomer giving a Δ-Λ-Δ-Λ-Δ-Λ-Δ-Λ repeating pattern (the continuous line – represents a BEDT-TTF layer) and an overall racemic lattice. In contrast, in the semiconducting salt, every anion layer contains a 50:50 mix of Δ and Λ enantiomers in alternating rows, giving a ΔΛ-ΔΛ-ΔΛ-ΔΛ pattern and thus an overall racemic lattice. A successful approach used to synthesize chiral crystals in the series of BEDT-TTF salts with tris(oxalato)metallate is to incorporate chiral guest molecules in the anion layer, which can lead to different donor packing motifs and thus to very different conducting properties. As an example, when the guest molecule is racemic (R/S)- or chiral (S)-sec-phenethyl alcohol, two BEDT-TTF salts are formed, which are isostructural and show differences in the metal-insulating properties due to the enantiomeric disorder observed in the (R/S)-salt vs. the (S)-salt.[37] By using *chiral induction*, i.e., by crystallizing from a chiral solvent, two polymorphs of a chiral conductor formulated as $(BEDT\text{-}TTF)_3NaCr(C_2O_4)_3\cdot CH_3NO_2$ **(9)** have been obtained by electrocrystallization of the racemic $Na_3Cr(ox)_3$ with BEDT-TTF and nitromethane in the presence of (R)-carvone as supporting electrolyte.[38] Attempts to grow crystals containing the opposite enantiomer of tris(oxalato)chromiate(III) by using (S)- instead of (R)-carvone by the same method did not produce crystals. The structure of $(BEDT\text{-}TTF)_3Na[\Delta\text{-}Cr(ox)_3]_{0.64}[\Lambda\text{-}Cr(ox)_3]_{0.36}\cdot CH_3NO_2$**(9a)** consists of alternating layers of BEDT-TTF radical organic cations and anion layers (Fig. 7.9a). The anion layer (Fig. 7.9b) contains $Cr(ox)_3^{3-}$, Na^+, and nitromethane and adopts the usual hexagonal packing found in the $(BEDT\text{-}TTF)_4[(A)M(ox)_3]\cdot$

S[39] series. In all the previous examples in the series, there has been a 50:50 mix of Δ and Λ enantiomers in an overall racemic lattice. The use of the chiral electrolyte clearly has an effect upon the crystal growth favoring a chiral structure.[38a] There is a disorder between the Cr and Na positions (64:36) clearly indicating that the structure contains predominantly one enantiomer.

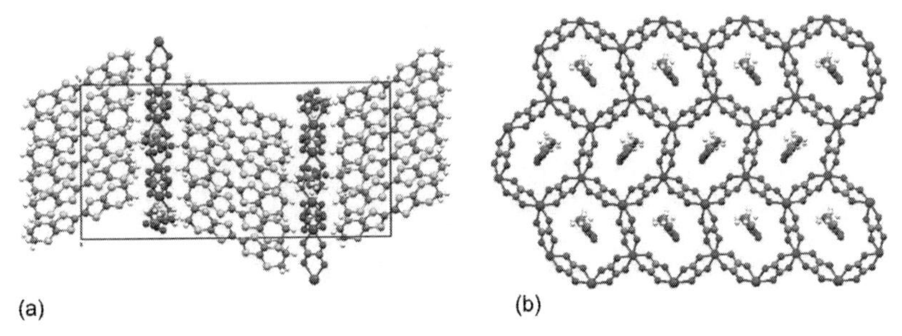

Figure 7.9 (a) Layered structure of **9a** viewed down the *a*-axis. (b) Inorganic layer of **9a** viewed down the *c*-axis. Reprinted from Ref. 38a. Copyright 2008, Royal Society of Chemistry (RSC).

The BEDT-TTF molecules are arranged in a unique packing motif (Fig. 7.10). Two of the molecules are face-to-face (A and B) while the TTF plane of the third molecule (C) is at 90° with respect to the other molecules. The face-to-face A and B BEDT-TTFs have a twisted ethylene conformation at one end and a boat conformation at the other, whilst the C molecule has a twisted conformation at both ends, which are eclipsed. There are several side-to-side contacts: (i) between A and B molecules in adjacent pairs (3.49(1)–3.58(1) Å; (ii) between C and A [the shortest, 3.46(1) Å], and adjacent C BEDT-TTFs [3.50(1) Å]. Compound **9b** instead is a racemic twin as indicated by structural parameters (Flack parameter of **9b**, 0.48(7); **9a**, 0.02(5), chiral structure). The packing motif of the BEDT-TTF is also novel— two BEDT-TTF molecules (A and B) are packed parallel to one another, but unlike in **9a**, A and B are not face-to-face (Fig. 7.10). The third molecule, C, is twisted at 45° with respect to the others. There are no face-to-face close S···S contacts below the sum of the van der Waals radii (<3.6 Å) between the A and B molecules but there are two side-to-side close contacts [3.59(1); 3.57(1) Å]. Similar to **9a**, the shortest S···S contacts are between BEDT-TTF C and either A or B [3.48(1)–3.58(1) Å].

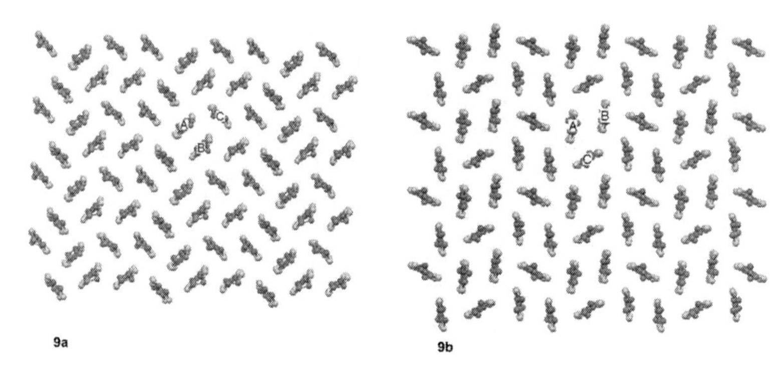

Figure 7.10 BEDT-TTF molecular packing of **9a** and **9b** viewed down the *c*-axis. Reprinted from Ref. 38a. Copyright 2008, RSC.

Conductivity measurements show that both **9a** and **9b** polymorphs are semiconductors with activation energies of 80 and 79 meV, respectively. Room temperature resistivities were $\rho_{RT} = 2$ and 22 Ω·cm, respectively. These systems represent the first chiral examples in this large family of materials. While the synthesis of chiral organic donor molecules has proved to be time consuming and expensive, chiral molecular conductors can be prepared by reacting achiral organic donors such as the BEDT-TTF molecule with inexpensive electrolytes and racemic anions. For the first time in the BEDT-TTF tris(oxalato)metallate series of charge transfer salts, it has been demonstrated that the use of a chiral solvent has been crucial for inducing chirality starting from a racemic precursor. This opens the route to the synthesis of chiral conducting materials and to study the effects of chirality upon metallic and even superconducting materials.

7.2.2.1 Ferromagnetic conductors

Several ferromagnetic metals[14] have been reported so far. Recently Coronado *et al.*[40] reported on the synthesis of the first molecular material showing coexistence of ferromagnetism, metal-like conductivity, and chirality by using the organic/inorganic molecular approach. A chiral BEDT-TTF (ET) derivative, (*S,S,S,S*)-tetramethyl-ET (TM-ET)[41], and the polymeric bimetallic oxalate (ox)-bridged complex, $[MnCr(ox)_3]^-$, which is responsible for the appearance of ferromagnetic ordering, has been used for preparing a new ferromagnetic molecular conductor with the double goal of increasing the functionality of the system and avoiding the disorder of the ethylenic groups in the organic layer. $[TM\text{-}ET]_x[MnCr(ox)_3]\cdot CH_2Cl_2$**(10)** consists of alternating layers of chiral organic radical cations, arranged in the typical β packing motif (Fig. 7.11), which are responsible for both the electrical conductivity and the optical activity, and the inorganic bimetallic oxalate-bridged honeycomb complexes, which are responsible for the occurrence of ferromagnetic properties.

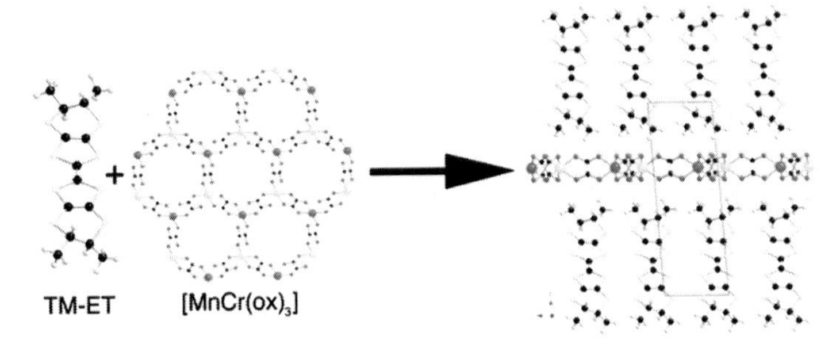

Figure 7.11 Crystal packing of $[TM\text{-}ET]_x[MnCr(ox)_3]\cdot CH_2Cl_2$. Reprinted from Ref. 40. Copyright 2010, American Chemical Society.

The room temperature conductivity value was 65 S cm^{-1}, and the temperature dependence of the resistance showed metal-like behavior down to ~190 K, where a minimum was reached (Fig. 7.12a).

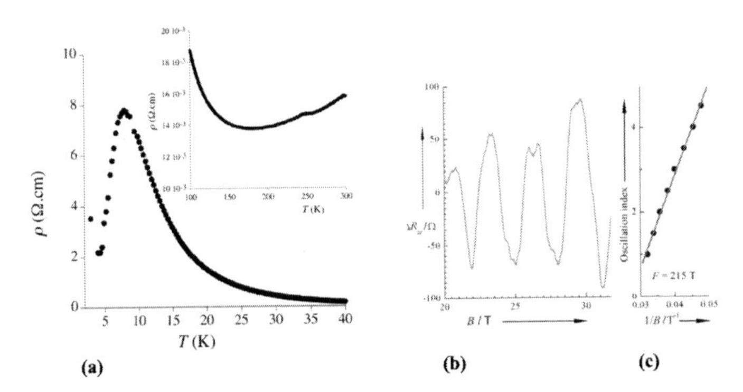

Figure 7.12 (a) Typical thermal dependence of the in-plane resistivity for crystals of **(10)**. (b) Shubnikov-de Haas oscillations at T = 550 mK. (c) Plot of the oscillation index vs. inverse field position, whose slope gives the frequency (i.e., cross-sectional area of the Fermi surface). Adapted from Ref. 40. Copyright 2010, American Chemical Society.

This change in slope was not abrupt, suggesting that no phase transition into a semiconducting state was reached but that a charge localization process appeared. Below 10 K, re-entrance into a metal-like regime occurred. Near 5 K, at the onset of ferromagnetic ordering, a second minimum was reached, and the resistivity increased again very rapidly. However, weak Shubnikov-de Haas (SdH) quantum oscillations were observed (Fig. 12b,c), suggesting the presence of small metallic regions exhibiting Fermi liquid-like behavior in coexistence with insulating domains at very low temperatures.

7.2.3 Single-Component Paramagnetic Conductors

SCMMs should greatly extend the development of new types of molecular conductors. All molecular metals developed before the year 2000 were, in fact, composed of two components: (i) the molecules forming the electronic band (designated by A) and (ii) a chemical species (designated by B); the formation of a conduction band and the generation of charge carriers by the intermolecular charge-transfer between A and B seemed two essential requirements to ensure a metallic state in these systems.[12] To generate a metallic or semimetallic band in single-component molecules, the energy separation between HOMO and LUMO should be small enough to make the HOMO and LUMO bands overlap each other through 2-D or 3-D van der Waals intermolecular interactions and to form partially filled bands. Therefore, the molecules must have a very small HOMO–LUMO gap (ΔE) and large intermolecular interactions, especially transverse S···S contacts. The role of the van der Waals interactions is therefore crucial in determining the solid state physical properties as well as the choice of the molecular building blocks. Square-planar neutral metal bisdithiolene complexes with ligands containing TTF-moiety seem to be the best candidates to be investigated since the first evidence of a very small ΔE from *ab initio* MO calculations performed on the $[Ni(ptdt)_2]$ **(11)** (ptdt = bis(propylenedithio)tetrathiafulvalene dithiolate) dithiolene complex.[42] The square–planar coordination geometry of metal bisdithiolenes is expected to favor π–π interactions while extended ligands are also expected to make more accessible different oxidation states. Theoretical calculations have shown that ΔE will tend to be small with increasing size on extended-TTF ligand system. Moreover, the TTF-like structure is crucial for obtaining large transverse intermolecular interactions because the HOMO of the TTF-like donor has the same sign on every sulfur atom, and all intermolecular contacts through sulfur atoms enhance the intermolecular interaction by contributing additively, as already mentioned, in the Section 7.1.[12]

7.2.3.1 SCMMs containing a TTF moiety

Tanaka and coworkers[43] have succeeded in preparing $[Ni(tmdt)_2]$ **(12)** (tmdt = trimethylenetetrathiafulvalene dithiolate), the first SCMM, which exhibits a stable metallic state down to 0.6 K and σ_{rt} = 400 S cm^{-1} on single crystal. The metallic properties have been explained on the basis of a closely packed structure and a very small HOMO–LUMO gap. The structural, electrical, and magnetic properties of $[Cu(dmdt)_2]^0$ **(13)** (dmdt = dimethyl-tetrathiafulvalenedithiolate),[44] have also been reported with the aim to clarify the possibility of obtaining single-component molecular conductors with magnetic properties. The molecular structure of the neutral complex **13** is nonplanar, while the dmdt ligand moiety is almost ideally planar. As shown in Fig. 7.13, the

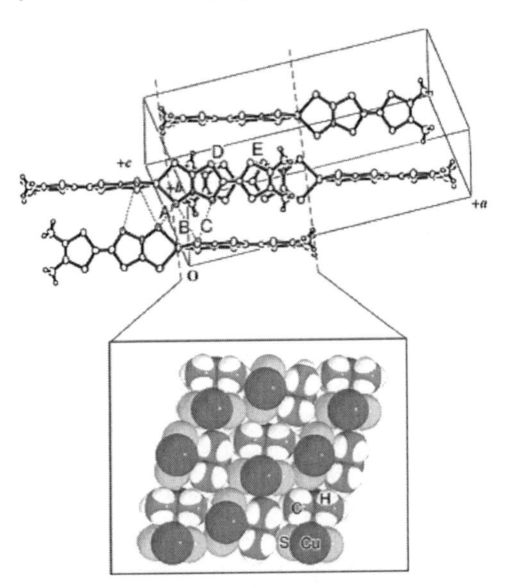

Figure 7.13 Molecular arrangement of **13**. The lower part of the figure presents the arrangement of ligand moieties projected along the molecular long axis. The S···S contacts are shown as dotted lines. Adapted from Ref. 44. Copyright 2002, American Chemical Society.

neutral molecules take an unprecedented molecular arrangement, which is completely different from the molecular arrangement in $[Ni(tmdt)_2]^0$, where, ideally, planar molecules are closely packed to form a 3-D metallic state.[43] One of the ligands of **13** overlaps, face-to-face, with the ligand of the adjacent molecule, and the opposite-side ligand overlaps with the ligand of the third molecule. Thus, the dmdt ligands show a packing motif similar to that of "κ-type organic superconductors".[45] S···S short contacts exist between adjacent molecules, as shown in Fig. 7.13, not only in the layer with a κ-type ligand arrangement, but also between the adjacent layers. These structural features suggest the possibility of a 3-D electrical conduction pathway. The resistivity measurements on the crystals of **13** showed semiconducting behavior with a room temperature conductivity value of about 3 S cm^{-1} and a very small activation energy (~40 meV).

The temperature dependence of the susceptibility of **13** was fitted by a Curie–Weiss plot, and the Curie constant value suggests the existence of 84% of $S = 1/2$ spin moments, which have been estimated on the basis of a g-value (2.035) obtained by EPR experiments. Therefore, a single-component molecular semiconductor incorporating magnetic moments has been obtained by reacting Cu(II) with the dmdt ligand.

Recently $[Au(tmdt)_2]$ **(14)**[46], which is isostructural with $[Ni(tmdt)_2]$ and highly conductive (50 S cm^{-1} at room temperature) down to 10 K, has been investigated by means of 1 H NMR for its unexpected magnetic behavior. The spectral broadening and relaxation-rate enhancement at 110 K provide microscopic evidence that $[Au(tmdt)_2]$ undergoes a magnetic phase transition at around 110 K, an extraordinarily high temperature among organic conductors (the quasi-two-dimensional system κ-$(BEDT\text{-}TTF)_2Cu[N(CN)_2]Cl$ shows a T_N at around 27 K) and comparable indeed to that of the typical inorganic antiferromagnetic Mn metal ($T_N \approx 100$ K). As the susceptibility does not show a sudden increase below this temperature,[47] the material cannot be ferromagnetic. The magnitude and temperature dependence of relaxation rate indicates that $[Au(tmdt)_2]$ is an unconventional metal with an antiferromagnetic order. It is well-known that low-dimensionality depresses a phase transition and causes spin contraction due to quantum fluctuations. $[Au(tmdt)_2]$ compound is a 3-D system, as shown by the Fermi surface topology,[48] and it is possible that 3-D causes the high transition temperature and the large ordered moment.[49] The crystal structure of [Au(tmdt)$_2$] was recently investigated[50] by powder X-ray diffraction experiments in the temperature range of 9–300 K using a synchrotron radiation source. Structural anomalies associated with the antiferromagnetic transition were observed around the transition temperature T_N = 110 K. The shortest intermolecular S···S distance along the a-axis shows a sharp decrease at around T_N, while distinct structure anomalies were not observed along the direction perpendicular to the (021) plane. These results suggest that the molecular arrangement in only the (021) plane changes significantly at T_N. Thus, the intermolecular spacing shows anomalous temperature dependence at around T_N only along that direction where the neighboring tmdt ligands have opposite spins in the antiferromagnetic spin structure model recently derived from *ab initio* band structure calculations (Fig. 7.14). The results of single-crystal four-probe resistance measurements on extremely small crystals (~25 μm) did not show a distinct resistance anomaly at T_N. The Au–S bond length decreases sharply at around 110 K, and this is consistent with the proposed antiferromagnetic spin distribution model, where the left and right ligands of the same molecule possess opposite spin polarizations.

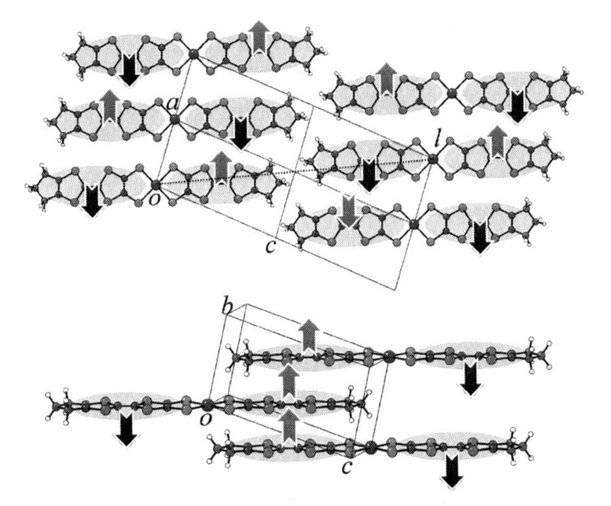

Figure 7.14 Antiferromagnetic spin-structure model derived from *ab initio* electronic band structure calculations.[50] The arrows indicate up and down spin distributions. Reprinted from Ref. 50. Copyright 2009, American Chemical Society.

New types of TTF-ligands fused with six-member rings such as cyclohexene and 1,4-dioxene, which have structural flexibility like the BEDT-TTF donor, have been used to obtain novel single-component highly conducting systems, $(R_4N)_n[Ni(chdt)_2]$ [R = Me, n = 2 **(15a)**; R = ^{n}Bu, n = 1 **(15b)**; n = 0 **(15c)**; chdt = cyclohexenotetrathiafulvalenedithiolate] and $(R_4N)_n[Ni(eodt)_2]$ [R = Me, n = 2 **(16a)**; R = ^{n}Bu, n = 1 **(16b)**; n = 0 **(16c)**; eodt = ethylenedioxytetrathiafulvalenedithiolate][51] (Chart 7.3).

Chart 7.3

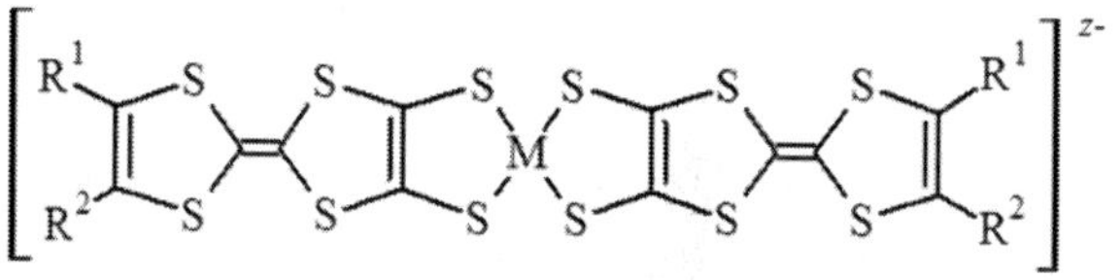

R^1^R^2 = –S(CH$_2$)$_3$S–; $[M(ptdt)_2]^{z-}$, M=Ni **(11)**

R^1^R^2 = –CH$_2$CH$_2$CH$_2$–; $[M(tmdt)_2]^{z-}$, M=Ni **(12)**, Au **(14)**

R^1 = R^2 = Me; $[M(dmdt)_2]^{z-}$, M=Cu **(13)**

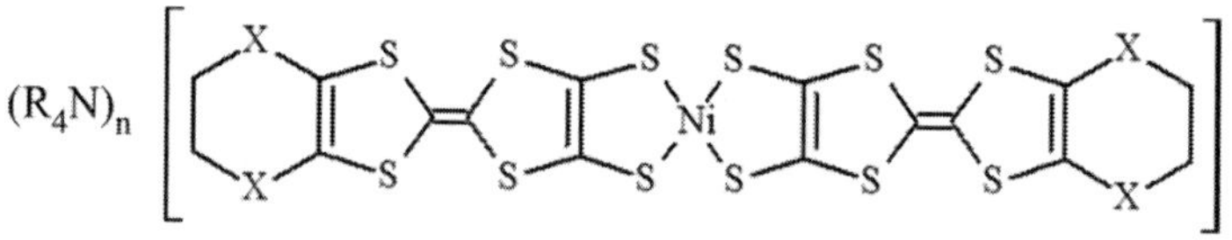

$(R_4N)_n[Ni(chdt)_2]$ **(15)**
a R = Me; X = CH$_2$; n = 2
b R = Bu; X = CH$_2$; n = 1
c X = CH$_2$; n = 0

$(R_4N)_n[Ni(eodt)_2]$ **(16)**
a R = Me; X = O; n = 2
b R = Bu; X = O; n = 1
c X = O; n = 0

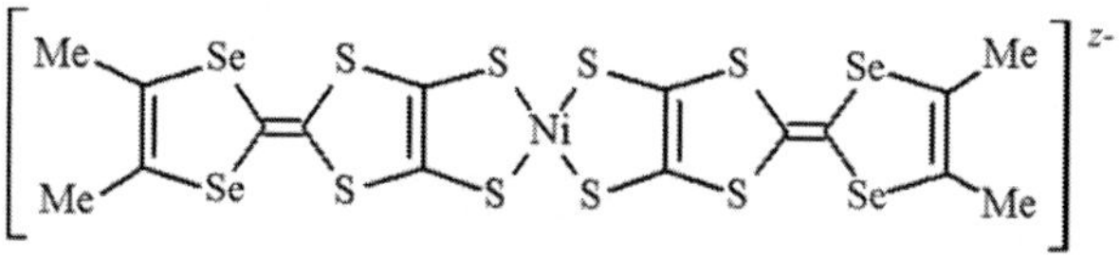

$[Ni(dmstfdt)_2]^-$ **(17)**

X-ray data are available only for **15b** and **16b,** which show sandwiched structures in which the chains or layers of the nickel complexes and cations are arranged alternately. The anions form a 1-D array along the c-axis, and between adjacent anions, there is one intermolecular S···S contact [3.646(1)Å] indicating side-by-side interactions along the *c*-axis. Interchain or interlayer S···S contacts less than the van der Waals radii were observed only in the transverse direction. The neutral species **15c** and **16c** showed large room temperature conductivity (σ_{rt} = 1–10 S cm^{-1}) measured on pressed pellets of samples. Complex **16c** showed metallic temperature dependence down to 120 K and retained high conductivity down to 0.6 K, thus being considered a new SCMM. The magnetic behavior of **15b**, shown in Fig. 7.15a, is in good agreement with the Bonner-Fisher model with J/kB = 28 K, showing the existence of an antiferromagnetic 1-D Heisenberg chain[52] in agreement with the arrangement of anions with regular distance along the 1-D chain (*c*-axis). Complex **16b** shows Curie–Weiss paramagnetism with antiferromagnetic interactions between the S = 1/2 states (Fig. 7.15b).

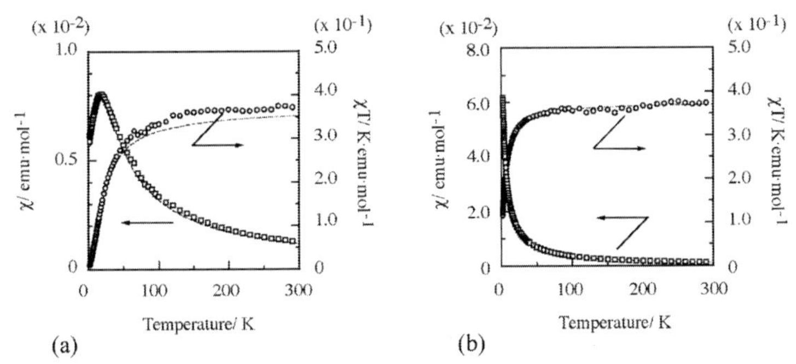

Figure 7.15 Temperature dependence of magnetic susceptibilities (χ) of polycrystals of the monoanionic species (a) (^{n}Bu$_4$N)[Ni(chdt)$_2$] and (b) (^{n}Bu$_4$N)[Ni(eodt)$_2$] at 5 kOe. The dotted lines indicate fitting curves of the Bonner–Fisher model (J/kB)-28 K for (a) and the Curie–Weiss law (C) 0.376 K·emu·mol^{-1} and θ = –4.6 K for (b), respectively. Reprinted from Ref. 51. Copyright 2004, American Chemical Society.

Recently, the new (^{n}Bu$_4$N)[Ni(dmstfdt)$_2$] **(17)** (dmstfdt = dimethyldiselenadithiafulvalene-dithiolate; Scheme 7.2) Ni complex has been synthesized and fully characterized.[53] This compound is a unique ambivalent molecular system exhibiting weakly metallic behavior above room temperature and weak ferromagnetism of localized spins at low temperature, despite the 1:1 stoichiometry similar to **15b** and **16b**. The crystallographically independent

$[Ni(dmstfdt)_2]^-$ anions (A and B) are arranged along the stacking *a*-axis alternately and the dihedral angle of 42.6° between the molecular planes of adjacent anions produce a weakly dimerized zigzag array, which form an anion layer parallel to the *ac*-plane. There are several intermolecular S···S(Se) contacts between the anions of the neighboring zigzag arrays in addition to many contacts between the adjacent anions in the zigzag array, leading to a 2-D network of intermolecular interactions in the anion layer (Fig. 7.16).

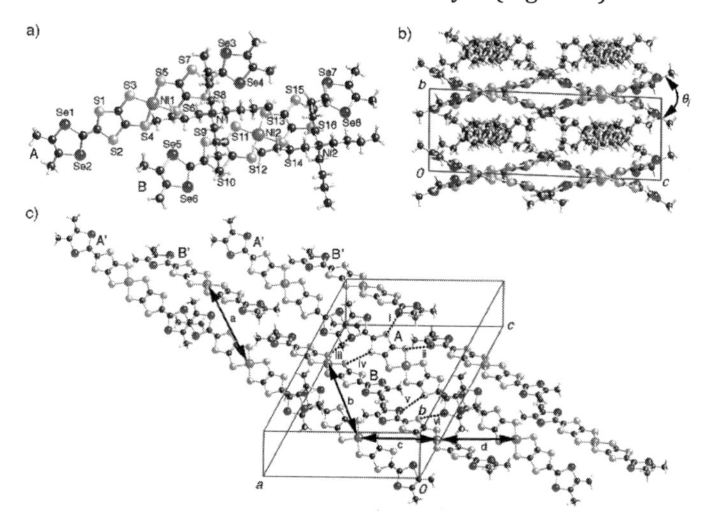

Figure 7.16 Crystal structure of $(^nBu_4N)[Ni(dmstfdt)_2]$ **(17)** at room temperature. (a) Crystallographically independent molecules with labeling of the atoms. (b) Sandwiched structure of anion and cation layers; two kinds of anions (A and B) form a zigzag –ABA′B′– array with dihedral angle (θ) of 42.6(1)° between the adjacent anions (A and B) along the *a*-axis. (c) Anion layer in **17**. Intermolecular contacts (Å): (i) S7···Se7, 3.723(6); (ii) S5···Se8, 3.795(5); (iii) Se4···S13, 3.638(6); (iv) S8···S11, 3.643(7); (v) S2···Se6, 3.693(6); and (vi) Se2···S10, 3.689(6). The Ni···Ni distances (Å) are (a) 14.436; (b) 12.562; (c) 9.747; and (d) 9.813. Reprinted from Ref. 53. Copyright 2007, American Chemical Society.

Despite the 1:1 stoichiometry of the complex, the metallic behavior is relevant. The room temperature conductivity is σ_{rt} = 0.2 S cm^{-1} and a weakly metallic behavior was observed above room temperature (inset of Fig. 7.17a). Below 280 K, the resistivity increased very slowly with a small, estimated activation energy of 17 meV down to 147 K a temperature at which a sharp transition to an insulating state occurred. This behavior is supported by simple tight-binding band structure calculation results.

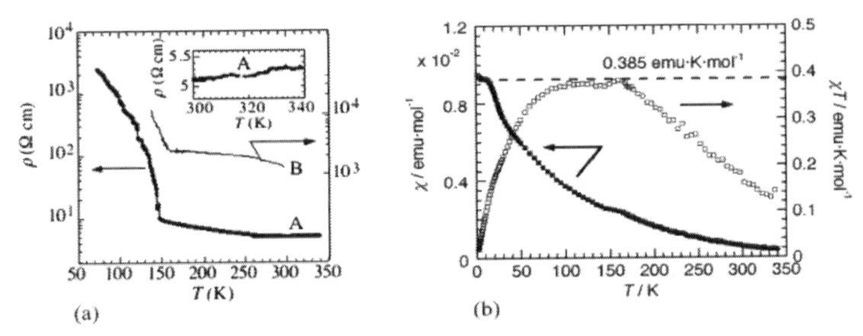

Figure 7.17 (a) Resistivities (ρ) along the long axis and perpendicular to the long axis (B) of **17** vs. the temperature (T). The inset is the temperature dependence of the resistivity (A) (right). (b) Temperature (T) dependence of magnetic susceptibility (χ) of **17** under the magnetic field of H = 10 kOe; the χT vs. T plot is also shown. The dotted line (0.385 emu K mol^{-1}) is the theoretical χT value estimated for a paramagnet with $S = 1/2$ and $g = 2.027$. Reprinted from Ref. 53. Copyright 2007, American Chemical Society.

The temperature dependence of the magnetic susceptibility is also very unique, and the increase of the χT product vs. T suggests a localization of the conduction electrons with lowering temperature in the 160–280 K range, where the resistivity gradually increases. The χT value was constant in the 80–150 K range and this fits well to the Curie–Weiss law, thus indicating that 17 undergoes a transition to an insulating state as a result of localization of the conduction electrons. In conclusion, 17 is a new molecular conductor that shows the characteristic susceptibility changes suggesting the gradual electron localization from the high-temperature weakly metallic state to the low-temperature localized spin state and further exhibits long-range magnetic ordering with a Néel ground state below 20 K, and it is found to be dependent on the applied field as expected for a canted antiferromagnet.

7.2.3.2 SCMMs containing a thiophene moiety

Novel extended dithiolene ligands incorporating not only a TTF[54a] but also thiophenic moieties were also investigated following the first report of Almeida and coworkers, on the metallic state in [Au (α-tpdt)$_2$]**(18)**, (α-tpdt = 3,4-thiophenedithiolate),[54b–c] a member of a novel class of gold complexes with thiophenedithiolate ligands, where the peripheral thiophene sulfur atoms promote additional intermolecular S–S contacts that control the packing pattern and thus the electronic properties.

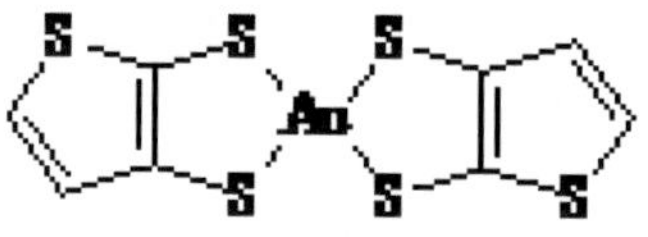

$[Au(\sigma\text{-tpdt})_2]$

The conductivity measurements, performed on a polycrystalline sample, show that $[Au(\alpha\text{-tpdt})_2]^\circ$ is a metal in the 15–300 K range with high room temperature conductivity, σ_{rt} = 6 S cm^{-1}, which is exceptionally large considering that the powder data are typically 100- to 1000-fold weaker than those observed in single crystals along their most conductive axis. This compound shows paramagnetic behavior almost independent of the temperature down to 10 K, reminiscent of the Pauli paramagnetism typical of metallic systems. Unfortunately no crystal data were available to further investigate the origin of the metallic state.[54] The extra thiophenic units were expected to enhance the degree of π delocalization over the ligand extremity, while the presence of the additional thiophenic sulfur atom were expected to favor the side intermolecular interactions at the ligand periphery, as found in other thiophene-type ligand-based compounds.[54,55] A summary of the used thiophenedithiolate ligands is reported in Chart 7.4.

Chart 7.4

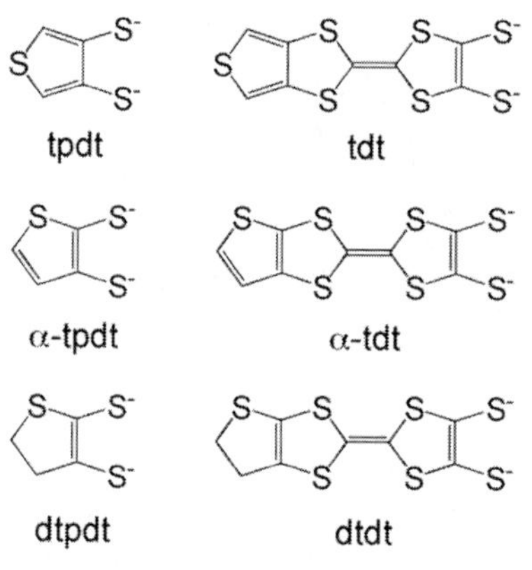

tpdt tdt

α-tpdt α-tdt

dtpdt dtdt

The importance of these complexes as useful building blocks to prepare molecular materials with very interesting magnetic and transport properties, ranging from metamagnets to SCMMs, has recently been discussed by Belo and Almeida.[56] Among these complexes, the most explored and studied have been the $[M(\alpha\text{-tpdt})_2]^-$ (M = Au, Ni) anions. In all their salts, although with different crystal packing motifs, the $[M(\alpha\text{-tpdt})_2]^-$ anions always pack in a layered structure composed of anionic zigzag type chains (Fig. 7.18).[57]

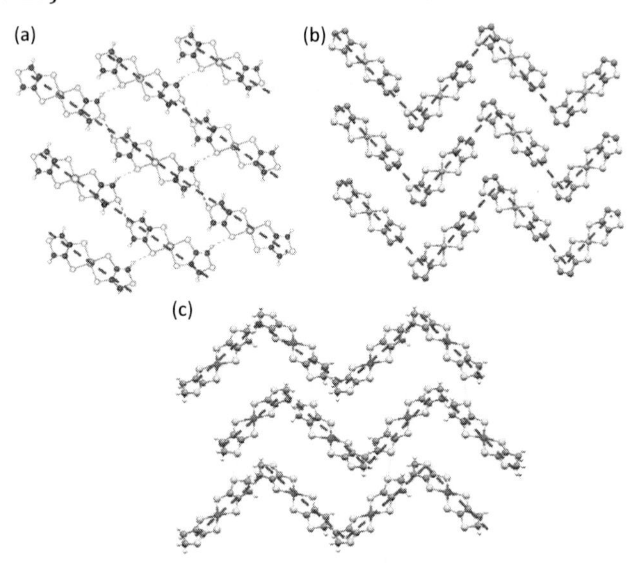

Figure 7.18 Anionic layers in the crystal structures of (a) $[Au(tpdt)_2]^-$, (b) $[Ni(\alpha\text{-tpdt})_2]^-$, and (c) $[Au(\alpha\text{-tpdt})_2]^-$. The red dashed line show the direction of the shorter interactions between anions, defining the anionic chains within the layers. Reprinted from Ref. 56. Copyright 2010, Elsevier. See also Color Insert.

In these chains, the key crystal engineering parameters concerning anion–anion interactions are the C–S interactions and the charge-assisted hydrogen bonds, mediated by the peripheral thiophenic sulfur atom. Another important feature seems to be the degree of *cis–trans*/orientation disorder in the position of the terminal thiophenic sulfur atom. This degree of disorder is correlated with the strength of the interactions with the surrounding cations. These contacts can be of two types: (i) hydrogen bonds involving the anion sulfur atoms or halogen atoms from the cations and (ii) aromatic π–π interactions. Since the terminal thiophenic ring is aromatic, the molecules in the crystal packing tend to arrange in order to maximize the π–π interactions when the cations contain an aromatic system.

This cation–anion interaction is important for stabilizing the anion in the most favorable trans geometry. Significant differences on their crystal structures were also observed in these salts depending on the cation nature. In the Et_4N salt, a clear segregation between the two species is observed, leading to an alternated cationic–anionic layered crystal structure. The $n\mathrm{Bu}_4N$ cation leads to a packing where isolated anions are surrounded by cations while substituted benzopyridine cations, such as BrBzPy, perfectly match the widths of the anion, leading to a crystal structure made of parallel alternated columns of cations and anions. Three new salts with these substituted benzopyridine cations, (RBzPy)[Ni(α-tpdt)$_2$] (R = H, Br, F) **(19a, b, c)** have been prepared and characterized,[57b] showing interesting structural features and magnetic properties. Previous studies with other dithiolene complexes[58] have shown that this type of cations favor segregated stacking of cations and anions leading to different kind of magnetic interactions and ordering, which depend on the overlap modes of the anions in the crystal structure. The cation substitution plays a role in determining structural differences and variable amounts of *cis–trans* disorder in the anionic dithiolene complexes related to the relative positions of the sulfur atoms of the thiophene rings. This is the first time that the *cis* configuration has been observed in a [Ni(α-tpdt)$_2$]$^{-\bullet}$dithiolene complex, probably as a consequence of preferential S···S interactions. The structure of all the compounds consist of alternated layers of anions and cations: the anions are arranged with thiophenic sulfur atoms connecting to a coordinating sulfur atom of a neighboring anion, placing the complexes almost perpendicular to their next neighbors, in zigzag chains. The cations are positioned with the pyridine rings inserted between thiophenic rings of anions, maximizing π–π interactions but failing in promoting segregated anion stacking, the common stacking pattern observed in several salts with other dithiolene complexes. There are several anion–cation interactions: C···C and C···S π–π interactions between the pyridine ring of the cations and the thiophenic ring of the anions, S···H–C hydrogen bonds, C–Br···S short contacts through a coordinating and thiophenic sulfur atom, and C–F···S interactions between anions and cations, which is responsible for the prevalence of the *cis* configuration. The *cis–trans* disorder affects the magnetic behavior of these compounds: (HBzPy)[Ni(α-tpdt)$_2$] **(19a)** shows dominant ferromagnetic interactions and, at low temperatures, typical behavior of a cluster glass as a consequence of disorder effects as shown in Fig. 7.19a;

the magnetic behavior of this compound can be essentially related to the intrachain arrangement, where the chains, in fact, consist of a random sequence of AFM coupled zigzag fragments, composed of a variable number of FM coupled $[Ni(\text{-tpdt})_2]_2$ units.

$(BrBzPy)[Ni(\alpha\text{-tpdt})_2]$ **(19b)** shows dominant antiferromagnetic interactions with a magnetic anomaly at $T \approx 6$ K (Fig. 7.19b) and $(FBzPy)[Ni(\alpha\text{-tpdt})_2]$ **(19c)** behaves like a paramagnet down to 1.5 K (weak ferromagnetic interactions are observed without magnetic ordering). This class of compounds shows how small changes within the building blocks can significantly affect the supramolecular interactions and thus the magnetic properties.

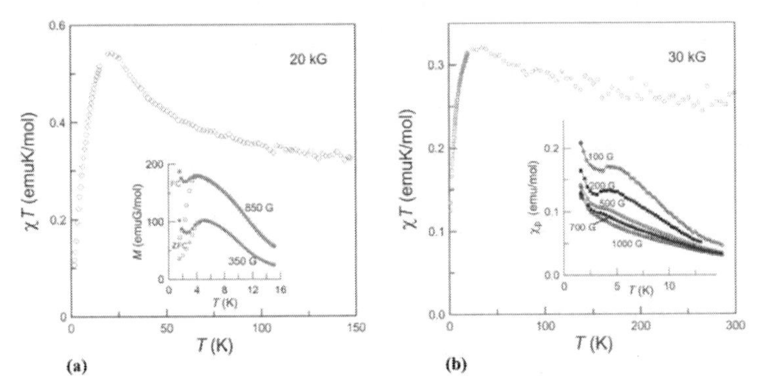

Figure 7.19 (a) Temperature dependence of χT for $(HBzPy)[Ni(\alpha\text{-tpdt})_2]$ under an applied field of 20 kG. The inset shows the ZFC (open symbols) and FC (closed symbols) magnetization curves for applied fields of 350 and 850 G, represented by circles and squares, respectively. (b) Temperature dependence of χT for compound **19b**. The inset shows the paramagnetic susceptibility behavior for various applied fields. Reprinted from Ref. 57b. Copyright 2006, RSC.

Moreover, the $[Ni(\alpha\text{-tpdt})_2]^-$ paramagnetic anion, in combination with a suitable counterion, has proven to be a challenging building block for materials showing unusual magnetic properties. For instance, when using decamethyl metallocenium cations, $(Cp^*_2Fe)[Ni(\alpha\text{-tpdt})_2]$ shows metamagnetic behavior below 2.56 K with a critical field of 600 G, while the corresponding manganese salt, $(Cp^*_2Mn)[Ni(\alpha\text{-tpdt})_2]$, is a frustrated magnet.[57c]

The gold and nickel bisdithiolene complexes of the new highly extended ligands, α-tdt (2,3-tiophenetetrathia-fulvalenedithiolate) and dtdt (dihydro-thiophenetetrathiafulvalenedithiolate), shown in Chart 7.5, were synthesized and fully characterized.[57f] These complexes were obtained under anaerobic conditions as mono-anionic [^{n}Bu$_4$N][Au(α-tdt)$_2$] **(20a)**, [^{n}Bu$_4$N][Au(dtdt)$_2$] **(21a)** and dianionic species [nBu$_4$N]$_2$[Ni(α-tdt)$_2$] **(20b)**, [nBu$_4$N]$_2$[Ni(dtdt)$_2$] **(21b)** and subsequently oxidized by air or iodine to the stable neutral state. The monoanionic complexes crystallize in at least two polymorphs, showing a similar packing motif with a strong segregation of anions and cations in alternating layers which are connected through several hydrogen bonds. The anionic layers are composed of tightly packed parallel chains of the nickel or gold complexes, which are almost parallel to the chain axis and form domino-like chains through the overlapping of their terminal thiophenic rings. In the case of **21b,** the overlapping between anions occurs through both the thiophenic ring and part of the TTF moiety, with all of the sulfur atoms connected by S···S short contacts. Several side-by-side short S···S contacts are observed between parallel chains which give rise to extended 2-D networks of contacts.

Chart 7.5

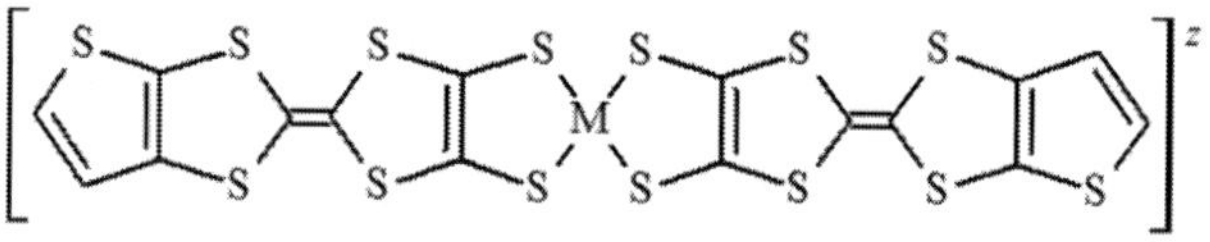

(20) [M(α-tdt)$_2$]^z z = -1, M = Au **(a)**;
z = -2, M = Ni **(b)**;
z = 0, M = Ni **(c)**;
z = 0, M = Au **(d)**

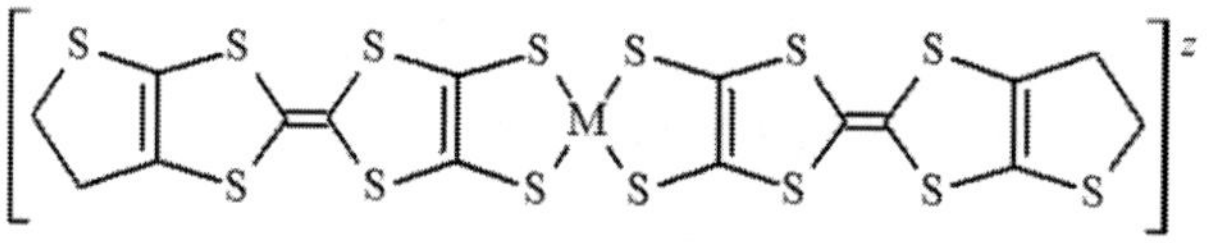

(21) [M(dtdt)$_2$]^z z = -1, M = Au **(a)**;
z = -2, M = Ni **(b)**;
z = 0, M = Ni **(c)**;
z = 0, M = Au **(d)**

All of the neutral complexes, obtained by means of iodine oxidation, show room temperature electrical conductivity of the order of 2, 2.5, 5, and 8 S cm^{-1} for [Ni(α-tdt)$_2$]**(20c)**, [Ni(dtdt)$_2$] **(21c)**, [Au(α-tdt)$_2$] **(20d)**, and [Au(dtdt)$_2$] **(21d)**, respectively. The more crystalline Ni samples, obtained by slow oxidation by air exposure, show higher electrical conductivity values of 200 S cm^{-1} for $[Ni(dtdt)_2]$ with a clear metallic behavior down to 80 K and 24 S cm^{-1} for $[Ni(\alpha\text{-tdt})_2]$, respectively. These complexes show also relatively large magnetic susceptibility values that correspond, beside the Pauli contribution typical of conducting systems, to effective magnetic moments in the range 1–3 μ_B, indicating that, in addition to delocalized conduction π-electrons, there are unpaired *d* electrons. These compounds are SCMMs where a 2-D network of intermolecular interactions is related to high electrical conductivity. Other examples of these SCMMs are mentioned in Ref. 59.

The intermolecular interactions as well as of the size of the dithiolene ligands play a crucial role on the physical properties of these systems. These multi-sulfurated molecules tend to stack side-by-side, in order to maximize the sulfur–sulfur interactions, allowing the neighboring stacks to interact through several S...S contacts. Along the stacks, the thiophenic ring lays over the TTF ring, allow contact between the sulfur atoms and the π–π interactions between the aromatic rings.

7.2.4 SMMs-Based Conductors

Since the discovery of the first dodecanuclear Mn cluster (Mn_{12}) exhibiting slow relaxation of the single-cluster magnetization early in the 1990s,[60] such size-defined clusters, now called *single-molecule magnets (SMMs)*, have been designed and synthesized with the aim of developing a new class of nanosized magnets,[61] the elucidation of quantum phenomena[62], and the quest for new potential of SMM-containing hybrid materials.[63–65] In a SMM, the cluster itself behaves as a general "magnet" at low temperatures; because of this the combination of a large spin ground state (ST) and large easy axis type anisotropy (axial zero-field splitting (ZFS) parameter, $D_{ST} < 0$) of the ST unit creates an energy barrier between the spin-up and spin-down magnetization states (±ms). In other words, in order to flip the spin of a Mn_{12} molecule from along the +*z* axis to along the –*z* axis of the disc-like $Mn_{12}O_{12}$ core, it takes some energy (the barrier) to reorient the spin (easy axis type of anisotropy). Indeed, they exhibit slow relaxation of the magnetization, which results in a barrier for spin reversal and in a hysteresis loop, below a blocking temperature, T_B. Therefore, the slow magnetization relaxation shown by an SMM is due to an individual molecule rather than a long-range ordering as observed in nanoscale magnetic domains of bulk magnets. SMM clusters are really peculiar magnetic particles dependent on intrinsic molecular characteristics, and so, have potential uses as ultimate devices for the information storage or as quantum computers using magnetization state levels and their mutual transfers.[66] Another appealing aspect of SMM, which has especially attracted much attention since the beginning of this century, is that SMM clusters are appealing as magnetic building blocks for further molecular assemblies.[67] Recent studies on SMM-based supramolecular oligomers[64] and network compounds[65–69] have demonstrated that the inter-SMM interaction via conducting electrons, albeit small, has a mutual influence on both SMM properties and conductivity.[69] Therefore, one of the goals in this field is the design of superparamagnetic/conducting hybrid materials combining SMM properties with high conductivity in order to tune the SMM properties using interactions with the itinerant electrons of the conducting sublattice. Even if the inter-SMM interaction is as weak as ~ -0.1 K, such an interaction crucially affects the original SMM properties (Scheme 7.2). Therefore, the SMM would also act as a sensitive magnetic sensor in this kind of materials. Another interesting point to be highlighted is the strong anisotropic nature of SMM clusters, in addition to the high spin state, and these characteristics make them potential candidates as spintronics materials. More details on this topic will be found in Chapter 3.

Scheme 7.2

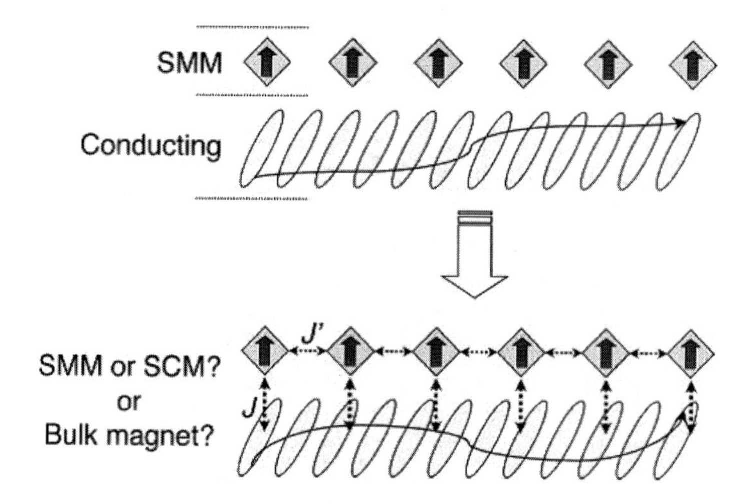

Yamashita and coworkers have reported on the synthesis of a unique hybrid material based on Mn_4SMM clusters, having $S_T = 9$ spin state and $[Pt(mnt)_2]^{n-}$ dithiolene complexes, $[Mn^{II}_2Mn^{III}_2(hmp)_6(MeCN)_2]$ $[Pt(mnt)_2]_4[Pt(mnt)_2]_2$ **(22)** (hmp^- = 2-hydroxymethyl pyridinate)[71] showing SMM/semi conducting behavior (Scheme 7.3).

Scheme 7.3 Reprinted from Ref. 71. Copyright 2007, American Chemical Society.

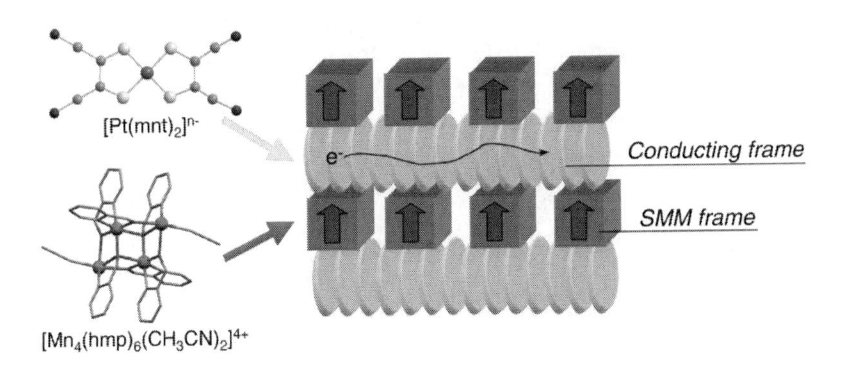

The structure consists of $[Mn^{II}_2Mn^{III}_2(hmp)_6(MeCN)_2]^{4+}[(Mn_4)^{4+}]$ units separated by 1-D double columns of $[Pt(mnt)_2]^{n-}$ complexes. There are two $[Pt(mnt)_2]^{n-}$ units (**A** and **B**) coordinated with the Mn^{II} ion of the $[Mn_4]^{4+}$ unit forming a discrete unit, $\{[Pt(mnt)_2]_2\text{-}[Mn_4]\text{-}[Pt(mnt)_2]_2\}$. The uncoordinated $[Pt(mnt)_2]^{n-}$ (**C**) and coordinated **A** and **B** are mutually stacked along the a-axis (Fig. 7.20a) to form a segregated 1-D double column possessing an [···**A**···**B**···**C**···] repeating unit (Fig. 7.20b). Considering the charge balance and the structures of **A**, **B**, and **C**, the $[Pt(mnt)_2]$ units show a fractional charge of −0.66. The electron transport properties of this salt, measured along the $[Pt(mnt)_2]^{0.66-}$ columns (stacking a-axis), show that it is a semiconductor down to 110 K (insulator at $T < 110$ K) with high room temperature conductivity ($\sigma_{rt} = 0.2$ S cm^{-1}) and activation energy $E_a = 136$ meV. Partially oxidized $[Pt(mnt)_2]^{n-}$ subunits allowing conductivity are rare and only Li salts of $[Pt(mnt)_2]^{n-}$ have been reported so far.[72]

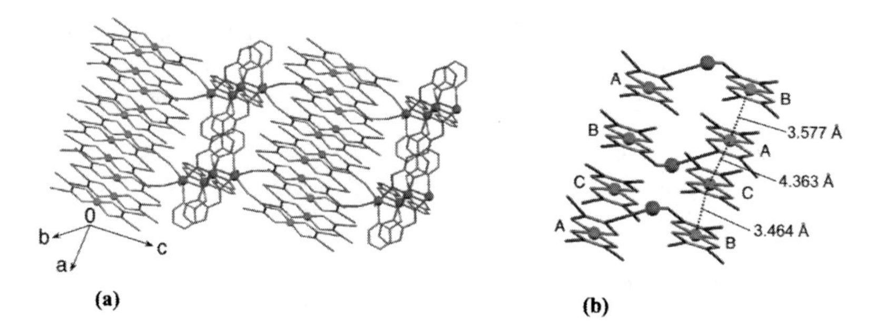

Figure 7.20 Packing view of $\{[Mn^{II}_2Mn^{III}_2(hmp)_6\ (MeCN)_2][Pt(mnt)_2]_4\}$ $[Pt(mnt)_2]_2$**(22)**, (a) and arrangement of part of $[Pt(mnt)_2]^{n-}$ moieties (b). Reprinted from Ref. 71. Copyright 2007, American Chemical Society.

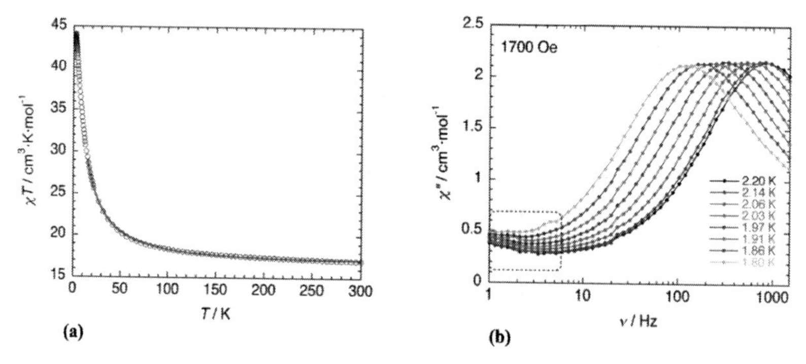

Figure 7.21 (a) Temperature dependence of χT for $\{Mn^{II}_2Mn^{III}_2(hmp)_6$ $(MeCN)_2][Pt(mnt)_2]_4\}[Pt(mnt)_2]_2$**(22)** measured at 1 kOe. Solid lines represent the best fit obtained with a tetranuclear model for which the contribution of four noninteracting paramagnetic species with $S = 1/2$ as $g = 2$ is taken into account. (b) Frequency dependence of χ'' for **22** at 1700 Oe as a function of temperature. An anomaly could be seen in the low-frequency region (dotted square). Reprinted from Ref. 71. Copyright 2007, American Chemical Society. See also Color Insert.

The temperature dependence of χT vs. T(K) (Fig. 7.21a) shows a typical intracluster ferromagnetic behavior, already seen in related $[Mn_4]$ clusters.[73] The larger χT value at 300 K compared to that of only a Mn_4 cluster (15–16 $cm^3 \cdot K \cdot mol^{-1}$) is due to the contribution of four $[Pt(mnt)_2]^-$ units with $S = ½$. Examination of the magnetic behavior at low temperatures show the SMM character of the $[Mn_4]^{4+}$ unit and the possible interunit interactions. The temperature dependence of *ac* susceptibilities measured at several *ac* frequencies (3 Oe *ac* field and zero *dc* field) exhibited frequency dependence of both in-phase (χ') and out-of-phase (χ'') components below 3 K. A small anomaly of χ'' in the low frequency region, in addition to the frequency-/field-dependent main signals in the high-frequency region (Fig. 7.21b), is observed; this anomaly might be due to interactions between the $[Mn_4]^{4+}$ units via the coordinated $[Pt(mnt)_2]^-$ units, although it should be very weak, it is not certain, because this effect does not influence the $[Mn_4]^{4+}$ SMM behavior, as confirmed by other magnetic data. To determine the correlation between $[Mn_4]^{4+}$ and $[Pt(mnt)_2]^{n-}$ units, high-field and high-frequency EPR were performed and they confirmed that the interunit interaction by which $[Mn_4]^{4+}$ units are coordinated (superexchange interaction) is negligible, since any significant inter-$[Mn_4]^{4+}$ interaction have shifted the main signals, as already demonstrated in SCM compounds.[74] However, the conductivity was present only in the temperature region where the SMM unit behaved as a general paramagnet and both SMM and conducting network acted independently in the entire temperature range. The $[Mn_4]^{4+}$ SMM has been also reacted with $(NBu_4)[Ni(dmit)_2]$, because the use of $[Ni(dmit)_2]^{n-}$ complex is more likely to lead to higher conductivity than the $[Pt(mnt)_2]^{n-}$ because of its smaller HOMO–LUMO gap and its tendency to a 2-D molecular arrangement.[9–10,75] Unfortunately, any attempts to obtain the $[Ni(dmit)_2]$-based desired hybrid material failed due to a decomposition of the SMM core. These results clearly indicate that the stability, i.e., structural stability and redox stability, of the used SMM clusters not only vs. the reaction solvent but also vs. the counterions (anionic dithiolene complexes) is a crucial factor for selecting the SMM molecular building block. The SMM family of Mn(III) salen-type out-of-plane dimers is one of the smallest nuclear SMMs[76] and is relatively redox stable at general conditions.[77] Hence, this SMM family is a good candidate of the SMM building block for this type of reactions. Very recently Yamashita and coworkers[70] reported on the second example of structurally hybridized compounds: $[Mn_2(5\text{-MeOsaltmen})(S)]_2[Ni(dmit)_2]_7\cdot 4(S)$ **(23)** (S = Acetone, **a**, Acetonitrile, **b**) (5-MeOsaltmen^{2-}=*N,N′*-(1,1,2,2-tetramethylethylene)bis(5-methoxysalicylideneiminate) consisting of ½$[Mn_2]^{2+}$ SMM layers and $[Ni(dmit)_2]^{n-}$ molecular conducting layers stacking along the *c* direction. In the $[Ni(dmit)_2]^{n-}$ layer (Fig. 7.22a), the $[Ni(dmit)_2]^{n-}$ units are connected through a 2-D network of S···S contacts. This layer is arranged in a *b*-type mode, in which four face-to-face overlaps (along the *a*–*b* direction; P, Q, R, S) and eight side-by-side contacts (along the *a*- or *b*-axis direction; p, q, r, s, t, u, v, w) are found. Two kinds of Ni···Ni distances are found because of the difference of the face-to-face stacking mode of $[Ni(dmit)_2]^{n-}$, regular mode and sliding mode, as shown in Fig. 7.22b. The R and S contacts give rise to the regular mode, while the P and Q contacts to the sliding mode. The amount of S···S contacts strongly depends on the stacking modes (regular or sliding) and therefore strongly affect the electronic transport properties of these compounds. Nevertheless, conducting phases are expected along the *ab* plane.

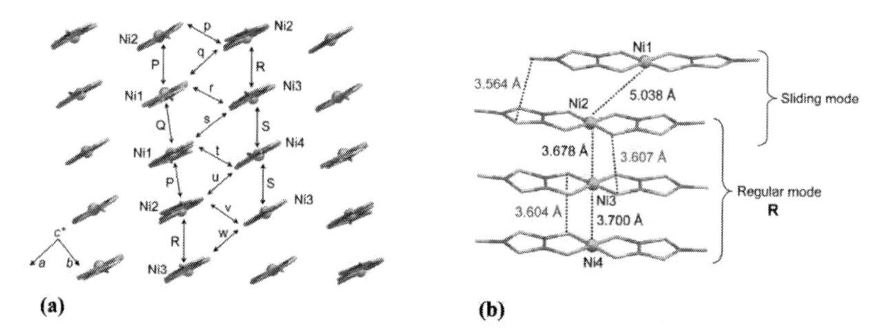

Figure 7.22 (a) End-on projection of a $[Ni(dmit)_2]^{n-}$ layer of **23a**; (b) Stacking features (regular mode and sliding mode) of $[Ni(dmit)_2]^{n-}$-complexes of **23a** and **23b** where dotted lines represent Ni···Ni contacts and the shortest S···S contacts. Reprinted from Ref. 70. Copyright 2007, Elsevier.

The conductivity at room temperature of **23a** and **23b** is σ_{rt} = 1.6 S cm^{-1} and 2.8 S cm^{-1}, respectively, which decreases gradually down to approximately 80 K (Fig. 7.23), indicating their semiconducting behavior. The observed conductivities are considerably higher than those observed in compounds having integer-valent $[Ni(dmit)_2]^-$ columns[78] and are 10 times higher than those observed in the first hybrid compound of $[Mn_4]^{4+}$ SMMs and $[Pt(mnt)_2]^{n-}$ columns (σ_{rt} = 0.22 S cm^{-1}).[71] Nevertheless, the semiconducting character of **23a** and **23b** may be attributed to the discontinuous stacking modes found in their packing structures (see Fig. 7.22). The Arrhenius plot of the resistivity (ρ) leads to an activation energy of 88 meV (50–200 K) and 143 meV (200–300 K) for **23a** and 119 meV (50–300 K) for **23b** (Fig. 7.23). Nevertheless, such a nonlinear conductivity may be, in the present case, due to local structural changes and/or change of carrier concentration resulting from the modification of the electronic band structures, depending on the temperature. The *dc* magnetic behavior of **23a** and **23b** is very similar to that of isolated $[Mn_2]^{2+}$ SMM in the temperature range of 1.8–300 K. The susceptibility obeys the Curie–Weiss and the increase of χT is purely due to the ferromagnetic coupling between Mn(III) ions via the bis-phenolate bridge in the $[Mn_2]^{2+}$ dimer, and the decrease at low temperatures is due to the ZFS attributed to Mn(III) ions and/or intermolecular interactions. Therefore, the χT behavior of **23a** and **23b** were fitted using a Heisenberg dimer model of $S = 2$ taking into account ZFS of Mn(III) ion. In conclusion, **23a** and **23b** exhibit a SMM/semiconductor character in which both characteristic components, however, act independently in the whole temperature range.

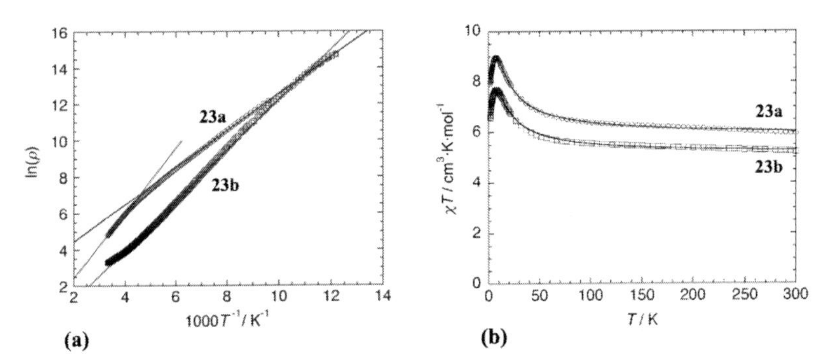

Figure 7.23 (a) ln(ρ) vs. $1/T$ plots for a single crystal of **23a** and **23b** (left side); the solid lines represent least-squares linear fits. (b) Temperature dependence of χT of **23a** and **23b** measured at 1 kOe. The solid lines represent best-fit curves obtained with a dimer model of $S = 2$. Adapted from Ref. 70. Copyright 2008, Elsevier.

Stoichiometric neutral assemblies with a formula of [Mn (5-Rsaltmen){M(dmit)$_2$}]$_2$ (R = MeO, M = Ni, **24a**; M = Au, **24b**; R = Me, M = Ni, **25a**; M = Au, **25b**) were obtained by reacting [Mn$_2$ (5-Rsaltmen)$_2$(H$_2$O)$_2$](PF$_6$)$_2$ (R = MeO, Me) with (NBu$_4$) [M(dmit)$_2$] (M = Ni, Au) under aerobic conditions, as reported in Scheme 7.4.[79]

Scheme 7.4 Adapted from Ref. 79. Copyright 2009, American Chemical Society.

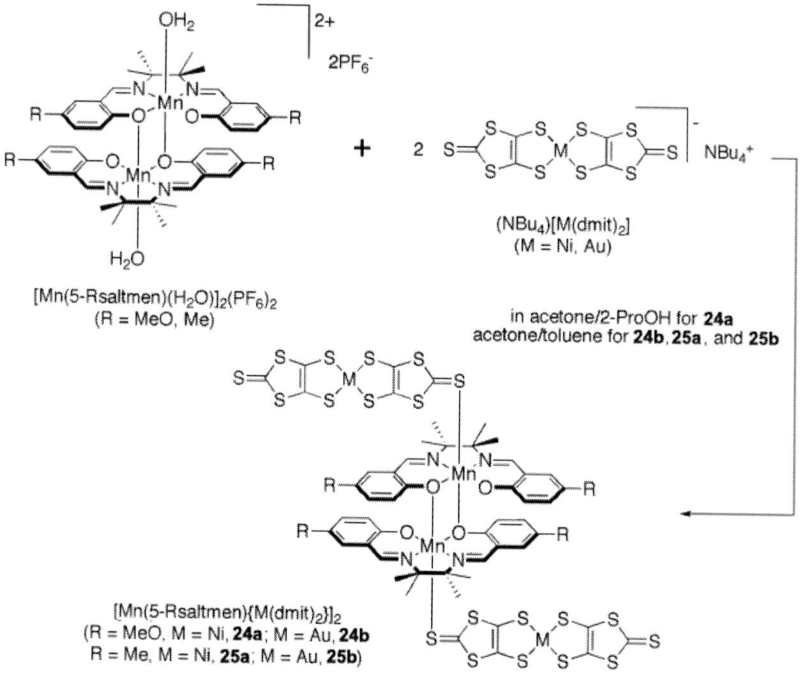

$[Mn(5\text{-}Rsaltmen)(H_2O)]_2(PF_6)_2$
(R = MeO, Me)

+ 2 $(NBu_4)[M(dmit)_2]$
(M = Ni, Au)

in acetone/2-ProOH for **24a**
acetone/toluene for **24b**, **25a**, and **25b**

$[Mn(5\text{-}Rsaltmen)\{M(dmit)_2\}]_2$
(R = MeO, M = Ni, **24a**; M = Au, **24b**
R = Me, M = Ni, **25a**; M = Au, **25b**)

These systems are interesting materials consisting of neutral single-component/neutral SMM-based molecular complexes that exhibit two major solid-state properties, that is, electrical conductivity and magnetism. They are Mn(III) saltmen out-of-plane dimers coordinated to $[M(dmit)_2]^-$ anions in apical positions, forming a linear type tetranuclear complex, $[M\text{-}(dmit)\text{-}Mn\text{-}(OPh)_2\text{-}Mn\text{-}(dmit)\text{-}M]$. The Mn ions are trivalent from the structural features, and thus the $[M(dmit)_2]^-$ complexes are −1-charged. The packing arrangements of the respective sets of **24a/24b** and **25a/25b** compounds show that they are isomorphous. Intermolecular π–π/S···S contacts are observed between the coordinating $[M(dmit)_2]^-$ moieties to form zigzag stair-like columns along the *a*-axis direction (Fig. 7.24).

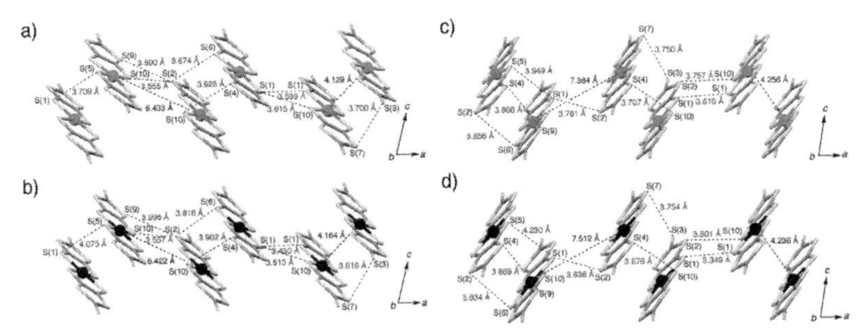

Figure 7.24 Short S–S contacts (blue dots) and M–M distances (black dots) in the stair-like stacking column of $[M(dmit)_2]^-$ moieties of **24a** (a), **24b** (b), **25a** (c), and **25b** (d). Reprinted from Ref. 79. Copyright 2009, American Chemical Society. See also Color Insert.

Hückel calculations confirm a strong intermolecular dimerization within the columns, leading to singlet $[M(dmit)_2]^-$ pairs that make them magnetically silent, even at room temperature. Nevertheless, **24a** and **25a** with M = Ni behave as semiconductors with σ_{rt} = 7 × 10^{-4} S cm^{-1} and E_a = 182 meV for **24a** and σ_{rt} = 1 × 10^{-4} S cm^{-1} and E_a = 292 meV for **24a**, while **24b** and **25b** with M = Au are insulators. This conductivity results from the 1-D columnar arrangement of the $[M(dmit)_2]^-$ anions with π–π and S···S contacts along the a-axis direction and the low value is due the absence of mixed valence and the presence of strong $[Ni(dmit)_2]_2^{2-}$ dimers along the columns. As a result of the strong dimerization of the $[M(dmit)_2]^-$ anions, the magnetic properties of **24a** are essentially identical to those of **24b** and **25b**, which can be described as isolated Mn(III) dimers, acting as SMMs. These compounds exhibit SMM properties with slow magnetization relaxation at low temperatures and fast zero-field quantum tunnel magnetization (QTM). Meanwhile, the magnetic properties of **25a** are dominated by the intermolecular Mn···Mn antiferromagnetic interactions via the singlet $[Ni(dmit)_2]_2^{2-}$ dimer ($J_{Mn\cdots Mn}/k$B = −2.85 K), inducing a long-range antiferromagnetic order at T_N = 6.4 K. The relatively strong antiferromagnetic interactions via this singlet pair (despite a long Mn–Mn distance) are probably associated with the singlet-triplet excitation of the dimeric $[Ni(dmit)_2]^-$ moiety. The magnitude of this intermolecular Mn···Mn interaction is indeed mainly dependent on the Mn/$[Ni(dmit)_2]^-$ orbital orientations and the strength of the σ-type orbital overlap on the Mn–S bond (Fig. 7.25).

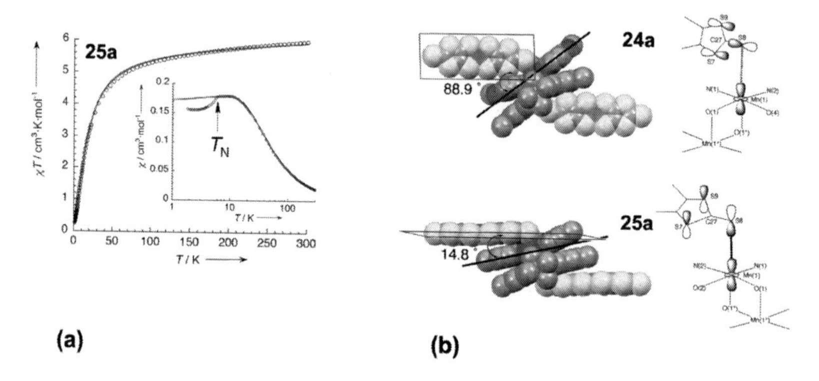

Figure 7.25 (a) Temperature dependence of χT measured at 1 kOe for **25a**; solid red lines represent the best fits to the experimental data. (b) View of coordination modes between the $[Ni(dmit)_2]^-$ moiety and the Mn(III) metal ion in **24a** and **25a**, where the given angles are the dihedral angles between least-square planes defined by S(7)-S(8)-S(9)-C(27) (for dmit) and O(1)-O(2)-N(2)-N(1)-Mn(1) [for Mn(saltmen)] and on the right, schematic orbital configurations of the d_{z2} orbital of Mn ion and a p orbital of S(8) in **24a** (a) and **25a** (b) and M–M distances (black dots) in the stair-like stacking column of $[M(dmit)_2]^-$ moieties of **24a** (a), **24b** (b), **25a** (c), and **25b** (d). Adapted from Ref. 79. Copyright 2009, American Chemical Society. See also Color Insert.

It must be highlighted that in these materials the bonding between the magnetic component and the conducting component play a key role in tuning the magnetic properties without significant changes of the conducting properties as shown in **24a** and **25a** whose behavior varies from semiconductor/SMM to semiconductor/antiferromagnet, respectively. In conclusion, the impact of such systems, showing synergic properties,[80] in the fields of molecular electronics and molecular devices will open challenging perspectives.

The physical properties of the compounds discussed in this chapter are summarized in Table 7.1.

Table 7.1 Physical properties of chalcogenolene-based molecular materials

Compound	Magnetic properties	Conducting properties	Ref.
κ-$BETS_2[Fe^{III}\text{-}(ox)Cl_2]$ **(1)**	Weakly FM	Conductor ($\sigma_{300K} = 10^2$ S cm^{-1}), retains the metallic state down to 2 K	20
$(TTF)_2[Fe(tdas)_2]$ **(2)**	AFM $[Fe(tdas)_2]_2^{2-}$ dimers	Semiconductor ($\sigma_{rt} = 0.03$ S cm^{-1})	22
$(BEDT\text{-}TTF)_2[Fe(tdas)_2]$ **(3)**	PM	Semiconductor ($\sigma_{rt} = 1$ S cm^{-1})	24
$(BETS)_2[Fe(tdas)_2]_2$ **(4)**	PM	$\sigma_{rt} = 0.2$ S cm^{-1}, retains the metallic state down to 200 K	25
$(BEDT\text{-}TTF)[Ni(tdas)_2]$ **(5)**	AFM coupling (dimer model)	Semiconductor ($\sigma_{rt} = 0.018$ S cm^{-1}), semiconductor–semiconductor transition (200 K)	26
$(Me\text{-}3,5\text{-}DIP)[Ni(dmit)_2]_2$ **(6)**	PM with AFM interactions	2-D metallic conductivity	28
α-$(BEDT\text{-}TTF)_5[Fe(croc)_3]\cdot 5H_2O$ **(7a)**	PM	Semiconductor ($\sigma_{rt} = 6$ S cm^{-1})	34
β-$(BEDT\text{-}TTF)_5[Fe(C_5O_5)_3]\cdot C_6H_5CN$ **(7b)**	PM	Metal down to 120 K	34
$[BEDT\text{-}TTF]_4[(H_3O)Cr(ox)_3]\cdot C_6H_5CN$ **(8)**	PM	Superconductor (Δ-Λ-Δ-Λ-Δ-Λ-Δ-Λ anion pattern) Semiconductor (ΔΛ-ΔΛ-ΔΛ-ΔΛ anion-layer pattern)	36
$(BEDT\text{-}TTF)_3NaCr(C_2O_4)_3\cdot CH_3NO_2$ **(9)**	PM	Semiconductor	38
$[TM\text{-}ET]_x[MnCr(ox)_3]\cdot CH_2Cl_2$**(10)**	FM	Conductor ($\sigma_{rt} = 65$ S cm^{-1}), metal-like behavior down to 190 K	40
$[Cu(dmdt)_2]$ **(13)**	Curie–Weiss PM (84% of $S = 1/2$ spin moments)	Semiconductor ($\sigma_{rt} = 3$ S cm^{-1})	44
$[Au(tmdt)_2]$ **(14)**	AFM phase transition at 110 K	Conductor ($\sigma_{rt} = 200$ S cm^{-1})	46
$(^nBu_4N)[Ni(chdt)_2]$ **(15b)**	AFM 1-D Heisenberg chain	-	51
$(^nBu_4N)[Ni(eodt)_2]$ **(16b)**	Curie–Weiss PM	-	51
$(^nBu_4N)[Ni(dmstfdt)_2]$ **(17)**	PM (Curie–Weiss), weak FM coupling ($T_N = 20$ K)	Weak metal behavior down to 147 K ($\sigma_{rt} = 0.2$ S cm^{-1})	53
$[Au(\alpha\text{-}tpdt)_2]$ **(18)**	PM	Metal (15–300 K, $\sigma_{rt} = 6$ S cm^{-1})	54
$(HBzPy)[Ni(\alpha\text{-}tpdt)_2]$ **(19a)**	FM coupling	-	57b
$(BrBzPy)[Ni(\alpha\text{-}tpdt)_2]$ **(19b)**	AFM coupling (anomaly at 6 K)	-	57b
$(FBzPy)[Ni(\alpha\text{-}tpdt)_2]$ **(19c)**	PM down to 1.5 K	-	57b
$[Ni(\alpha\text{-}tdt)_2]$ **(20c)**; $[Ni(dtdt)_2]$ **(21c)**	Large χ (1–3 μ_B)	Conductors ($\sigma_{rt} = 24$ S cm^{-1} **20c**; $\sigma_{rt} = 200$ S cm^{-1}(**21c**), metallic behavior down to 80 K)	57f
$\{[Mn_4]^{4+}\}\{Pt(mnt)_2\}_4][Pt(mnt)_2]_2$ **(22)**	SMM character of the $[Mn_4]^{4+}$ unit	Semiconductor down to 110 K ($\sigma_{rt} = 0.2$ S cm^{-1}, insulator at $T < 110$ K)	71

(Continued)

Table 7.1 (*Continued*)

Compound	Magnetic properties	Conducting properties	Ref.
$[Mn_2(5\text{-MeOsaltmen})(S)]_2$ $[Ni(dmit)_2]_7$ **(23)**	SMM character	Semiconducting behavior (σ_{rt} = 1.6 S cm^{-1}) **23a**, σ_{rt} = 2.8 S cm^{-1} (**23b**)	70
[Mn(5-MeOsaltmen) $\{M(dmit)_2\}]_2$ **(24)** (M = Ni, **a**, M = Au, **b**)	SMM character of Mn(III) dimers	Semiconductor **(24a)** Insulator **(24b)**	79
{Mn(5-Mesaltmen) $[M(dmit)_2]\}_2$ **(25)** (M = Ni, **a**, M = Au, **b**)	Mn···MnAFM interactions **(25a)** SMM character of Mn(III) dimers **(25b)**	Semiconductor **(25a)** Insulator **(25b)**	79

PM = paramagnetic; FM = ferromagnetic; AFM = antiferromagnetic.

7.3 Processing of Mono/Multifunctional Molecular Materials

In this chapter, we have shown that playing with molecular chemistry, it is possible to design functional materials with the desired physical properties. However, it is difficult to envision a technological breakthrough by using molecular materials in the usual single crystal form, a morphology in which they can be hardly integrated in molecular devices. Preparation as thin films offers an attractive alternative to overcome this difficulty, and encouraging results have actually been emerging during the past 20 years. Several efforts to prepare thin films of metallic, semiconducting, and magnetic materials have been reported by using different techniques, such as thermal sublimation in high vacuum, chemical vapor deposition (CVD), Langmuir–Blodgett (LB) techniques, spin coating, halide evaporation on an organic donor-treated polymer film, adsorption in solution, and electrodeposition on intrinsic micro-rough silicon (001) wafers. For gas-phase techniques, such as sublimation in high vacuum and CVD, precursors should fit specific physical and chemical criteria: volatility, transport in the gas phase without decomposition, etc., and these methods, involving a gas–surface reaction, lead to a better homogeneity of the film. In techniques based on the use of solutions of the molecular building blocks, such as adsorption in solution and electrodeposition, the chemical criteria such as solubility, stability in organic solution, etc. are more familiar to chemists. These techniques have been successfully applied to prepare thin films of molecule-based compounds containing some of the molecular building blocks reported in Chart 7.1, in particular, bis(dithiolene) transition metal complexes and tetrathiafulvalene derivatives for organic field-effect transistors (OFETs). Conductors based on dithiolene complexes have been intensively studied as thin films and a survey of their potential electronic device applications and patents on electrical properties has been reported by Cassoux and Faulmann.[9] Conducting LB films based on $Au(dmit)_2$ dithiolene complex, electrical sensors such as electrodeposited films of $[n\text{-}Bu_4N][Ni(dmit)_2]$ used for detecting SO_2 or NO, electrical switching and memory phenomena on devices formed by uniform thin films based on $Ni(dmit)_2$ and $Ni(dmid)_2$ dithiolene complexes have been reported. When prepared as thin films, intrinsic molecular conductors become semiconductors in most cases because of the formation of segregated domains (grains) separated by boundaries; the transport properties are thus determined by the morphology of the films and studies on the effects of grain boundaries on the intrinsic physical properties of the parent material are essential when targeting technological applications. Few examples of thin films showing metallic behavior down to liquid helium temperatures have been reported such as the ϑ-$(BET\text{-}TTF)_2Br{\cdot}3H_2O$ (BET-TTF = bis(ethylenethio)-tetrathiafulvalene) salt, grown on polycarbonate substrates under exposure to bromine vapor, with room temperature conductivity σ_{rt} = 120 S cm^{-1};[81] the $[TTF][Ni(dmit)_2]_2$ salt, prepared as polycrystalline thin film on silicon wafers by electrocrystallization technique, which show a grain-like morphology[82] and a MI transition at T = 12 K and the single-component neutral molecular metal $Ni(tmdt)_2$, grown as polycrystalline thin film on silicon wafers by electrocrystallization technique,[83] which exhibits metallic behavior down to T = 6 K, with $\sigma_{rt} \approx 100$ S cm^{-1} and $\sigma_{6K} \approx 135$ S cm^{-1}. The SEM image reported in Fig. 7.26 clearly evidences the polycrystalline morphology of the films and XRD studies reveal that the films are formed by $Ni(tmdt)_2$ molecules which crystallize in the triclinic group, the same crystallographic phase as single crystals. As expected, σ_{rt} is clearly lower than the conductivity values found in single crystals (σ_{300K} = 400 S cm^{-1}) because of grain boundary effects. de Caro *et al.* have reported on the preparation of $[Per]_2[Au(mnt)_2]$films, grown electrochemically on a silicon wafer used as anode, which show a nanowire-like morphology (Fig. 7.26), never observed to date.[84]

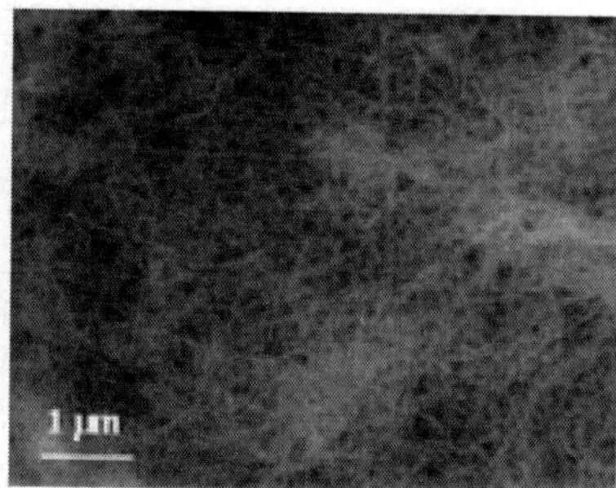

Figure 7.26 SEM image of a neutral thin film of $Ni(tmdt)_2$ (thickness $\approx 4\,\mu m$) grown on a passivated silicon wafer (left). Reprinted from Ref. 83. Copyright 2006, American Chemical Society. SEM image of $[Per]_2[Au(mnt)_2]$nanowire films (right). Reprinted from Ref. 84. Copyright 2007, RSC.

The choice to study $[Au(mnt)_2]^-$ complexes, though diamagnetic, in association with perylene molecules was mainly driven by the high room temperature conductivity value found when growing as α-phase on the silicon wafer. Nanowire films of $[Per]_2[Au(mnt)_2$ show room temperature conductivity, $\sigma_{rt} = 0.02$ S cm^{-1}, 10^4 times lower than that of single crystals and thermally activated semiconducting behavior with activation energy of 0.88 eV in the 100–298 K range. In this context, the most recent and significant advance is represented by the observation by Valade *et al.* of a superconducting transition in a fiber-like film of $[TTF][Ni(dmit)_2]_2$, electrodeposited on silicon substrates.[85] The fiber-like morphology of the films mimics the needle-like morphology of the single crystals and the SEM results suggest that the crystallographic *b*-axis related to the longer dimensions of the fibers is mostly oriented in the plane of the film while both *a*- and *c*-axes are randomly oriented. A broad decrease of the resistance below 0.8 K has been observed under a hydrostatic pressure of 7.7 kbar, and this incomplete superconducting transition is shifted to lower temperatures by increasing the magnetic field. These results evidence that the electrodeposition technique affords a promising way to produce films with the same properties as single crystals but, in comparison to vapor phase methods, it is more versatile and it can be used for all kind of molecular conductors. The very recent use of tetrathiafulvalene derivatives for fabricating organic field- OFETs either from vacuum deposition or from solution has been highlighted by Rovira[86] owing to the easy processability and high device performance of these materials for applications in modern microelectronics. TTF derivatives seem to be good candidates for the preparation of OFETs, due to (i) the possibilities of synthesizing tailored derivatives; (ii) the crystallization of TTF derivatives is controlled by the π–π stacking, which, together with the S···S interactions allows for an intermolecular electronic transfer responsible for their transport properties; (iii) TTF derivatives are generally soluble in various solvents and are easily modified by chemical methods. The electrical measurements, performed on the TTF derivatives reported in Chart 7.6 were in agreement with theoretical calculations, that is, the materials showing higher mobility have lower reorganization energy (which needs to be small for efficient charge transport) and larger transfer integral (the electronic coupling between adjacent molecules).[87] In addition, further calculations demonstrated that the reorganization energy values strongly depend on the crystal structure and, thus, investigating the role of the intermolecular interactions on TTF crystals is crucial to understand their transport properties.[88] These results are very promising for the design of new materials for organic field-effect transistors.

Chart 7.6 Reprinted from Ref. 86. Copyright 2006, RSC.

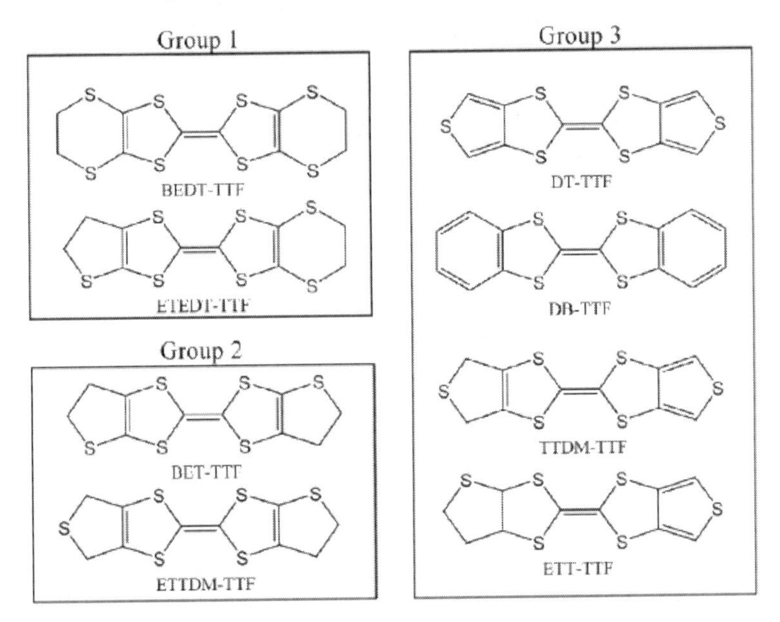

Also SMMs, for their quantum physical properties, have been proposed as potential candidates for several technological applications such as high-density magnetic storage, sensors, quantum computing, and spintronics that require highly controlled thin films and artificially organized micro- and nanopatterns. Preparation of thin films of SMMs, especially based on Mn_{12} size-defined cluster, have been reviewed by Coronado[89] along with the different methods used for the deposition and organization of SMMs on surfaces. The noncovalent approach successfully followed by Cavallini *et al.*, should be highlighted here.[90] Using stamp-assisted techniques, they demonstrated how the patterning over native silicon oxide of aggregates formed by a few hundred molecules of Mn_{12}-(phenylbenzoate) could be achieved. This technique has provided size and distance control on multiple-lengths scale (Fig. 7.27).

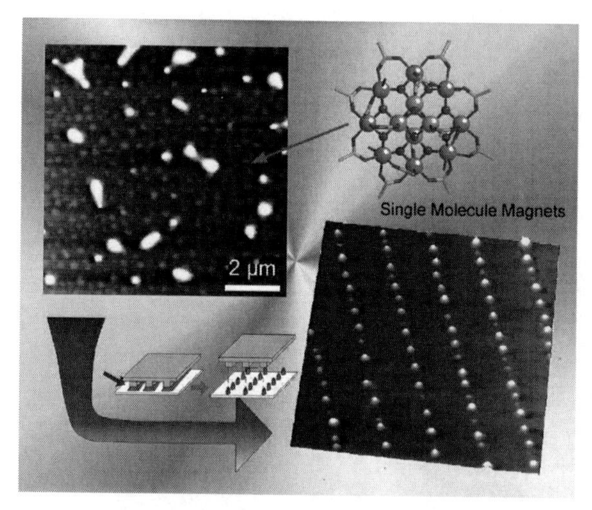

Figure 7.27 AFM image of printed Mn_{12} SMM on Si(100) with native SiO_x. Adapted from Ref. 90. Copyright 2008, RSC.

Although these methods are developed on a laboratory scale, some of them such as the patterning by modeling of homogeneous films and lithographically controlled de-mixing, can be easily integrated in traditional industrial nanofabrication methods for future industrial developments.

In conclusion, the processing of molecular materials into thin films, nanowires or embedded in polymeric matrices represent the forefront of the molecular electronics and other nanotechnologies. The preparation of thin films of multifunctional materials still remain an unexplored field, which is highly appealing for their potential applications as "smart" materials in "smart" switches, sensors, etc.

7.4 Conclusion

The compounds described in this chapter, summarized in Table 7.1, clearly show that the biggest challenge of molecule-based materials can be envisaged in the design and synthesis of the molecular building blocks, some of them reported in Chart 7.1. In this context, the molecular approach plays a key role since it enables their chemical design allowing for fine-tuning of the physical properties. In fact, it is possible to control the packing of the organic network and thus the conducting properties by playing with the size, shape, symmetry, and charge of the inorganic counterions, which determine the intermolecular interactions, while the strength and directionality of these interactions yield to the supramolecular structure of the material, targeting for its desired properties. The use of noncovalent interactions is absolutely critical in this regard and the molecular–supramolecular balance has to be strictly controlled. These interactions such as van der Waals interactions, π–π, and π–d interactions, H-bonding, etc. play a crucial role in self-assembling these predesigned molecular units and may provide a powerful way to afford layered molecular materials with new or unknown physical properties. As an example, the intermolecular chalcogen–chalcogen S(O)···S(O, Se) interactions, in addition to other contacts such as S···I, S···O, S···N, etc. between adjacent molecules have yielded to 2-D or 3-D networks and have been shown to be responsible for the occurrence of the metallic state observed in SCMMs and molecular magnetic conductors up to low temperatures. Future work must take the opportunity to build on the current understanding of the way these molecules assemble. Thus, new studies must be increasingly target driven and involve preparation of particular material types by rational analysis of the known behavior of the component building blocks or by synthesis of new building blocks expected to interact in the solid state through the above-mentioned inter molecular interactions in a predetermined way. However, BEDT-TTT, BETS, dithiolene complexes, and polymeric metal oxalate complexes still remain the most successful molecular bricks to play with, since their assemblage through the above-mentioned interactions, might yield materials with exotic properties out of SCMM, SMMs, and chiral molecular conductors, which are of utmost importance for the development of applications in molecular electronics, spintronics, and other related areas of current interest in material science. In particular, the magneto-chiral effects in such multifunctional systems are worth to be investigated at different levels.[91] Crystal engineering of these hybrid molecule-based materials is in its infancy right now and many appealing combinations of the above-mentioned molecular bricks can still obtained. However, it is difficult to envision a technological breakthrough by using them in the usual single-crystal form and their processing as thin films, nanowires, or embedded in polymeric matrices offers an attractive alternative to overcome this difficulty and to represent the forefront of the molecular electronics and other related nanotechnologies.

Acknowledgments

This work was supported by the University of Cagliari and the COST Action D35, WG11 "Multifunctional and Switchable Molecular Materials." Special thanks to the guest editor, Lahcène Ouahab, for his kind invitation to give a contribution to this book. Finally, I'd like to dedicate this chapter "to my aunty Maria Rosaria" – Maria Laura.

References

1. Cambi, L., Gagnasso, A. (1931). Iron dithiocarbamates and nitrosodithiocarbamates, *Atti. Accad. Naz. Lincei,* **13**, 809–813.
2. Gütlich, P., Ksenofontov, V, Gaspar, A. B. (2005). Pressure effect studies on spin crossover systems, *Coord. Chem. Rev.,* **249**, 1811–1829.
3. Williams, J. M., Ferraro, J. R., Thorn, R. J., Carlson, K. D., Geiser, U., Wang, H. H., Kini, A. M, Whangbo M. H. (1992). *Organic Superconductors,* Prentice Hall, Englewood Cliffs.
4. Miller, J. S, Drillon, M. (2004). *Magnetism: Molecules to Materials,* Wiley-VCHs, Weinheim, Vol. 1–4.
5. (a) Prasad, P. N, Williams, D. J. (1991). *Introduction of Nonlinear Optical Effects In Molecules And Polymers,* J. Willey and Sons, New York; (b) Chemla, D. S, Zyss, J. (1987). *Nonlinear Optical Properties of Organic Molecules and Crystals,* Academic Press, New York, Vol. 1, 2.

6. (a) Curry, R. J, Gillin, W. P. (1999). 1.54 μm electroluminescence from erbium (III) tris(8-hydroxyquinoline) ErQ.-based organic light-emitting diodes, *Appl. Phys. Lett.*, **75**, 1380–1382; (b) Gillin, W. P, Curry, R. J. (1999). Erbium (III) tris(8-hydroxyquinoline) (ErQ): A potential material for silicon compatible 1.5 μm emitters, *Appl. Phys. Lett.*, **74**, 798–799; (b) Mitscke, U, Bauerle, P. (2000). The electroluminescence of organic materials, *J. Mater. Chem.*, **10**, 1471–1507; (c) Kelley, T. W., Baude, P. F., Gerlach, C., Ender, D. E., Muyres, D., Haase, M. A., Vogel, D. E, Theiss, S. D. (2004). Recent progress in organic electronics: Materials, devices, and processes, *Chem. Mater.*, **16**, 4413–4422.

7. Wolf, S. A., Awschalom, D. D., Burham, R. A., Daughton, J. M., von Molnár, S., Roukes, M. L., Chtchelkanova, A. Y, Treger, D. M. (2001). Spintronics: A spin-based electronics vision for the future, *Science*, **294**, 1488–1495, and references therein.

8. Fraxedas, J. (2002). Perspectives on thin molecular organic films, *Adv. Mater.*, **14**, 1603–1614, and references therein.

9. Faulmann, C, Cassoux, P. (2004), Dithiolene Chemistry: Synthesis, Properties and Applications, In: *Progress in Inorganic Chemistry*, 52nd edn. (Stiefel, E. I., ed.), Wiley, Chichester, pp. 399–489.

10. Robertson, N, Cronin, L. (2002). Metal bis-1,2-dithiolene complexes in conducting or magnetic crystalline assembles, *Coord. Chem. Rev.*, **227**, 93–127.

11. Kato, R. (2004). Conducting metal dithiolene complexes: Structural and electronic properties, *Chem. Rev.*,**104**, 5319–5346.

12. Kobayashi, A., Fujiwara, E, Kobayashi, H. (2004). Single-component molecular metals with extended-TTF dithiolate ligands, *Chem. Rev.*, **104**, 5243–5264.

13. Coronado, E, Day, P. (2004). Magnetic molecular conductors, *Chem. Rev.*, **104**, 5419–5448.

14. Coronado, E, Galán-Mascarós, J.-R. (2008). Molecule-based ferromagnetic conductors: Strategy and design, *C. R. Chimie*, **11**, 1110–1116.

15. Mercuri, M. L., Deplano, P., Pilia, L., Serpe A, Artizzu F. (2010). Interactions modes and physical properties in transition metal chalcogenolene-based molecular materials, *Coord. Chem. Rev.*, **254**, 1419–1433.

16. (a) Graham, A. W., Kurmoo, M, Day, P. (1995). β''-(bedt-ttf)$_4$[(H$_2$O) Fe(C$_2$O$_4$)$_3$]·PhCN: The first molecular superconductor containing paramagnetic metal ions, *J. Chem. Soc., Chem. Commun.*, 2061–2062; (b) Enoki, T, Miyazaki, A. (2004). Magnetic TTF-based charge-transfer

complexes, *Chem. Rev.*, **104**, 5449–5478; (c) Kobayashi, H., Cui, H. B, Kobayashi, A. (2004). Organic metals and superconductors based on BETS (BETS = Bis(ethylenedithio)tetraselenafulvalene), *Chem. Rev.*, **104**, 5265–5288; (d) Kurmoo, M., Graham, A. W., Day, P., Coles, S. J., Hursthouse, M. B., Caulfield, J. L., Singleton, J., Pratt, F. L., Hayes, W., Ducasse, L, Guionneau, P. (1995). Superconducting and semiconducting magnetic charge transfer salts: (BEDT-TTF)$_4$AFe(C_2O_4)$_3$·cntdot.C_6H_5CN (A = H_2O, K, NH_4), *J. Am. Chem. Soc.*, **117**, 12209–12217; (e) Martin, L., Turner, S. S., Day, P., Mabbs, F. E, McInnes, E. J. L. (1997). New molecular superconductor containing paramagnetic chromium(III) ions, *Chem. Commun.*, 1367–1368; (f) Rashid, S., Turner, S. S., Day, P., Howard, J. A. K., Guionneau, P., McInnes, E. J. L., Mabbs, F. E., Clark, R. J. H., Firth, S, Biggs, T. (2001). New superconducting charge-transfer salts (BEDT-TTF)$_4$[A·M(C_2O_4)$_3$]·$C_6H_5NO_2$ (A = H_3O or NH_4, M = Cr or Fe, BEDT-TTF = bis(ethylenedithio)tetrathiafulvalene), *J. Mater. Chem.*, **11**, 2095–2101.

17. Fujiwara, H., Fujiwara, E., Nakazawa, Y., Narymbetov, B. Zh., Kato, K., Kobayashi, H., Kobayashi, A., Tokumoto, M, Cassoux, P. (2001). A novel antiferromagnetic organic superconductor κ-(BETS)$_2$FeBr$_4$ [where BETS = Bis(ethylenedithio)tetraselenafulvalene], *J. Am. Chem. Soc.*, **123**, 306–314.

18. Coronado, E., Galán-Mascarós, J. R., Gómez-García, C. J, Laukhin, V. N. (2000). Coexistence of ferromagnetism and metallic conductivity in a molecule-based layered compound, *Nature*, **408**, 447–449.

19. Uij, S., Shinagawa, H., Terashima, T., Yakabe, T., Terai, Y., Tokumoto, M., Kobayashi, A., Tanaka, H, Kobayashi, H. (2001). Magnetic-field-induced superconductivity in a two-dimensional organic conductor, *Nature*, **410**, 908–910.

20. (a) Zhang, B., Wang, Z., Zhang, Y., Kobayashi, H., Kurmoo, M., Mori, T., Pratt, F. L., Inoue, K, Zhu, D. (2007). Dual-function molecular crystal containing the magnetic chain anion [Fe^{III}(C_2O_4)Cl_2^-]$_n$, *Polyhedron*, **26**, 1800–1804; (b) Fujiwara, H., Wada, K., Hiraoka, T., Hayashi, T., Sugimoto, T., Nakazumi, H., Yokogawa, K., Teramura, M., Yasuzuka, S., Murata, K., Mori, T. (2005). Stable Metallic Behavior and Antiferromagnetic Ordering of Fe(III) *d* Spins in (EDO-TTFVO)$_2$·FeCl$_4$, *J. Am. Chem. Soc.*, **127**, 14166–14167 and references therein.

21. Zhang, B., Wang, Z. M., Zhang, Y., Takahashi, K., Okano, Y., Cui, H. B., Kobayashi, H., Inoue, K., Kurmoo, M., Pratt, F. L, Zhu, D. B. (2006). Hybrid organic-inorganic conductor with a magnetic chain anion: -(BETS)$_2$[Fe^{III}-(ox)Cl_2] [BETS = Bis(ethylenedithio)tetraselenafulvalene], *Inorg. Chem.*, **45**, 3275–3280.

22. Robertson, N., Awaga, K., Parsons, S., Kobayashi, A, Underhill, A. E. (1998). $[TTF]_2[Fe(tdas)_2]$: A molecular conductor containing magnetic counter-ions, *Adv. Mater. Opt. Electron.*, **8**, p. 93–96.

23. Awaga, K., Okuno, T., Maruyama, Y., Kobayashi, A., Kobayashi, H., Schenk, S, Underhill, A. E. (1994). Possible "reentrant" behavior in magnetic properties of $TBA[Fe(tdas)_2]$ Complex, *Inorg. Chem.*, **33**, 5598–5600.

24. Mercuri, M. L., Deplano, P., Leoni, L., Schlueter, J. A., Geiser, U., Wang, H. H., Kini, A. M., Manson, J. L., Gómez-García, C., Coronado, E, Whangbo, M. H. (2002). A two-dimensional radical salt based upon BEDT-TTF and the dimeric, magnetic anion $[Fe(tdas)_2]_2$ 2-: $(BEDT\text{-}TTF)_2$ $[fe(tdas)_2]$ (tdas = 1,2,5-thiadiazole-3,4-dithiolate), *J. Mater. Chem.*, **12**, 3570–3577.

25. Pilia, L., Faulmann, C., Malfant, I., Collière, V., Mercuri, M. L., Deplano, P, Cassoux, P. (2002). $(BETS)_2[Fe(tdas)_2]_2$: A new metal in the molecular conductor family, *Acta Cryst. C*, **58**, 240–242.

26. Curreli, S., Deplano, P., Mercuri, M. L., Pilia, L., Serpe, A., , Schlueter, J. A., Whited, M. A., Geiser, U., Coronado, E., Goméz-García, C. J, Canadell, E. (2004). Synthesis, crystal structure, and physical properties of (BEDT-TTF)$[Ni(tdas)_2]$ (BEDT-TTF = Bis(ethylenedithio)tetrathiafulvalene; tdas = 1,2,5-thiadiazole-3,4-dithiolate): First monomeric $[Ni(tdas)_2]$-monoanion, *Inorg. Chem.*, **43**, 2049–2056.

27. Akutagawa, T., Shitagami, K., Aonuma, M., Noro, S, Nakamura, T. (2009). Ferromagnetic and antiferromagnetic coupling of $[Ni(dmit)_2]^-$ anion layers induced by Cs_2^+(benzo[18]crown-6)$_3$ supramolecule, *Inorg. Chem.*, **48**, 4454–4461.

28. Kosaka, Y., Yamamoto, H. M., Nakao, A., Tamura, M, Kato, R. (2007). Coexistence of conducting and magnetic electrons based on molecular π-electrons in the supramolecular conductor (Me-3,5-DIP)$[Ni(dmit)_2]_2$, *J. Am. Chem. Soc.*, **129**, 3054–3055.

29. Coomber, A. T., Beljonne, D., Friend, R. H., Brédas, J. L., Charlton, A., Robertson, N., Underhill, A. E., Kurmoo, M, Day, P. (1996). Intermolecular interactions in the molecular ferromagnetic $NH_4Ni(mnt)_2 \cdot H_2O$, *Nature*, **380**, 144–146.

30. Avarvari, N, Wallis, J. D. (2009). Strategies towards chiral molecular conductors, *J. Mater. Chem.*, **19**, 4061–4076.

31. (a) Coronado E, Galán-Mascarós, J. R. (2005). Hybrid molecular conductors, *J. Mater. Chem.*, **15**, 66–74; (b) Day, P, Kurmoo, M. (1997). Molecular magnetic semiconductors, metals and superconductors: BEDT-TTF salts with magnetic anions, *J. Mater. Chem.*, **7**, 1291–1295.

32. Madalan, A. M., Canadell, E., Auban-Senzier, P., Brânzea, D., Avarvari N, Andruh, M. (2008). Conducting mixed-valence salt of bis(ethylenedithio) tetrathiafulvalene (BEDT-TTF) with the paramagnetic heteroleptic anion $[Cr^{III}(oxalate)_2(2,2'\text{-bipyridine})]^-$, *New J. Chem.*, **32**, 333–339.

33. Clemente-León, M., Coronado, E., Gómez-García, C. J., Soriano- Portillo, A., Constant, S., Frantz, R, Lacour, J. (2007). Unusual packing of ET molecules caused by π–π stacking interactions with TRISPHAT molecules in two [ET][TRISPHAT] salts (ET = bis(ethylenedithio)tetrathiafulva lene, TRISPHAT = (tris(tetrachlorobenzenediolato)phosphate(V))), *Inorg. Chim. Acta*, **360**, 955–960.

34. Coronado, E., Curreli, S., Giménez-Saiz, C., Gómez-García, C. J., Deplano, P., Mercuri, M. L., Serpe, A., Pilia, L., Faulmann, C, Canadell, E. (2007). New BEDT-TTF/$[Fe(C_5O_5)_3]^{3-}$ hybrid system: Synthesis, crystal structure, and physical properties of a chirality-induced α phase and a novel magnetic molecular metal, *Inorg. Chem.*, **46**,4446–4457.

35. Castro, I., Calatayud, M. L., Lloret, F., Sletten, J, Julve, M. (2002). Syntheses, crystal structures and magnetic properties of di- and trinuclear croconato-bridged copper(II) complexes, *J. Chem. Soc., Dalton Trans.*, 2397–2403.

36. Martin, L., Turner, S. S., Day, P., Abdul Malik, K. M., Coles, S. J, Hursthouse, M. B. (1999). Polymorphism based on molecular stereoisomerism in tris(oxalato) Cr(III) salts of bedt-ttf [bis(ethylenedithio)tetrathiafulval ene], *Chem. Commun.*, 513–514.

37. Martin, L., Day, P., Akutsu, H., Yamada, J.-I., Nakatsuji, S.-I., Clegg, W., Harrington, R. W., Horton, P. N., Hursthouse, M. B., McMillan, P, Firth, S. (2007). Metallic molecular crystals containing chiral or racemic guest molecules, *Cryst. Eng. Comm.*, **9**, 865–867.

38. (a) Martin, L., Day, P., Horton, P., Nakatsuji, S., Yamada, J, Akutsu, H. (2010). Chiral conducting salts of BEDT-TTF containing a single enantiomer of tris(oxalato)chromate(III) crystallised from a chiral solvent, *J. Mater. Chem.*, **20**, 2738–2742; (b) Martin, L., Day, P., Nakatsuji, S., Yamada, J., Akutsu, H, Horton, P. (2010). A molecular charge transfer salt of BEDT-TTF containing a single enantiomer of tris(oxalato) chromate(III) crystallised from a chiral solvent, *Cryst. Eng. Comm.* **12**, 1369–1372.

39. (a) Martin, L., Turner, S. S., Day, P., Guionneau, P., Howard, J. A. K., Uruichi, M, Yakushi, K. (1999). Synthesis, crystal structure and properties of the semiconducting molecular charge-transfer salt $(bedt\text{-}ttf)_2Ge(C_2O_4)_3{\cdot}PhCN$ [bedt-ttf = bis(ethylenedithio)tetrathiafulv alene], *J. Mater. Chem.*, **9**, 2731–2736; (b) Prokhorova, T. G., Khasanov,

S. S., Zorina, L. V., Buravov, L. I., Tkacheva, V. A., Baskakov, A. A., Morgunov, R. B., Gener, M., Canadell, E., Shibaeva, R. P, Yagubskii, E. B. (2003). Molecular Metals Based on BEDT-TTF Radical Cation Salts with Magnetic Metal Oxalates as Counterions: β''-(BEDT-TTF)$_4$A[M(C_2O_4)$_3$]·DMF (A = NH^{4+}, K^+; M = Cr^{III}, Fe^{III}), *Adv. Funct. Mater.*, **13**, 403–411; (c) Coronado, E., Curreli, S., Giménez-Saiz, C., Gómez-García, C. J. (2005). A novel paramagnetic molecular superconductor formed by bis(ethylenedithio) tetrathiafulvalene, tris(oxalato)ferrate(III) anions and bromobenzene as guest molecule: $ET_4[(H_3O)Fe(C_2O_4)_3]\cdot C_6H_5Br$, *J. Mater. Chem.*, **15**, 1429–1436.

40. Galán-Mascarós, J.-R., Coronado, E., Goddard, P. A., Singleton, J., Coldea, A. I., Wallis, J. D., Coles, S. J, Alberola, A. (2010). A chiral ferromagnetic molecular metal, *J. Am. Chem. Soc.*, **132**, 9271–9273.

41. (a) Wallis, J. D., Karrer, A., Dunitz, J. D. (1986). Chiral metals? A chiral substrate for organic conductors and superconductors, *Helv. Chim. Acta*, **69**, 69–70; (b) Karrer, A., Wallis, J. D., Dunitz, J. D., Hilti, B., Mayer, C. W., Bürkle, M., Pfeiffer, J. (1987). Structures and Electrical Properties of Some New Organic Conductors Derived from the Donor Molecule TMET (*S,S,S,S*,-Bis(dimethylethylenedithio) tetrathiafulvalene), *Helv. Chim. Acta*, **70**, 942–953.

42. Kobayashi, A., Tanaka, H., Kumasaki, M., Torii, M., Narymbetov, B, Adachi, T. (1999). Origin of the High Electrical Conductivity of Neutral [Ni(ptdt)$_2$] (ptdt^{2-} = propylenedithiotetrathiafulvalenedithiolate): A Route to Neutral Molecular Metal, *J. Am. Chem. Soc.*, **121**, 10763–10771.

43. Tanaka, H., Okano, Y., Kobayashi, A., Suzuki, W, Kobayashi, H. (2001). A Three-Dimensional Synthetic Metallic Crystal Composed of Single-Component Molecules, *Science*, **291**, 285–287.

44. Tanaka, H., Kobayashi, A, Kobayashi, H. (2002). A Conducting Crystal Based on A Single-Component Paramagnetic Molecule, [Cu(dmdt)$_2$] (dmdt = Dimethyltetrathiafulvalenedithiolate), *J. Am. Chem. Soc.*, **124**, 10002–10003.

45. Harris, N. J, Underhill, A. E. (1987). Preparation, structure, and properties of $[Ni\{S_2C_2(CF_3)_2\}_2]^-$ and $[Ni\{Se_2C_2(CF_3)_2\}_2]^-$ salts with small cations, *J. Chem. Soc., Dalton Trans.*, 1683–1685.

46. Hara, Y., Miyagawa, K., Kanoda, K., Shimamura, M., Zhou, B., Kobayashi, A, Kobayashi, H. (2008). NMR Evidence for Antiferromagnetic Transition in the Single-Component Molecular Conductor, [Au(tmdt)$_2$] at 110 K, *J. Phys. Soc. Jpn.*, **77**, 053706 (4 pages).

47. Zhou, B., Shimamura, M., Fujiwara, E., Kobayashi, A., Higashi, T., Nishibori, E., Sakata, M., Cui, H., Takahashi, K, Kobayashi, H. (2006). Magnetic

Transitions of Single-Component Molecular Metal [Au(tmdt)$_2$] and Its Alloy Systems, *J. Am. Chem. Soc.*, **128**, 3872–3873.

48. Ishibashi, S., Tanaka, H., Kohyama, M., Tokumoto, M., Kobayashi, A., Kobayashi, H, Terakura, K. (2005). *Ab Initio* Electronic Structure Calculation for Single-Component Molecular Conductor Au(tmdt)$_2$ (tmdt = Trimethylenetetrathiafulvalenedithiolate), *J. Phys. Soc. Jpn.*, **74**, 843–846.

49. Miyagawa, K., Kawamoto, A., Nakazawa, Y. and Kanoda, K. (1995). Antiferromagnetic Ordering and Spin Structure in the Organic Conductor, κ-(BEDT-TTF)$_2$Cu[N(CN)$_2$]Cl, *Phys. Rev. Lett.*, **75**, 1174–1177.

50. Zhou, B., Kobayashi, A., Okano, Y., Cui, H., Graf, D., Brooks, J. S., Nakashima, T., Aoyagi, S., Nishibori, E., Sakata, M, Kobayashi, H. (2009). Structural Anomalies Associated with Antiferromagnetic Transition of Single-Component Molecular Metal [Au(tmdt)$_2$], *Inorg. Chem.*, **48**, 10151–10157.

51. Fujiwara, E., Kobayashi, A., Fujiwara, H, Kobayashi, H. (2004). Syntheses, Structures, and Physical Properties of Nickel Bis(dithiolene) Complexes Containing Tetrathiafulvalene (TTF) Units, *Inorg. Chem.*, **43**, 1122–1129.

52. Estes, W. E., Gavel, D. P., Hatfield, W. E, Hodgson, D. (1978). Magnetic and structural characterization of dibromo- and dichlorobis(thiazole) copper(II), *Inorg. Chem.*, **17**, 1415–1421.

53. Fujiwara, E., Yamamoto, K., Shimamura, M., Zhou, B., Kobayashi, A., Takahashi, K., Okano, Y., Cui, H, Kobayashi, H. (2007). (^{n}Bu$_4$N)[Ni(dmstfdt)$_2$]: A Planar Nickel Coordination Complex with an Extended-TTF Ligand Exhibiting Metallic Conduction, Metal–Insulator Transition, and Weak Ferromagnetism, *Chem. Mater.*, **19**, 553–558.

54. (a) Le Narvor, N., Robertson, N., Weyland, T., Kilburn, J. D., Underhill, A. E., Webster, M., Svenstrup, N, Becher, J. (1996). Synthesis, structure and properties of nickel complexes of 4,5-tetrathiafulvalene dithiolates: high conductivity in neutral dithiolate complexes, *J. Chem. Soc., Chem. Commun.*, 1363–1364; (b) Belo, D., Alves, H., Lopes, E. B., Duarte, M. T., Gama, V., Henriques, R. T., Almeida, M., Pérez-Benítez, A., Rovira, C, Veciana, J. (2001). Gold Complexes with Dithiothiophene Ligands: A Metal Based on a Neutral Molecule, *Chem. Eur. J.*, **7**, 511–519; (c) Belo, D., Alves, H., Lopes, E. B., Duarte, M. T., Gama, V., Henriques, R. T., Almeida, M., Pérez-Benítez, A., Rovira, C, Veciana, J. (2001). New dithiothiophene complexes for conducting and magnetic materials, *Synth. Met.*, **120**, 699–702.

55. Belo, D., Figueira, M. J., Mendonça J., Santos, I. C., Almeida, M., Henriques, R. T., Duarte, M. T., Rovira, C, Veciana, J. (2005). Copper, cobalt and

platinum complexes with dithiothiophene-based ligands, *Eur. J. Inorg. Chem.*, 3337–3345.

56. Belo, D, Almeida, M. (2010). Transition metal complexes based on thiophene-dithiolene ligands, *Coord. Chem. Rev.*, **254**, 1479–1492.

57. (a) Belo, D., Alves, H., Rabaça, S., Pereira, L. C., Duarte, M. T., Gama, V., Henriques, R. T., Almeida, M., Ribera, E., Rovira, C, Veciana, J. (2001). Nickel complexes based on thiophenedithiolate ligands: Magnetic properties of metallocenium salts, *Eur. J. Inorg. Chem.*, 3127–3133; (b) Belo, D., Figueira, M. J., Nunes, J. P. M., Santos, I. C., Pereira, L. C., Gama, V., Almeida, M, Rovira, C. (2006). Magnetic properties of RBzPy[Ni (α-tpdt)$_2$] (R = H, Br, F): effects of *cis–trans* disorder, *J. Mater. Chem.*, **16**, 2746–2756; (c) Belo, D. Mendonça, J., Santos, I. C., Pereira, L. C. J., Almeida, M., Novoa, J. J., Rovira, C., Veciana, J, Gama, V. (2008). Metallocenium salts of nickel Bis(α-thiophenedithiolate) [M(Cp*)$_2$][Ni(α-tpdt)$_2$] (M = Fe, Mn, Cr): Metamagnetism and magnetic frustration, *Eur. J. Inorg. Chem.*, 5327–5337; (d) Belo, D., Pereira, L. C., Almeida, M., Rovira, C., Veciana, J, Gama, V. (2009). Magnetization inverted hysteresis loops in the molecular magnets[M(Cp*)$_2$][Ni(α-tpdt)$_2$] (M = Fe, Mn). *Dalton Trans.*, 4176–4180; (e) Naves, A. I. S., Dias, J. C., Vieira, B. J. C., Bronco, M. B. C., Pereira, L. C. J., Santos, I. C., Waerenborgh, J. C., Almeida, M., Belo, D, Gama, V. (2009). A new hybrid material exhibiting room temperature spin-crossover and ferromagnetic cluster-glass behavior, *Cryst. Eng. Comm.*, **11**, 2160–2168; (f) Nunes, J. P. M., Figueira, M. J., Belo, D., Santos, I. C., Ribeiro, B., Lopes, E. B., Henriques, R. T., Vidal-Gancedo, J., Veciana, J., Rovira, C, Almeida, M. (2007). Transition metal bisdithiolene complexes based on extended ligands with fused tetrathiafulvalene and thiophene moieties: New single-component molecular metals, *Chem. Eur. J.*, **13**, 9841–9849.

58. Ni, Z., Ren, X. M., Ma, J., Xie, J., Ni, C., Chen, Z, Meng, Q. (2005). Theoretical studies on the magnetic switching controlled by stacking patterns of Bis(maleonitriledithiolato) nickelate(III) dimers, *J. Am. Chem. Soc.*, **127**, 14330–14338.

59. Belo, D., Figueira, M. J., Nunes, J. P. M., Santos, I. C., Almeida, M., Crivillers, N, Rovira, C. (2007). Synthesis and characterization of the novel extended TTF-type donors with thiophenic units, *Inorg. Chim. Acta*, **360**, 3909–3914.

60. Sessoli, R., Tsai, H.-L., Schake, A. R., Wang, S., Vincent, J. B., Folting, K., Gatteschi, D., Christou, G, Hendrickson, D. N. (1993). High-spin molecules: [$Mn_{12}O_{12}(O_2CR)_{16}(H_2O)_4$], *J. Am. Chem. Soc.*, **115**, 1804–1816.

61. (a) Gatteschi, D., Caneschi, A., Pardi, L, Sessoli, R. (1994). *Science*, **265**, 1054;(b) Christou, G., Gatteschi, D., Hendrickson, D. N, Sessoli, R. (2000). Single-molecule magnets, *MRS Bull.*, **25**, 66–71; (c) Gatteschi, D., Sessoli, R. (2003). Quantum Tunneling of Magnetization and Related Phenomena in Molecular Materials, *Angew. Chem., Int. Ed. Engl.*, **42**, 268–297; (d) Gatteschi, D., Sessoli, R, Villain, J. (2006). *Molecular Nanomagnets*, Oxford University Press.

62. (a) Thomas, L., Lionti, F., Ballou, R., Gatteschi, D., Sessoli, R, Barbara, B. (1996). Macroscopic quantum tunnelling of magnetization in a single crystal of nanomagnets, *Nature*, **383**, 145–147; (b) Friedman, J. R., Sarachik, M. P., Tejada, J., Ziolo, R. (1996). Macroscopic Measurement of Resonant Magnetization Tunneling in High-Spin Molecules, *Phys. Rev. Lett.*, **76**, 3830–3833; (c) Barra, A.-L., Brunel, L.-C., Gatteschi, D., Pardi, L, Sessoli, R. (1998). High-Frequency EPR Spectroscopy of Large Metal Ion Clusters: From Zero Field Splitting to Quantum Tunneling of the Magnetization, *Acc. Chem. Res.*, **31**, 460–466; (d) Friedman, J. R., Sarachik, M. P., Hernandez, J. M., Zhang, X. X., Tejada, J, Molins, E. (1997). Effect of a transverse magnetic field on resonant magnetization tunneling in high-spin molecules, *J. Appl. Phys.*, **81**, 3978-3981; (e) Barra, A. L., Gatteschi, D, Sessoli, R. (1997). High-frequency EPR spectra of a molecular nanomagnet: Understanding quantum tunneling of the magnetization, *Phys. Rev. B*, **56**, 8192–8198; (e) Wernsdorfer, W, Sessoli, R. (1999). Quantum Phase Interference and Parity Effects in Magnetic Molecular Clusters, *Science*, **284**, 133–135; (f) Wernsdorfer, W., Sessoli, R., Caneschi, A., Gatteschi, D., Cornia, A, Mailly, D. (2000). Landau–Zener method to study quantum phase interference of Fe_8 molecular nanomagnets,*J. Appl. Phys.*, **87**, 5481–5486; (g) Wernsdorfer, W., Bhaduri, S., Boskovic, C., Christou, G, Hendrickson, D. N. (2002). Spin-parity dependent tunneling of magnetization in single-molecule magnets, *Phys. Rev. B*, **65**, 180403 (4 pages); (h) Wernsdorfer, W., Aliaga-Alcalde, N., Hendrickson, D. N, Christou, G. (2002). Exchange-biased quantum tunnelling in a supramolecular dimer of single-molecule magnets, *Nature*, **416**, 406–409; (i) Solace, L., Wernsdorfer, W., Thereon, C. , Barra, A.-L., Puccini, M., Mailly, D, Barbara, B. (2003). In-plane magnetic reorientation in coupled Ferro- and antiferromagnetic thin films, *Phys. Rev. B*, **68**, 220407 (4 pages).

63. (a) Cornia, A., Barbette, A. C., Puccini, M., Bobbi, L., Bronchi, D., Caneschi, A., Gatteschi, D., Big, R., Del Penning, U., De Rienzi, V., Gurevich, L, van der Zant, H. S. J. (2003). Direct Observation of Single-Molecule Magnets Organized on Gold Surfaces,*Angew. Chem., Int. Ed. Engl.*, **42**, 1645–1648; (b) Clemente-León, M., Coronado, E., Forment-Aliaga, A., Amorós, P., Ramírez-Castellanos, J, González-Calbet, J. M. (2003). Incorporation of

Mn_{12} single molecule magnets into mesoporous silica, *J. Mater. Chem.*, **13**, 3089–3095; (c) Clemente-León, M., Coronado, E., Forment-Aliaga, A, Romero, F. M. (2003). Organized assemblies of magnetic clusters, *C. R. Chimie*, **6**, 683–688.

64. Forment-Aliaga, A., Coronado, E., Feliz, M., Gaita-Ariño, A., Llusar, R, Romero, F. M. (2003). Cationic Mn_{12} Single-Molecule Magnets and Their Polyoxometalate Hybrid Salts, *Inorg. Chem.*, **42**, 8019–8027.

65. Bolink, H. J., Coronado, E., Forment-Aliaga, A, Gómez-García, C. J. (2005). Conductive Hybrid Films of Polyarylamine Electrochemically Oxidized with the Molecular Nanomagnet $[Mn_{12}O_{12}(H_2O)_4(C_6F_5COO)_{16}]$, *Adv. Mater.*, **17**, 1018–1023.

66. Leuenberger, M. N, Loss, D. (2001). Quantum computing in molecular magnets, *Nature*, **410**, 789–793.

67. Miyasaka, H, Yamashita, M. (2007). A look at molecular nanosized magnets from the aspect of inter-molecular interactions, *Dalton Trans.*, 399–406.

68. (a) Clérac, R., Miyasaka, H., Yamashita, M, Coulon, C. (2002). Evidence for Single-Chain Magnet Behavior in a Mn^{III}–Ni^{II} Chain Designed with High Spin Magnetic Units: A Route to High Temperature Metastable Magnets, *J. Am. Chem. Soc.*, **124**, 12837–12844; (b) Miyasaka, H., Clérac, R., Mizushima, K., Sugiura, K., Yamashita, M., Wernsdorfer, W, Coulon, C. (2003). $[Mn_2(saltmen)_2Ni(pao)_2(L)_2](A)_2$ with L = Pyridine, 4-Picoline, 4-*tert*-Butylpyridine, *N*-Methylimidazole and A = ClO_4^-, BF_4^-, PF_6^-, ReO_4^-: A Family of Single-Chain Magnets, *Inorg. Chem.*, **42**, 8203–8213; (c) Ferbinteanu, M., Miyasaka, H., Wernsdorfer, W., Nakata, K., Sugiura, K., Yamashita, M., Coulon, C, Clérac, R. (2005). Single-Chain Magnet $(NEt_4)[Mn_2(5\text{-}MeOsalen)_2Fe(CN)_6]$ Made of Mn^{III}–Fe^{III}–Mn^{III} Trinuclear Single-Molecule Magnet with an $S_T = {}^9/_2$ Spin Ground State, *J. Am. Chem. Soc.*, **127**, 3090–3099.

69. (a) Miyasaka, H., Nakata, K., Lecren, L., Coulon, C., Nakazawa, Y., Fujisaki, T., Sugiura, K., Yamashita, M, Clérac, R. (2006). Two-Dimensional Networks Based on Mn_4 Complex Linked by Dicyanamide Anion: From Single-Molecule Magnet to Classical Magnet Behavior, *J. Am. Chem. Soc.*, **128**, 3770–3783; (b) Miyasaka, H., Nakata, K., Sugiura, K., Yamashita, M, Clérac, R. (2004). A Three-Dimensional Ferrimagnet Composed of Mixed-Valence Mn_4 Clusters Linked by an $\{Mn[N(CN)_2]_6\}^{4-}$, *Angew. Chem., Int. Ed. Engl.*, **43**, 707–711.

70. Hiraga, H., Miyasaka, H., Takaishi S., Kajiwara T, Yamashita, M. (2008). Hybridized complexes of $[Mn^{III}_2]$ single-molecule magnets and Ni dithiolate complexes, *Inorg. Chim. Acta*, **361**, 3863–3872.

71. Hiraga, H., Miyasaka, H., Nakata, K., Kajiwara, T., Takaishi, S., Oshima, Y., Nojiri, H, Yamashita, M. (2007). Hybrid molecular material exhibiting single-molecule magnet behavior and molecular conductivity, *Inorg. Chem.*, **46**, 9661–967.

72. Ahmad, M. M., Turner, D. J, Underhill, A. E. (1984). Physical properties and the Peierls instability of $Li_{0.82}[Pt(S_2C_2(CN)_2)_2]\cdot 2H_2O$, *Phys. Rev. B*, **29**, 4796–4799.

73. (a) Lecren, L., Wernsdorfer, W., Li, Y.-G., Roubeau, O., Miyasaka, H, Clérac, R. (2005). Quantum tunneling and quantum phase interference in a $[Mn^{II}_2Mn^{III}_2]$ single-molecule magnet, *J. Am. Chem. Soc.*, **127**, 11311–11317; (b) Lecren, L., Wernsdorfer, W., Li, Y.-G., Vindigni, A., Miyasaka, H, Clérac, R. (2007). One-dimensional supramolecular organization of single-molecule magnets, *J. Am. Chem. Soc.*, **129**, 5045–5051.

74. Oshima, Y., Nojiri, H., Asakura, K., Sakai, T., Yamashita, M., Miyasaka, H. (2006). Collective magnetic excitation in a single-chain magnet by electron spin resonance measurements, *Phys. Rev. B*, **73**, 214435 (5 pages).

75. (a) Kato, R. (2004). Conducting metal dithiolene complexes: Structural and electronic properties, *Chem. Rev.*, **104**, 5319–5346; (b) Cassoux, P. (1999). Complexes, *Coord. Chem. Rev.*, **185**, 213–232; (c) Cassoux, P, Valade, L. (1991). Molecular metals and superconductors derived from metal complexes of 1,3-dithiol-2-thione-4,5-dithiolate (dmit), *Coord. Chem. Rev.*, **110**, 115–160.

76. (a) Miyasaka, H., Clérac, R., Wernsdorfer, W., Lecren, L., Bonhomme, C., Sugiura, K, Yamashita, M. (2004). A Dimeric Manganese(III) Tetradentate Schiff Base Complex as a Single-Molecule Magnet, *Angew. Chem., Int. Ed. Engl.*, **43**, 2801–2805; (b) Li, Z., Yuan, M., Pan, F., Gao, S., Zhang, D, Zhu, D. (2006). Syntheses, crystal structures, and magnetic characterization of five new dimeric manganese(III) tetradentate Schiff base complexes exhibiting single-molecule-magnet behavior, *Inorg. Chem.*, **45**, 3538–3548.

77. Miyasaka, H., Saitoh, A, Abe, S. (2007). Magnetic assemblies based on Mn(III) salen analogues, *Coord. Chem. Rev.*, **251**, 2622–2664.

78. (a) Akutagawa, T, Nakamura, T. (2000). $[Ni(dmit)_2]$ salts with supramolecular cation structure, *Coord. Chem. Rev.*, **198**, 297–311; (b) Imai, H., Otsuka, T., Naito, T., Awaga, K, Inabe, T. (1999). $M(dmit)_2$ salts with nitronyl nitroxide radical cations (M = Ni and Au, dmit = 1,3-dithiol-2-thione-4,5-dithiolate). Nonmagnetic single-chain formation vs antiferromagnetic spin-ladder-chain formation of $M(dmit)_2$, *J. Am. Chem. Soc.*, **121**, 8098–8103.

79. Hiraga, H., Miyasaka, H., Clérac, R., Fourmigue´, M, Yamashita, M. (2009). $[M^{III}(dmit)_2]^-$-coordinated Mn^{III} salen-type dimers ($M^{III} = Ni^{III}$, Au^{III}; $dmit^{2-}$ = 1,3-dithiol-2-thione-4,5-dithiolate): Design of single-component conducting single-molecule magnet-based materials, *Inorg. Chem.*, **48**, 2887–2898.

80. Matsushita, M. M., Kawakami, H., Sugawara, T, Ogata, M. (2008). Molecule-based system with coexisting conductivity and magnetism and without magnetic inorganic ions, *Phys. Rev. B*, **77**, 195208 (6 pages).

81. Mas-Torrent, M., Laukhina, E., Rovira, C., Veciana, J., Tkacheva, V., Zorina, L, Khasanov, S. (2001). New transparent metal-like bilayer composite films with highly conducting layers of -$(BET\text{-}TTF)_2Br{\cdot}3H_2O$ nanocrystals, *Adv. Funct. Mat.*, **11**, 299–303.

82. de Caro, D., Malfant, I., Fraxedas, J., Faulmann, C., Valade, L., Milon, J., Lamère, J.-F, Collière, V. (2004). Metallic thin films of $TTF[Ni(dmit)_2]_2$ by electrodeposition on (001)-oriented silicon substrates, *Adv. Mat.*, **16**, 835–838.

83. Malfant, I., Rivasseau, K., Fraxedas, J., Faulmann, C., de Caro, D., Valade, L., Kaboub, L., Fabre, J. M, Senocq, F. (2006). Metallic polycrystalline thin films of the single-component neutral molecular solid $Ni(tmdt)_2$, *J. Am. Chem. Soc.*,**128**, 5612–5613.

84. Savy, J. P., de Caro, D., Malfant, I., Faulmann, C., Valade, L., Almeida, M., Koike, T., Fujiwara, H., Sugimoto, T., Fraxedas, J., Ondarçuhu, T, Pasquier, C. R. (2007). Nanowires of molecule-based charge-transfer salts, *New J. Chem.*, **31**, 519–527.

85. Savy, J. P., de Caro, D., Valade, L., Legros, J.-P., Auban-Senzier, P., Fraxedas, J., Pasquier, C. R, Senocq, F. (2007). Superconductivity in $TTF[Ni(dmit)_2]_2$ films, *EPL*,**78**, 37005.

86. Mas-Torrent, M, Rovira, C. (2006). Tetrathiafulvalene derivatives for organic field effect transistors, *J. Mater. Chem.*, **16**, 433–436. and references therein.

87. Mas-Torrent, M., Hadley, P., Bromley, S. T., Ribas, X., Tarrés, J., Mas, M., Molins, E., Veciana J, Rovira, C. (2004). Correlation between crystal structure and mobility in organic field-effect transistors based on single crystals of tetrathiafulvalene derivatives, *J. Am. Chem. Soc.*, **126**, 8546–8553.

88. Bromley, S. T., Mas-Torrent, M., Hadley, P, Rovira, C. (2004). Importance of intermolecular interactions in assessing hopping mobilities in organic field effect transistors: Pentacene versus dithiophene-tetrathiafulvalene, *J. Am. Chem. Soc.*, **126**,6544–6545.

89. Coronado, E., Martí-Gastaldo, C, Tatay, S. (2007). Magnetic molecular nanostructures: Design of magnetic molecular materials as monolayers, multilayers and thin films, *Appl. Surf. Sci.*, **254**, 225–235.
90. Cavallini, M., Facchini, M., Albonetti, M, Biscarini, F. (2008). Single molecule magnets: From thin films to nano-patterns, *Phys. Chem. Chem. Phys.*, **10**, 784–793.
91. Train, C., Gheorghe, R., Krstic, V., Chamoreau, L.-M., Ovanesyan, N. S., Rikken, G. L. J. A., Gruselle, M., and Verdaguer, M. (2008). Strong magneto-chiral dichroism in enantiopure chiral ferromagnets, *Nature Mater.*, **7**, 729–734.

Index

Color Insert

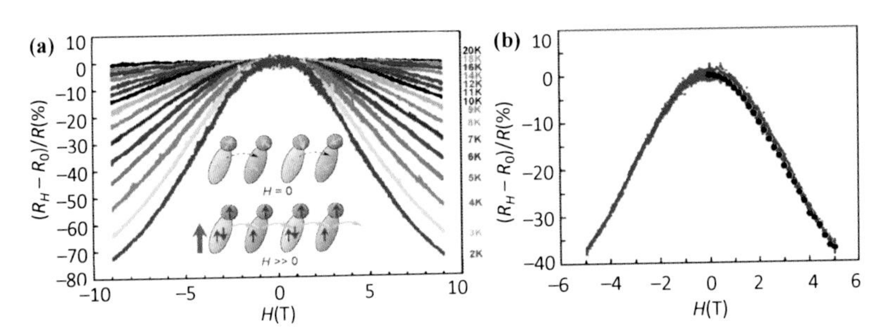

Figure 1.18

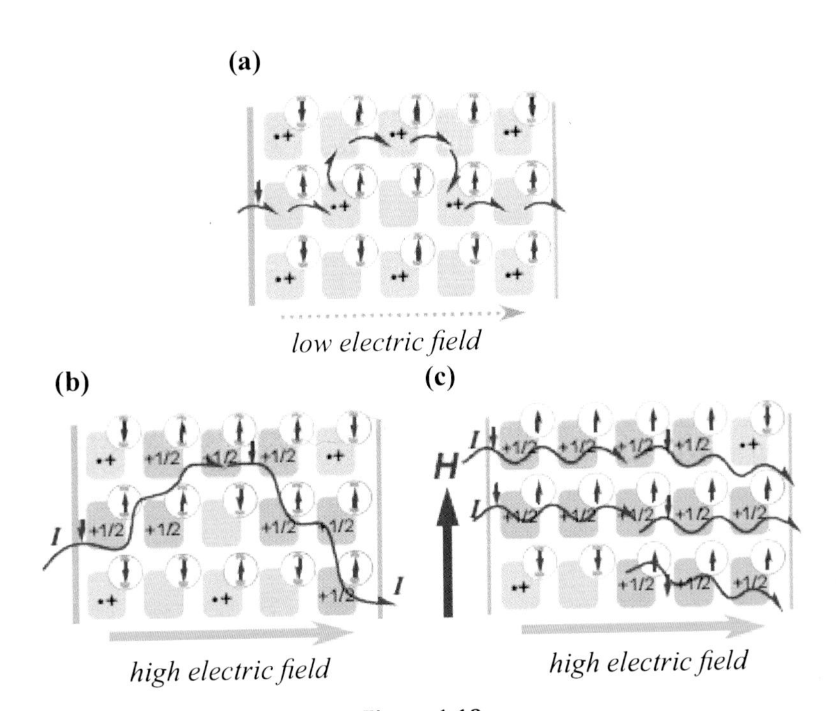

Figure 1.19

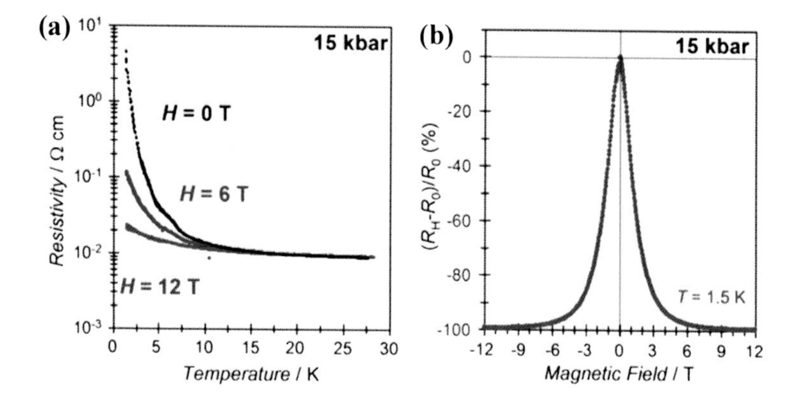

Figure 1.21

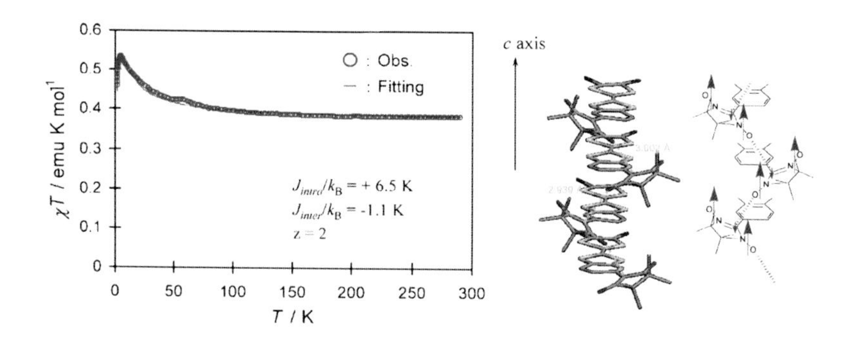

Figure 1.26

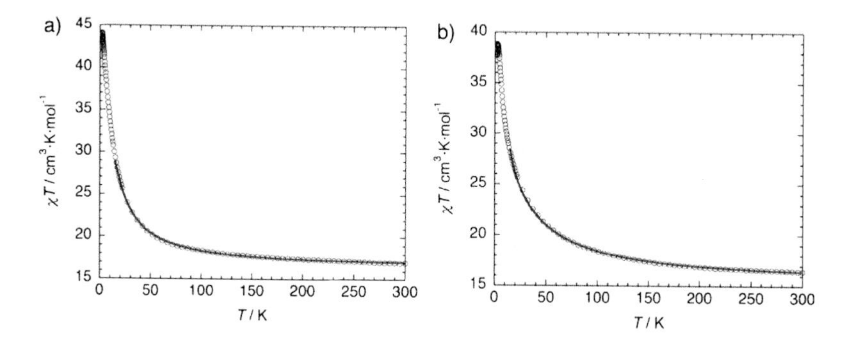

Figure 2.9

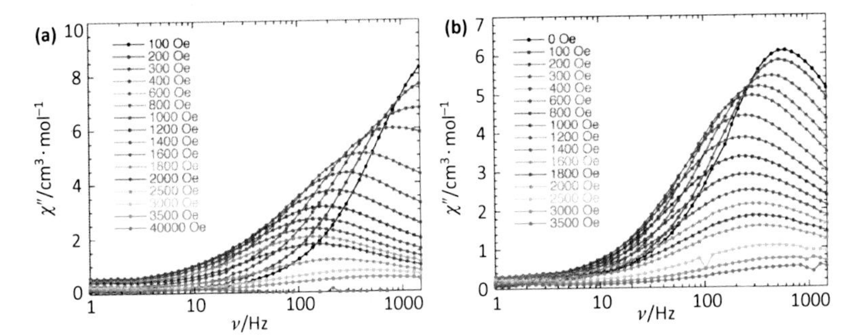

Figure 2.12

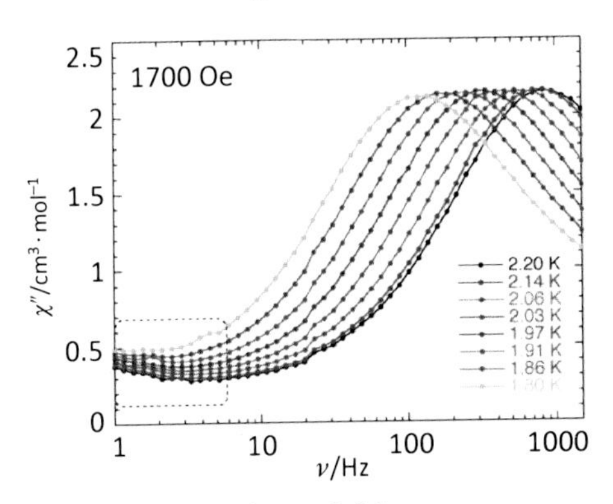

Figure 2.14

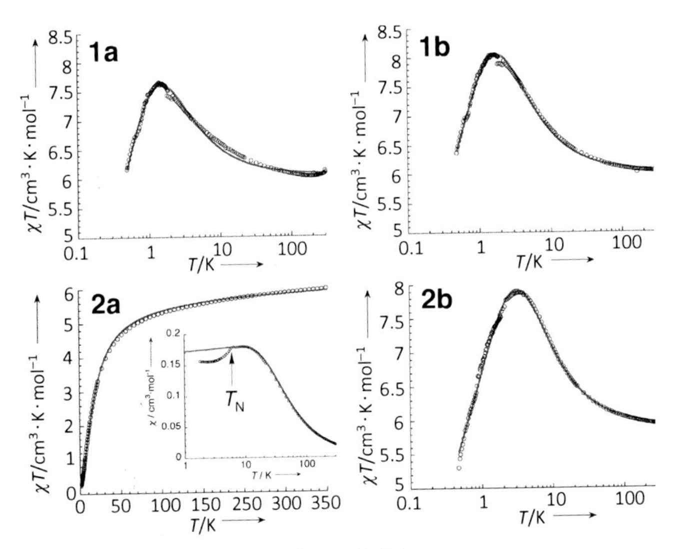

Figure 2.17

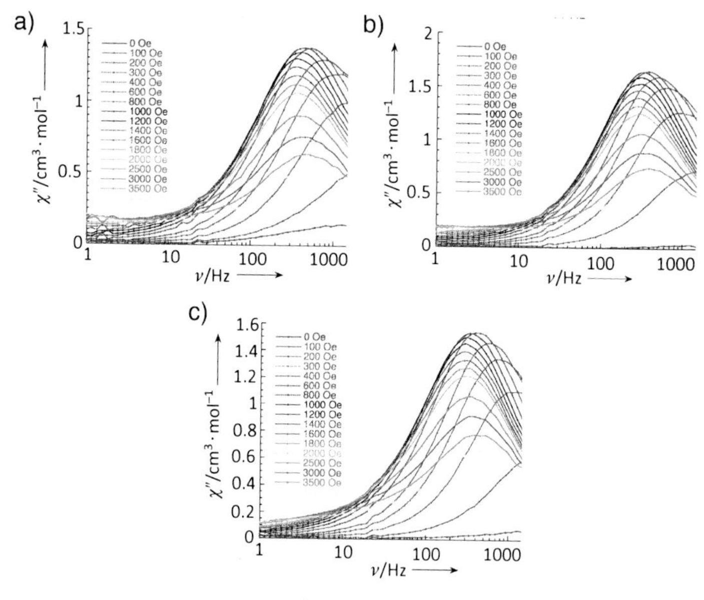

Figure 2.19

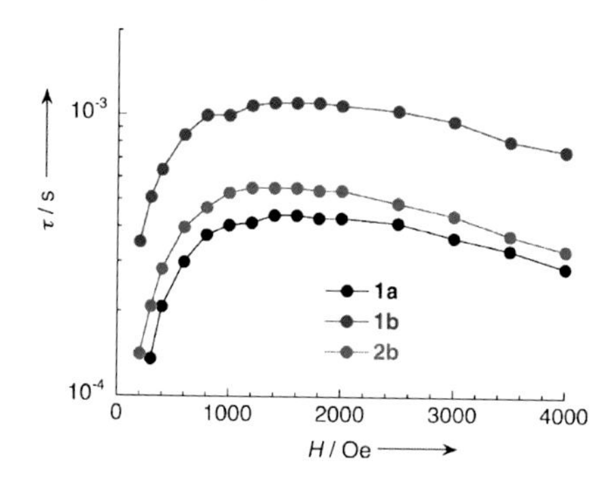

Figure 2.20

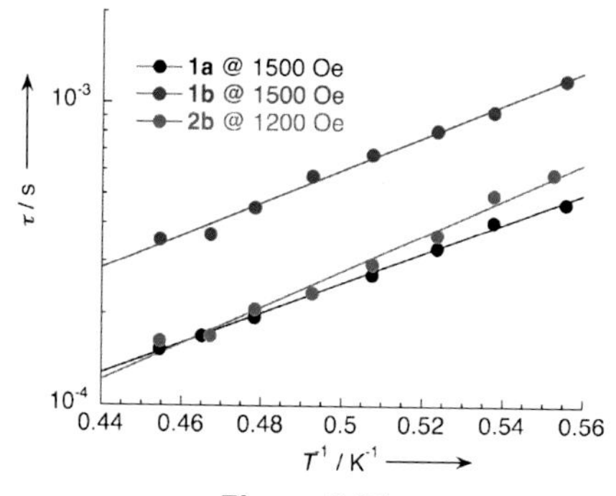

Figure 2.21

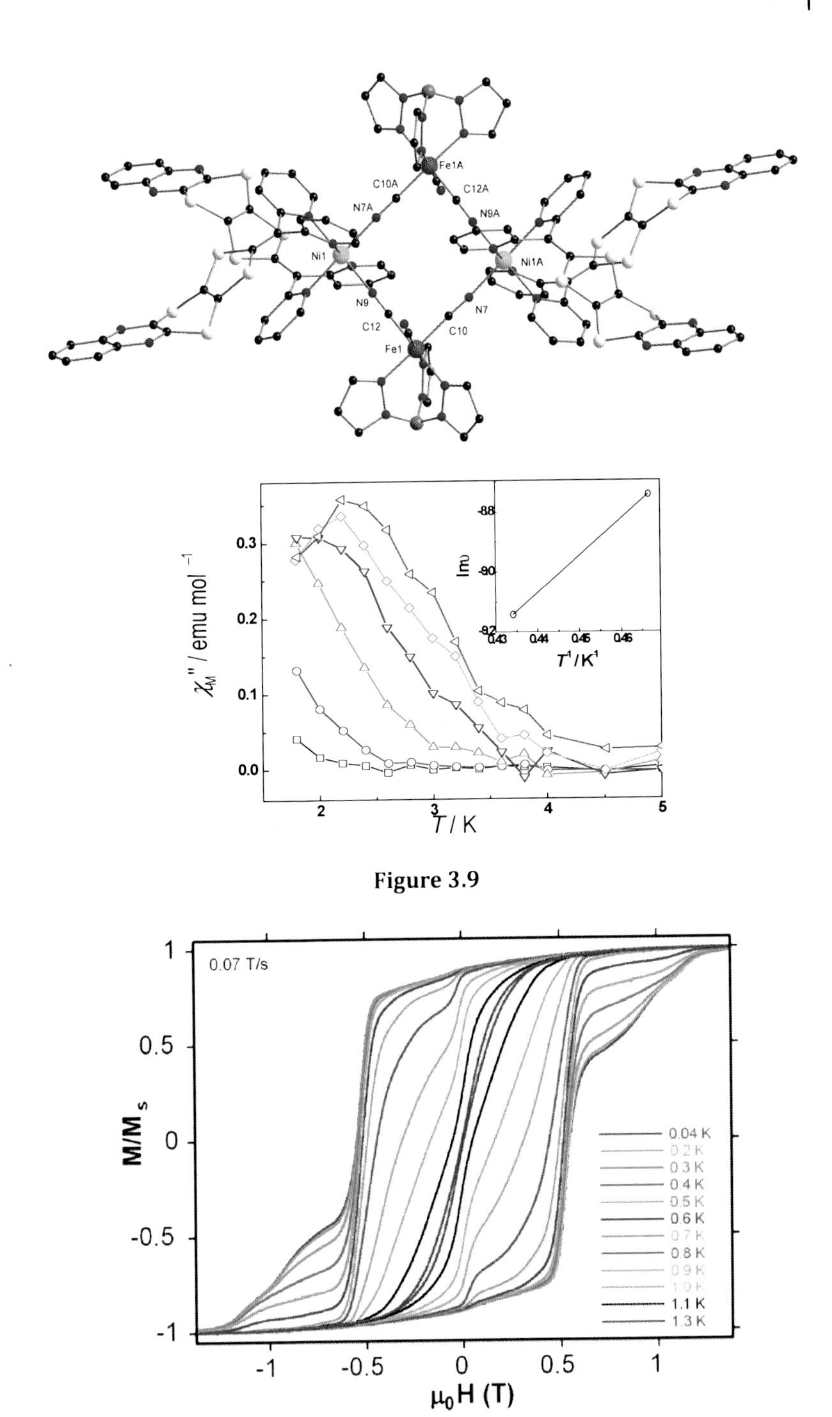

Figure 3.9

Figure 3.11

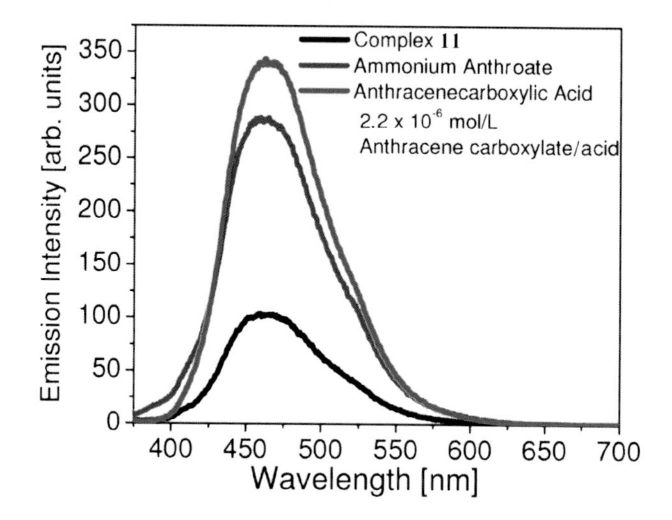

Figure 3.12

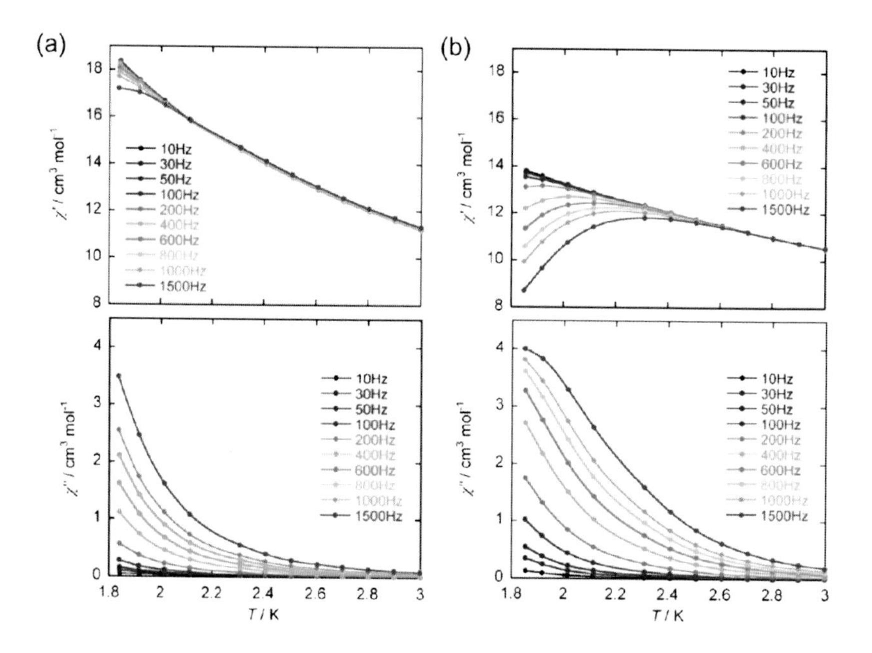

Figure 3.14

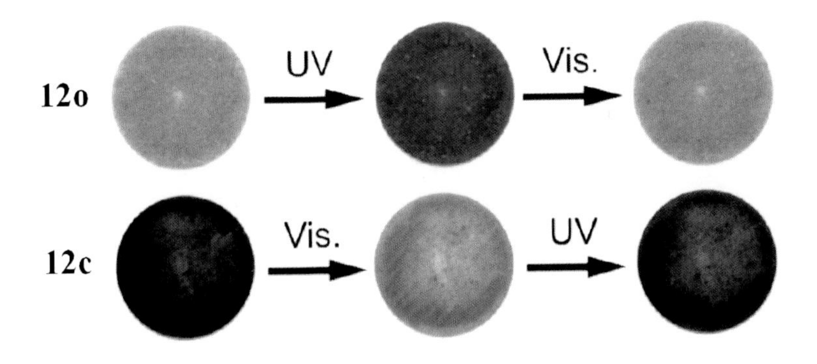

Figure 3.15

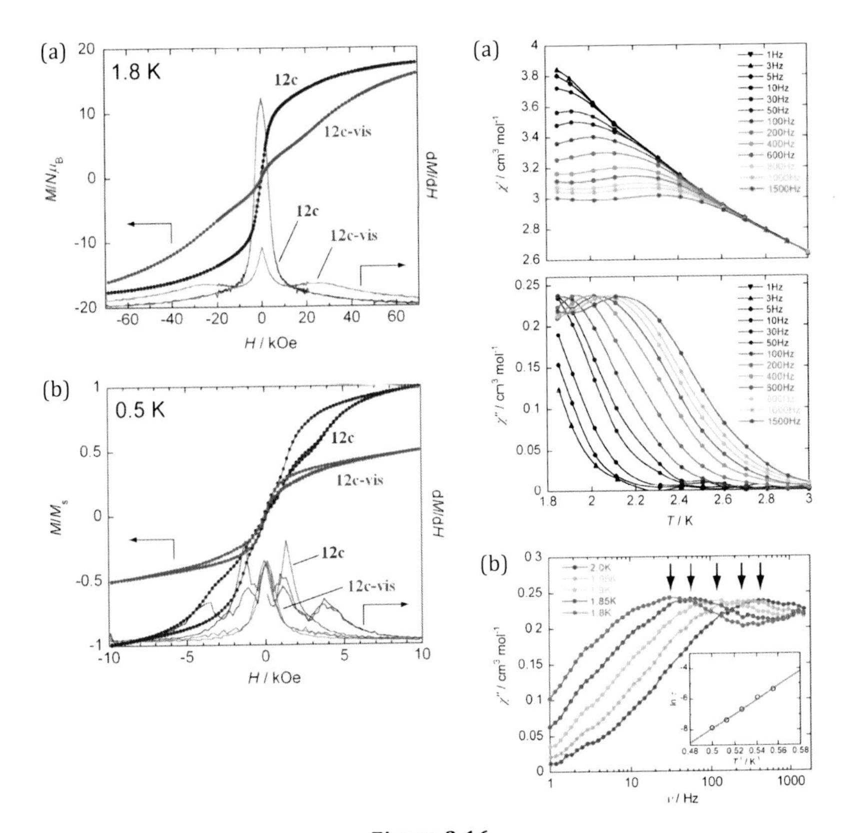

Figure 3.16

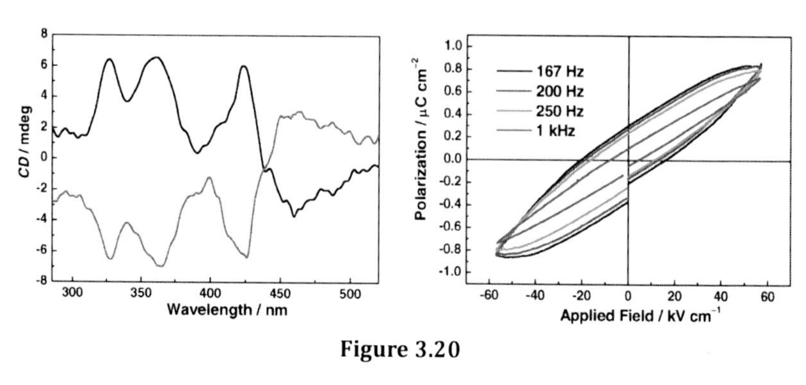

Figure 3.20

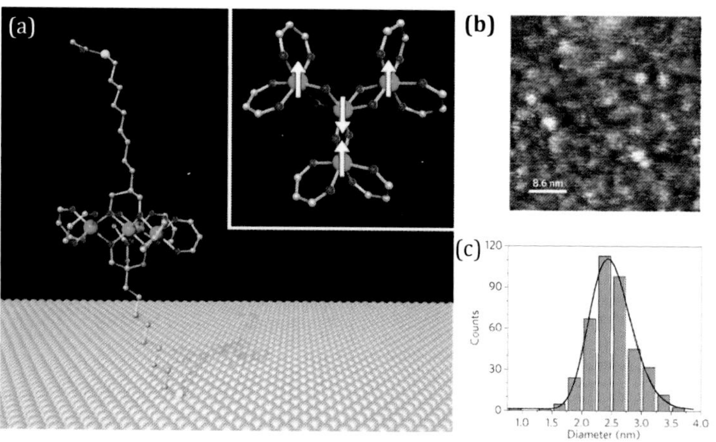

Figure 3.26

Figure 6.12

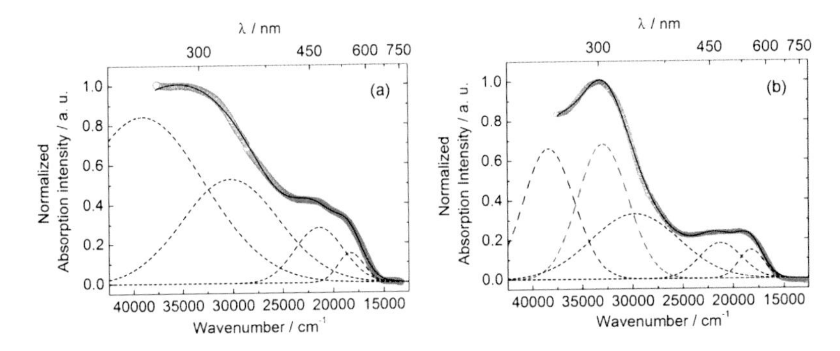

Figure 6.14

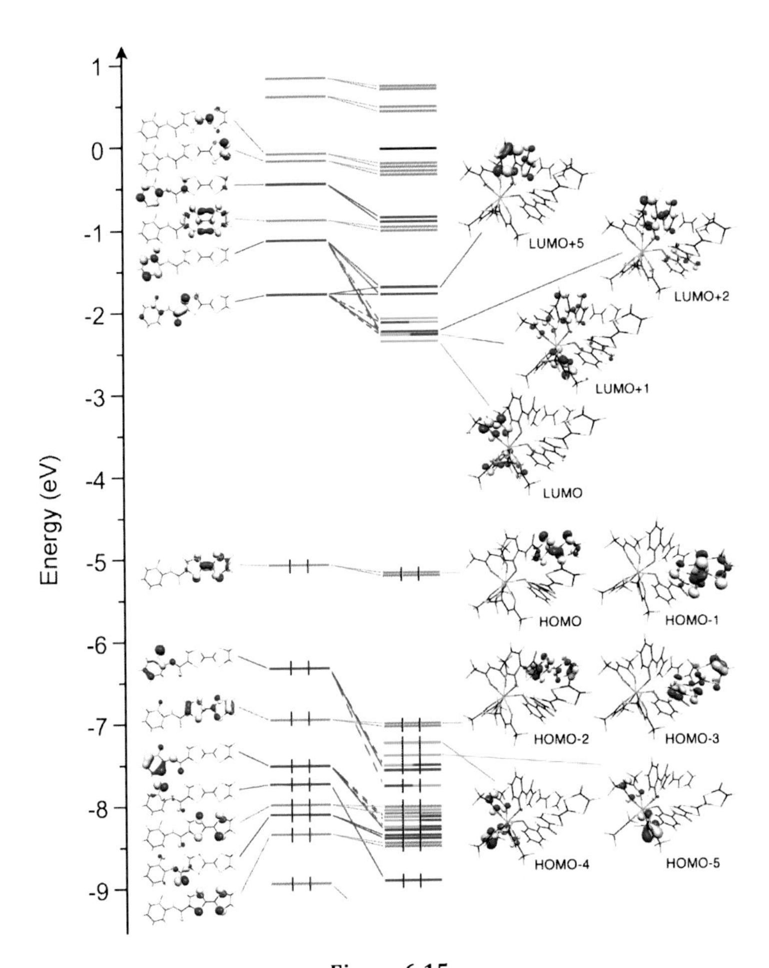

Figure 6.15

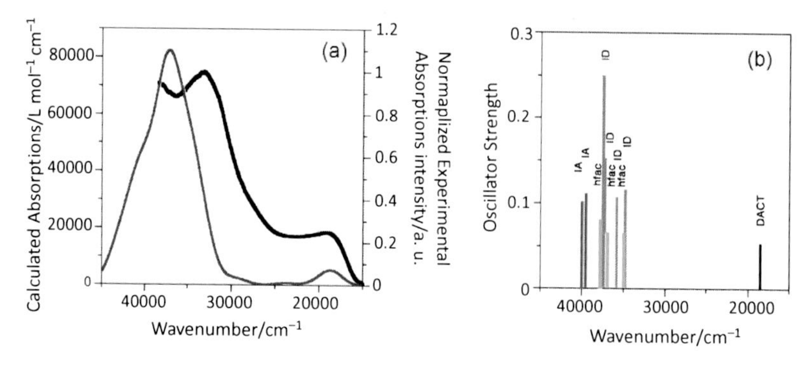

Figure 6.16

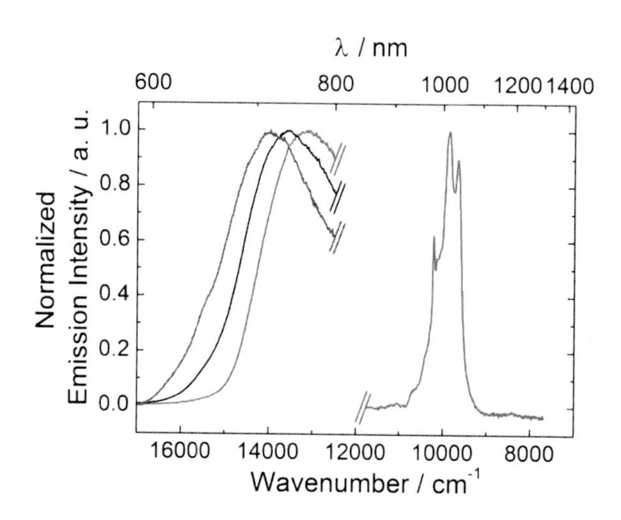

Figure 6.17

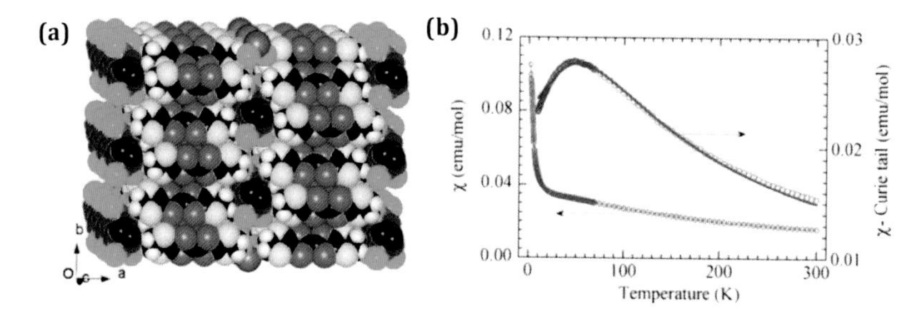

Figure 7.1

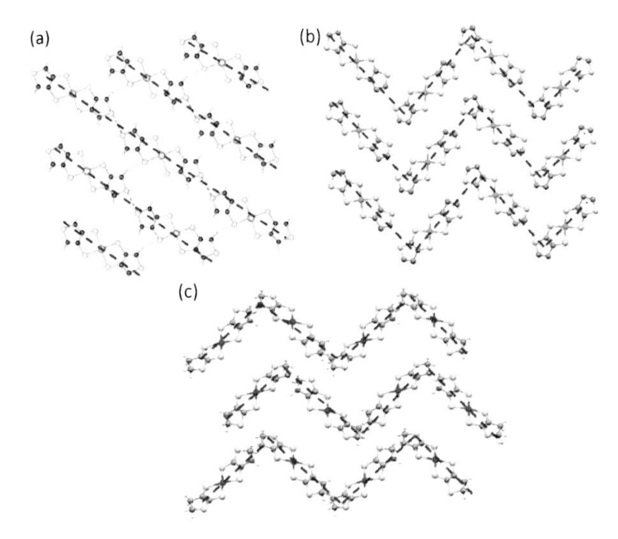

Figure 7.18

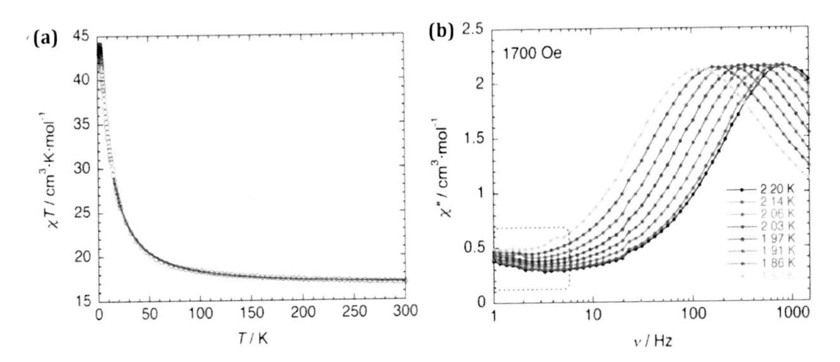

Figure 7.21

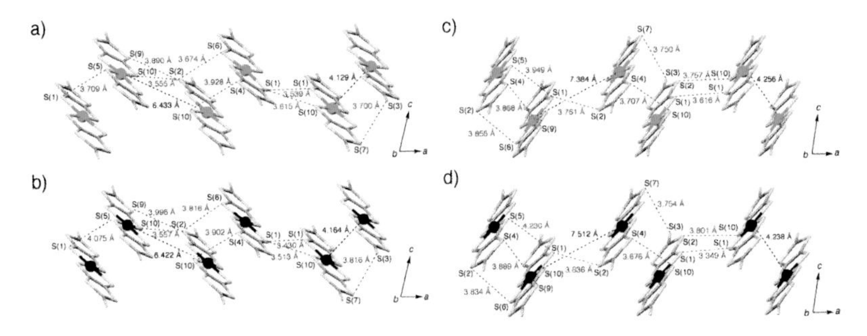

Figure 7.24

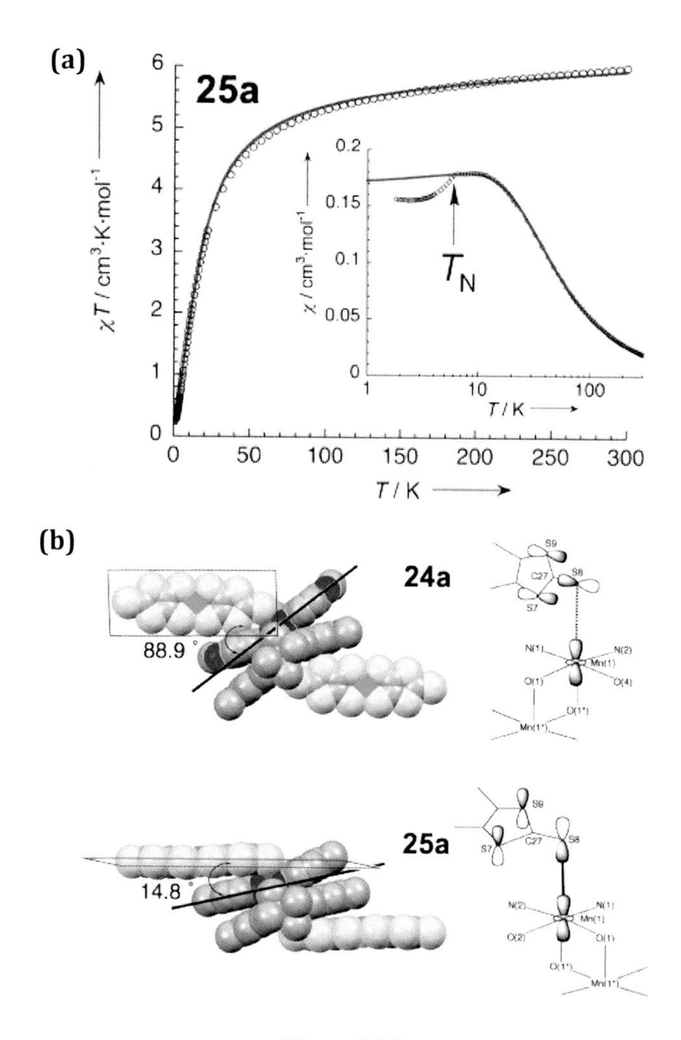
(a)
25a
χT / cm^3·K·mol^{-1}
T / K
χ / cm^3·mol^{-1}
T_N
(b)
24a
88.9 °
25a
14.8 °
S9
S8
C27
S7
N(1)
N(2)
Mn(1)
O(1)
O(4)
O(2)
O(1*)
Mn(1*)

Figure 7.25